SHENANDOAH UNIVERSITY LIBRARY
WINCHESTER, VA 22601

WITHDRAWN

D0205479

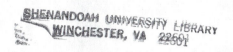

SHENANDOAH UNIVERSITY LIBRARY
WINCHESTER, VA 22601

Inventing Polymer Science

The Chemical Sciences in Society Series

Arnold Thackray, Editor

EDITORIAL BOARD

Arnold Thackray, Chemical Heritage Foundation
Otto Theodor Benfey, Chemical Heritage Foundation
Owen Hannaway, Johns Hopkins University
John E. Lesch, University of California, Berkeley
Mary Jo Nye, Oregon State University
John Servos, Amherst College
Jeffrey L. Sturchio, Merck & Co., Inc.

Sponsored by

CHEMICAL
HERITAGE
FOUNDATION

SHENANDOAH UNIVERSITY LIBRARY
WINCHESTER, VA 22601

Inventing Polymer Science

Staudinger, Carothers, and the
Emergence of Macromolecular Chemistry

Yasu Furukawa

PENN

University of Pennsylvania Press

Philadelphia

3.2-99

Copyright © 1998 University of Pennsylvania Press
All rights reserved
Printed in the United States of America on acid-free paper

10 9 8 7 6 5 4 3 2 1

Published by
University of Pennsylvania Press
Philadelphia, Pennsylvania 19104–4011

Library of Congress Cataloging-in-Publication Data
Furukawa, Yasu.
Inventing polymer science : Staudinger, Carothers, and the
emergence of macromolecular science / Yasu Furukawa.
     p.   cm. — (The chemical sciences in society series)
Includes bibliographical references (p.   –) and index.
ISBN 0-8122-3336-0 (acid-free paper)
1. Polymers — History.   2. Polymerization — History.
3. Macromolucules — History.   4. Staudinger, Hermann, 1881–1965.
5. Carothers, Wallace Hume, 1896–1937.   I. Title.   II. Series.
QD381.F87   1998
547′.7 — dc21   97-36390
                                                            CIP

Frontispiece: Wallace Hume Carothers, ca. 1930 (courtesy Hagley Museum and Library);
Hermann Staudinger, 1950s (courtesy Deutsches Museum)

QD   Furukawa, Yasu.
381
.F87   Inventing polymer science
1998

547.7 F984i

Contents

Acknowledgments

This project started when I was a graduate student in the History of Science Department at the University of Oklahoma. My prior experience as a practicing chemist, along with my interest in Hermann Staudinger and Wallace Hume Carothers, compelled me to write a dissertation on the early history of macromolecular chemistry in which the two scientists had played central roles. The book evolved from that Oklahoma dissertation, completed in 1983, by way of my extensive postdoctoral research conducted in my home country, Japan. I wish to express my sincere gratitude to my mentor, Mary Jo Nye, for her sound advice and continuous encouragement from the project's inception to the final revisions.

I am grateful to O. Theodor Benfey and Jeffrey L. Sturchio, who read the entire manuscript with care and offered thoughtful advice and moral support at crucial stages. Peter J. T. Morris and Teiji Tsuruta kindly gave of their time to read parts of the manuscript and provided helpful suggestions. I have also benefited from the criticisms offered by John K. Smith and Herbert Morawetz on my earlier draft. I owe a special debt to Elizabeth Sandager, who meticulously read the final manuscript and corrected errors in my English.

My study was greatly aided by the recollections of living witnesses. Magda Staudinger not only offered me extensive interviews in Freiburg but also answered many queries about her husband, Hermann, through subsequent correspondence until a few years before her death in 1997 at the age of 95. The information given by her greatly broadened my perspective on Staudinger's work and social life. The late Gerard J. Berchet, the late Yukichi Go, William E. Hanford, the late Julian W. Hill, Joseph Labovsky, the late Herman F. Mark, Seizo Okamura, Yutaka Sakurada, and Rudolph Signer all provided me with invaluable firsthand information (often not available in published sources) on Staudinger, Carothers, and other luminaries, as well as their own experiences, through interviews conducted between 1982 and 1993 in the United States, Germany, Switzerland, and Japan. I have also made use of the transcripts of other interviews, especially those done by the oral history project of the Chemical Heritage Foundation.

I would like to acknowledge the assistance I received at several archives and libraries while examining archival sources. Rudolf Heinrich assisted me in exploiting a rich store of the Hermann Staudinger Papers at the Deutsches Museum in Munich. Eckart Henning and Andreas K. Walter of the Archiv zur Geschichte der Max-Planck-Gesellschaft in Berlin were most helpful during my visit there. Between 1982 and 1993, I made several trips to the Hagley Museum and Library in Wilmington, Delaware, to work on the papers of Carothers and DuPont Company documents. Among the staff members there, I owe particular thanks to Richmond D. Williams, Michael Nash, Marjorie G. McNinch, John C. Rumm, and Barbara D. Hall for their kind help. When I first visited Wilmington in early 1982, Adeline B. C. Strange kindly loaned me her private collection of Carothers-John R. Johnson correspondence and transcripts of her interviews with Carothers's colleagues, which very much stimulated my interest. Her collection is now stored in the Hagley Library. I also want to thank archivists and librarians of the Firmenarchiv der Hoechst AG; the Chemical Heritage Foundation; the American Philosophical Society; the University of Illinois Archives; the University of Chicago Archives; the Special Collections and Archives, State University of New York, Albany; Leo Baeck Institute; the Pennsylvania Hospital Medical Library; Thomas Library, Wittenberg University, Springfield, Ohio; Bancroft Library, University of California, Berkeley; the Harvard University Archives; the Massachusetts Institute of Technology Archives; the National Museum of American History; and the Library of Congress, Washington, D.C.

A number of historians and chemists have provided me with valuable information, comments, and research material, to which this study has been indebted in various ways. For this, I wish to thank Clement H. Bamford, James J. Bohning, Michael Chayut, Harald Cherdron, Emo Chiellini, Eric Elliott, Wilhelm Füßl, John Heitmann, Koichi Hatada, Christiane Hess, David A. Hounshell, Hiroshi Inagaki, Susumu Iwabuchi, Jeffrey A. Johnson, Chikayoshi Kamatani, Hugo Ott, Claus Priesner, A. Truman Schwartz, Taro Tachibana, Takeshi Takahiko, Shigeya Takeuchi, and Takeshi Wada. I am also thankful to my colleagues, among them Mitsuo Shiraishi, at Tokyo Denki University, who sustained my research activity over the past years.

Part of the material in this book has appeared in different form in my articles in *Historia Scientiarum* and *Kagakushi*. A short version of some of the themes in this book has also appeared as an essay in *Science in the 20th Century*, edited by John Krige and Dominique Pestre (1997).

My research was supported by a University of Oklahoma Graduate College Travel Grant (1982); Eleutherian Mills-Hagley Foundation (now Hagley Foundation) Grant-in-Aid (1982); and Japan's Ministry of Education, Science, and Culture Grant-in-Aid for Scientific Research (1986–1987). The Tokyo Denki University Fund for Scholastic Research Advancement granted a subvention that bore a part of the publication cost of this book.

During my sabbatical from 1992 to 1993, I enjoyed continuing my research at the Chemical Heritage Foundation in Philadelphia. I am particularly grateful to Arnold Thackray and O. Theodor Benfey for their hospitality and their encouragement to publish this book as a volume of the Chemical Sciences in Society Series. Also invaluable was the professional assistance of Frances C. Kohler at the Chemical Heritage Foundation; and of Patricia Smith, Alison A. Anderson, and Kym Silvasy at the University of Pennsylvania Press.

Finally, I wish to thank my family members, Natsuko, Yoko, and Soh, for their understanding, patience, and warm encouragement.

Introduction

The emergence of macromolecular chemistry marks an epoch in the history of twentieth-century science. The field, termed more broadly polymer science, represents the study of polymers or large-molecular substances such as rubber, cellulose, starch, proteins, DNA, and plastics. The new concepts and methodologies developed in this field not only significantly broadened the intellectual horizons of chemical science, but also provided a foundation for the new growth of molecular biology and molecular physics. The practical applications of macromolecular chemistry are more conspicuous to the public. Indeed, it is difficult to envision a modern society without synthetic polymers — synthetic rubbers, synthetic fibers, and an enormous variety of useful plastics that are non-existent in nature.

Despite the significance of macromolecular chemistry in modern scientific development, general textbooks on the history of chemistry have failed to draw adequate attention to the field. Henry M. Leicester's *The Historical Background of Chemistry* (1956), James R. Partington's *A History of Chemistry* (1964), and Eduard Farber's *The Evolution of Chemistry* (1969), even though they were published after macromolecular chemistry had reached a certain maturity, make no mention of this field.[1] Aaron Ihde's *The Development of Modern Chemistry* (1964), Alexander Findlay's *A Hundred Years of Chemistry* (1965), and William H. Brock's more recent book, *The Fontana History of Chemistry* (1992), briefly cover the subject, but treat it as a part of industrial or applied chemistry.[2] There is no doubt that macromolecular chemistry has strong ties with industry. Yet to present this science as a mere branch of industrial chemistry is misleading.

This bent in historical textbooks affords a parallel with the tendency of general chemistry textbooks. In his posthumous autobiography, published in 1993, Herman F. Mark, a pioneer of the field, was puzzled:

It is difficult to understand why, although half of our professional chemists work with polymers, elementary textbooks on organic and physical chemistry hardly mention their existence. I have tried to suggest ways in which this omission might be remedied, but I am afraid we have a long way to go before polymers will receive reasonable treatment.[3]

Paul J. Flory, an American polymer chemist and Nobel laureate, shared this frustration. When he was professor in the Department of Chemistry at Stanford University in the 1960s and 1970s, his colleagues never allowed him to teach undergraduate courses on macromolecules. As he stated in a 1977 interview:

Unfortunately, most students graduating from college have virtually no exposure to macromolecules or to the basic principles of macromolecular science. They know it only superficially in terms of its applications in industry and in biology. . . . Some instructors pay lip service by injecting material on macromolecules in one or two lectures at the end of a course, or by adding a chapter at the end of the text. This is worse than nothing. It should be interwoven into the syllabus of the course or into the textbook.[4]

Macromolecular chemistry was then a well established scientific discipline that flourished in graduate programs at American universities. The lack of understanding among fellow chemists and authors of history textbooks is perhaps not explained merely by their ignorance. Rather, it is due largely to the widespread image of the field's conspicuous technological applications, which no doubt has obscured the relevancy of macromolecular chemistry as a pure science. It would be relatively easy in a historical narrative to place this science a priori into the great melting pot of industry. But it is more important and relevant to analyze and elucidate the historical relationship, tension, and dynamism between the scientific study of polymers and industrial practice.

The literature on the history of macromolecular chemistry is not meager. It has traditionally been dominated by the writings of practitioners in the field. Their autobiographies, memoirs, and scientific reviews provide numerous historical accounts.[5] In the past fifteen years, the literature has been augmented, especially in the United States, through the efforts of a group of the American Chemical Society, with Raymond S. Seymour as its most active leader. Through symposia, the group has collected firsthand information from polymer chemists on the history of their working subjects in polymer chemistry and technology. The results have been published in a number of monographs: *History of Polymer Science and Technology* (1982); *History of Polyolefins: The World's Most Widely Used Polymers* (1986); *High Performance Polymers: Their Origin and Development* (1986); *Pioneers in Polymer Science* (1989); *Organic Coatings: Their Origin and Development* (1990); *Manmade Fibers: Their Origin and Development* (1993).[6] Of uneven quality, these works comprise collections of essays, with the usual focus on precursors and technical detail, that are colored by scientists' values. The tradition of practitioners' history culminated in a book, *Polymers: The Origins and Growth of a Science*, published in 1985 by Herbert Morawetz, a student of Herman Mark. Perhaps comparable to Joseph S. Fruton's history of biochemistry in its depth and scope, this work presents an extensive general history of macromolecular chemistry with an admirable command of the vast chemical literature. Like Fruton's study, it is primarily an impersonal scientific review, describing the evolution of theories and experiments.[7]

Apart from insiders' contributions, there is also a sign of growing interest in the field among historians and biographical writers. Several books have shed light on certain aspects of the history of this specialty from various perspectives. Robert Olby has included a chapter on Staudinger's macromolecular concept in his history of molecular biology, *The Path to the Double Helix* (1974).[8] Frank M. McMillan's *The Chain Straighteners* (1979) portrays the polymer research of Karl Ziegler and Giulio Natta.[9] In his work, *H. Staudinger, H. Mark, und K. H. Meyer: Thesen zur Größe und Struktur der Makromoleküle* (1980), Claus Priesner has documented the controversy between Hermann Staudinger and Meyer-Mark in the 1920s and 1930s.[10] Peter J. T. Morris, working on a polymer project of the Beckman Center for the History of Chemistry, U.S.A., has published a concise booklet on the history of polymer chemistry, *Polymer Pioneers: A Popular History of the Science and Technology of Large Molecules* (1986) and a study on the U.S. emergency program for synthetic rubber during World War II: *The American Synthetic Rubber Research Program* (1989).[11] David A. Hounshell and John K. Smith have written a comprehensive work, *Science and Corporate Strategy: Du Pont R & D, 1902–1980* (1988), which includes chapters on polymer research at the DuPont Company.[12] More recently, Jeffrey I. Meikle has published *American Plastic: A Cultural History* (1995),[13] and Matthew E. Hermes a biography of Wallace H. Carothers, *Enough for One Lifetime: Wallace Carothers, Inventor of Nylon* (1996).[14]

Despite the relatively extensive literature on the subject, there are few analytical histories which deal with both the intellectual and the social setting of macromolecular chemistry in its formative period. This study seeks to fill that void in the historical scholarship. It aims to explore the emergence of macromolecular chemistry between the 1920s and 1940s in its intellectual, institutional, industrial, and political context. Special emphasis is placed on the scientific work and activities of the German Nobelist Hermann Staudinger (1881–1965) and the American chemist Wallace Hume Carothers (1896–1937) and their respective research groups. The roles of these two scientists were irrefutably central to the new field of chemical science during the interwar period. This study not only examines the origins and development of their scientific work, but seeks to illuminate their particular styles in chemistry and in their professional activities, and to contrast the peculiar institutional and social milieux in which they pursued their science.

The subject encompasses the formation of a scientific discipline as well. Macromolecular chemistry is the field of science which deals with a class of substances called polymers that have special properties, such as colloidal nature in solution and fibrousness or elasticity in the solid state, as exemplified by rubber, cellulose, proteins, starch, resins, and numerous synthetic polymers. This field arose not as a consequence of the cultivation of an unexplored area of ignorance. Rather, it emerged out of a fundamental reinterpretation of existing objects of inquiry. Staudinger was to play an important role in this reinterpretation during and after World War I. His controversial theory on the nature and properties of macromolecules stimulated a new wave of polymer studies, including the pathbreaking work of Carothers. In this respect, Flory's following statement is agreeably clear:

Polymer science dates from the recognition that polymers . . . consist of the very large molecules that we call macromolecules. More specifically, they consist of long chains of atoms linked by chemical bonds. The number of atoms in these chains usually runs into the thousands. This rudimentary, but fundamental, conception of the molecular constitution of polymeric substances is the cornerstone of modern polymer science. Without it, a *science* of polymers could not have been founded and elaborated.[15]

It is essential to explore, in the first place, the intellectual origins of the science. I examine in depth a major theme in the early development of the field, namely the establishment of the nature of macromolecules and their structures and properties. By the mid-1930s, after numerous stormy debates, the concept of macromolecules had been accepted by the majority of scientists. By the immediate post-World War II era, macromolecular chemistry had won recognition as a growing new scientific discipline. In this short time, we see a transition from theory formation to discipline formation. Although theory formation and discipline formation are often interwoven, they are not necessarily one and the same. How did this transformation take place? Here I examine not only the intellectual elements, but also the personal, institutional, industrial, and social factors that interacted in the shaping of macromolecular chemistry as a science.

Problems with terminology accompany the writing of the history of macromolecular chemistry. Well before the birth of macromolecular chemistry, there existed such jargon-laden terms as "polymers," "colloids," "high-molecular compounds" (*hochmolekulare Verbindungen*) — key words that would be taken over by macromolecular chemists. A careful treatment is in order when we confront these terms in the older literature. The meaning behind the terminology differs before and after the establishment of the macromolecular theory in the 1930s.

The field of macromolecular chemistry is now more often called "polymer chemistry" in English-speaking countries. However, in the context of nineteenth-century chemistry, "polymer" meant any compound that consisted of molecules having the same repeating atomic units, without regard to the molecular size. By Staudinger's time, a group of colloidal substances had become known as polymers. In the early decades of the twentieth century, chemists studied polymers such as rubber, cellulose, and resins, but many of them considered these polymers to be made up of small molecules. In other words, "the chemistry of polymers" had existed before macromolecular chemistry, and that "polymer chemistry" differed from what we now mean by the term. Likewise, practitioners of this area were not "polymer chemists" in the sense we use the term today. The chemistry of polymers became synonymous with the chemistry of macromolecules in scientists' view only after their reinterpretation of polymer structure.

The German word *hochmolekulare Verbindungen* (named after high molecular weight compounds) was also already widely used in the chemical community in the early decades of the twentieth century. Yet, again, *hochmolekulare Verbindungen* were not historically the same as compounds of large molecules. As will be seen in Chapter 1, many chemists did not regard the reported high values of molecular weights of polymers as evidence for big "chemical molecules," but considered them apparent values of physical aggregations of small molecules. In short, "high-molecular compounds" meant compounds of small molecules.

Based on his new concept, Staudinger introduced the term "*Makromolekül* (macromolecule)" in 1922.[16] He preferred to call his science *makromolekulare Chemie,* as indicated by the title of his journals: the *Journal für makromolekulare Chemie,* which ran from 1943 to 1944, and *Die Makromolekulare Chemie,* which was established in 1947.[17] Staudinger's coinage, which embraced a new concept and made a very clear distinction from previous usage, was remarkably effective in avoiding unnecessarily anachronistic terminology. *Makromolekulare Chemie* is still in wide use in Germany.

In America, "macromolecular chemistry" did not strike root as an official terminology. A fairly wide acceptance of "polymer chemistry" in English must be related to the history of its usage. Throughout his scientific career in the 1920s and 1930s, Carothers only occasionally called his field "the chemistry of macromolecular materials."[18] We find no mention of "macromolecular chemistry" or "polymer chemistry" even in

Roger Adams's 1939 obituary on Carothers, which described the latter's study of polymers and polymerization.[19] Far more influential in naming the field was perhaps Herman F. Mark, who spent his early professional career in Germany and Austria and emigrated to the United States in 1940. Mark, who had been involved in a bitter controversy with Staudinger over the priority of the concept of long chain molecules, was loath to call his field "macromolecular chemistry," the name associated with Staudinger. Instead, Mark favored "*hochpolymere Chemie*" or "high polymeric chemistry" in his publications in the late 1930s and in a well-circulated English monograph series, *High Polymers*, which he edited beginning in 1940.[20] It took little time before the word "high" was omitted in conventional usage. When he established the first English language periodical devoted exclusively to the field in 1946, Mark titled it the *Journal of Polymer Science*. In 1950 the American Chemical Society founded the Division of High Polymer Chemistry with Mark as its secretary-treasurer, but soon the name was shortened to the Division of Polymer Chemistry. By the 1950s, "polymer science" or "polymer chemistry" had taken hold on the American chemical community, as illustrated by Paul Flory's monumental 1953 textbook, *Principles of Polymer Chemistry*, and Fred W. Billmeyer's widely-read 1957 work, *Textbook of Polymer Chemistry* (its later editions were titled *Textbook of Polymer Science*).[21]

The name of the field is not without controversy even today. "The Glossary of Basic Terms in Polymer Science," prepared in 1994 by the Commission on Macromolecular Nomenclature of the International Union of Pure and Applied Chemistry, showed an officially predominant use of "polymers" and "polymer science" over "macromolecules" and "macromolecular science." It is not surprising that Magda Staudinger, the widow and closest collaborator of Hermann Staudinger, was uncomfortable about the action, and requested that the Committee replace the name "polymer science" with "macromolecular science," a protest that has an historical justification. As she urged:

Time has come to give careful consideration to the historical development of both terms: polymer and macromolecule. The decision should definitely be in favour of the well-defined term *macromolecule*. This term should be applied for the titles of publications, including journals and books, and particularly for the names of institutions for scientific research. They should be named "for macromolecular science" and no longer include the term polymer.[22]

Whether conventionality will outweigh an historical claim is less material to this present study. In the main, I have adopted Staudinger's original name, macromolecular chemistry, in this historical study, while I also use the words, polymer chemistry and polymer chemist, insofar as their meanings are unequivocal from the context.

Three major theses animate this study. These interrelate in certain ways. First, I depict chemists' differing approaches to polymers from the late nineteenth century through the 1930s in the light of two distinct traditions of chemical practice: the "organic-structural tradition" and the "physicalist tradition."[23] The former is rooted in the classical organic structural chemistry developed by the nineteenth-century chemist August Kekulé and his followers. Those who belonged to this tradition considered the molecule to be the chemical entity and viewed that properties of polymers could best be understood in terms of the internal structure of the organic molecule. The latter tradition, as represented by views of colloid chemists, physical chemists, and their sympathizers, minimized the role of molecules as the entity of scientific inquiries. They stressed that polymers were a physical state of aggregates of small molecules held together by physical forces, and that special properties of these compounds, such as colloidal nature, could be sufficiently explained in terms of matter, forces, and energy rather than molecules themselves.

Tension existed between the two traditions, reflecting not only theoretical and methodological differences but also disciplinary and institutional competition between classical organic chemistry and physical chemistry. As I explain in Chapter 1, the rise of the aggregate theory of polymers in the early part of the twentieth century meant a triumph for the physicalist school in the early decades of this century. It is irrelevant hindsight to see this triumph as an unfortunate obstruction or "the dark age" in the history of polymer-colloid studies.[24] For one thing, the physicalist theories served as the first coherent unitary views to explain the phenomena of a wide variety of polymeric substances. They also encouraged the further study of polymers. Chapter 2 portrays Staudinger, the organic structuralist, and analyzes the origins and development of his theory that polymers were macromolecules. The subsequent debate in the late 1920s over the existence of macromolecules largely exhibits an intellectual and disciplinary clash between the neo-organic structuralist Staudinger and his physicalist opponents. The "new micelle theory," proposed by Kurt H. Meyer and Herman Mark in 1928, embraced aspects of the macromolecular theory and the aggregate theory. Although Staudinger continued to criticize their eclectic theory, it acted as a bridge between the organic-structural and physicalist traditions. Chapter 3 focuses on yet another approach that arose within the organic-structural tradition, namely that of Carothers. There were no direct ties between Carothers and Staudinger or his German school. By exploring the American context of polymer studies and by tracing Carothers's educational background and initial contact with DuPont, I show how he first became interested in polymers, how he came to adopt the macromolecular theory, and how he developed his unique views of polymerization. By the mid-1930s, Staudinger and Carothers succeeded in establishing their macromolecular views by bringing the organic-structural approach back to the study of polymers, and by adding new concepts to classical organic chemistry. As the study of polymers expanded, however, it became increasingly apparent that there were certain limits to this approach. Chapter 5 discusses the restoration of the physicalist approach to polymer research. The period between the mid-1930s and late 1940s saw the coming of age of the physical chemistry of macromolecules. The two distinct chemical approaches thus proved to be necessary partners before macromolecular science could become fully established.

The second thesis concerns the political dimensions of macromolecular chemistry, especially in the case of Staudinger. Previous studies have discussed only sporadically some of the political aspects of Staudinger's life, and none have linked the technical debates over macromolecules to the broader political and institutional contexts. In Chapter 2, I argue that Staudinger's political activities during World War I, particularly his controversy with Fritz Haber over chemical warfare, affected the course of the macromolecular debate in the mid-1920s. The rise of Nazism further changed the picture of the macromolecular debate. While his Jewish scientific opponents fled Nazi Germany, Staudinger remained in the Third Reich but suffered oppression under Nazism due to his political past. In Chapter 4, by scrutinizing Staudinger's political struggles with his Nazi academic opponents (such as Kurt Hess and Wolfgang Ostwald), I show that the last stage of the macromolecular debate in Germany reflects not only a theoretical conflict but also the political tensions of the time. Staudinger succeeded in both surviving the Nazi threat and continuing his research by appealing rhetorically to the role he might play in German autarky. Chapter 6 also discusses the impact of World War II on the study of polymers, as illustrated by the American synthetic rubber program and the Japanese government's institutionalization of their fiber-oriented polymer studies. As these examples demonstrate, the science of macromolecules was by no means isolated from its social and political context.

Third, I analyze the historical interactions between polymer studies and industry. The polymer industries, which dealt with the manufacture of rubbers, cellulose, and resins, had sprung up just before the rising interest in polymer-colloid theories and shortly before the emergence of macromolecular chemistry. As I show in Chapter 1, advances in these early industries relied on experience rather than chemical theories, but industrial practice stimulated research into the chemistry of colloids and polymers. In Chapter 2, I show that Staudinger's primary interest lay in the pure science of polymers but, because of his research objectives, he was able to maintain a good relationship with the German chemical industry, especially the I. G. Farbenindustrie, which showed great interest in his study of polymers. This relationship helped to secure industrial posts for Staudinger's students as well as research funds at the University of Freiburg. Furthermore, as argued in Chapter 4, German industry played a significant role in rescuing Staudinger and his research school during the Nazi period. In America, the study of macromolecules first sprang from industry and then made a significant impact on university science. Chapter 3 expounds the nature of Carothers's pioneering study of macromolecular synthesis and his research activities at DuPont between 1928 and 1937. An analysis of the invention of nylon and neoprene by Carothers and his research group exemplifies how his theoretical study was transformed to practical applications within the industrial framework. The nylon venture, which would become a prototype for the science-based polymer industry, also turned out to be a large-scale test of the validity of macromolecular theory. Carothers's study of polymers represented a reciprocal approach in which theoretical principles and industrial innovations were advanced by interacting with each other. In Chapter 4, I discuss Carothers's dilemma as an industrial scientist and examine its relation to the growth of his mental depression, which eventually led to his suicide in 1937. Chapter 6 discusses the role of industrial development in the legitimization of macromolecular chemistry, sketches in outline the rapid expansion of macromolecular chemistry and the polymer industry during and after World War II, and epitomizes the legacy of Staudinger and Carothers.

The history of macromolecular chemistry provides us with a panorama of several captivating features of modern science: the rise and fall of scientific theories, the creation of new ideas and their epistemological grounds; scientific debate; the reception process of a new concept; the formation of a scientific discipline; the tension and interaction between disciplines; the relationship between science and technology; the strength and weakness of professional scientists; nature of the scientific community and schools; national style of science; and the institutional, industrial, and political intersections of modern science.

Chapter 1
Background, 1800–1920

The great experimental difficulties encountered in the investigation of colloidal bodies are responsible for the slow development of the chemistry of these substances.
　　　　—Carl O. Weber, "The Nature of India-Rubber," 1900.

I found myself in a swarm of organic chemists gathered about Emil Fischer, already regarded as the future leader of our science, since what was not organic chemistry was not recognized as chemistry. To his disparaging remark about our new direction [i.e., physical chemistry] I answered that the organic chemists owe us thanks for the possibility of determining the molecular weights of non-volatile substances. Fischer replied: "That was entirely unnecessary; I see directly the molecular weight of every new substance, and do not need your methods."
　　　　—Wilhelm Ostwald, *Lebenslinien*, 1927.

All those sticky, mucilaginous, resinous, tarry masses which refuse to crystallize, and which are the abomination of the normal organic chemist; those substances which he carefully sets toward the back of his cupboard and marks "not fit for further use," just these are the substances which are the delight of the colloid chemist.
　　　　—Wolfgang Ostwald, *Die Welt der vernachlässigten Dimensionen*, 1915.

The tendency of the organic chemist to discard the tarry residues or resinous products of a reaction and to explain them away by merely mentioning their formation is unfortunate.
　　　　—Roy H. Kienle and A. G. Hovey, "The Polyhydric Alcohol-Polybasic Acid Reaction," 1929.

Chemistry, a subject dealing with the nature and changes of matter, was itself both an intellectually and institutionally distinct and a rapidly evolving field of science in the nineteenth century. One of the most remarkable features of its development was the growth of organic chemistry. The outlook of organic chemistry underwent a series of significant changes throughout this period. During the first half of the nineteenth century, organic chemistry was essentially analytic. Practitioners of this field devoted themselves to three major activities: isolating organic substances by means of distillation and crystallization, determining their composition by burning the samples, and investigating their properties. In Germany, Justus von Liebig's school at the University of Giessen scored great success in research and teaching in organic chemistry in the 1830s. His program of research and laboratory instruction was an effective response to the demands of the field at that time. The core of Liebig's program was organic analysis. His school provided a large number of aspiring chemists with reliable techniques for the qualitative and quantitative analysis of organic compounds, an approach that quickly achieved a national and international reputation.[1]

Toward the latter half of the century, organic chemists were shifting their emphasis from the composition to the structure of molecules. Structural representation of molecules became a leading method to address the problem of substances associated with organic origins, the substances known to contain a large amount of the element carbon. By mid-century, chemical vitalism had lost its wide appeal, and chemists embarked on a new course to synthesize organic compounds artificially from inorganic or simpler organic substances, thereby ushering in the epoch of organic synthesis. Throughout this transition, organic chemistry expanded its meaning from the chemistry of products inherent in living things, such as plants and animals, to the chemistry of carbon compounds. Thus the German master of organic chemistry, August Kekulé, declared in his monumental *Lehrbuch der organischen Chemie*, published in 1859, "We define organic chemistry as the chemistry of carbon compounds. In doing this, we see no opposition between organic and inorganic compounds."[2] The successful rise of organic-chemical industries in the late nineteenth century was concurrent with the maturity of organic chemistry as a field of research.

Certain theoretical elements which were ultimately to shape the development of macromolecular chemistry can be seen in the context of the nineteenth century. During this period, such concepts as polymers, polymerization, and colloids emerged from various stages of scientific investigation. Chemists extended their inquiries into the elucidation of the structure and properties of a class of organic substances with a colloidal nature, now known as macromolecular compounds, such as rubber, cellulose, starch, proteins, and synthetic resins. Even though their investigations met with experimental difficulties, the possible existence of very large molecules for these compounds was suggested by a number of scientists working within the framework of organic structural chemistry. However, the concept of big molecules disappeared from mainstream scientific thinking in the early part of the twentieth century. Consequently, in chemical circles the interpretation of polymers and polymerization underwent a radical change.

This chapter will explore the background of macromolecular chemistry beginning in the early part of the nineteenth century through the 1920s — around the time of Hermann Staudinger and Wallace Carothers. It will illuminate the rise and decline of the large-molecular concept in light of changing views and approaches in chemistry, and also elucidate the industrial nexus of polymer studies during this period.

Polymers and Colloids

The words "polymer" and "colloid" were nineteenth-century inventions. In 1832 the Swedish chemist Jöns Jacob Berzelius recognized the existence of compounds with the same proportionate composition but different numbers of constituent atoms. He referred to such a phenomenon as "polymeric" (from the Greek πολύμερής, "consisting of many parts"), differentiating it from his notion of "isomeric" (from the Greek ἰσομερής, "consisting of equal parts"), which designated compounds with the same composition and the same number of atoms but different properties. The examples Berzelius provided for polymeric substances were "olefiant gas," or what we now call ethylene (as he put it, CH^2) and Faraday's volatile oil "Weinöl" (C^4H^8). These were polymeric, he reasoned, because the relative number of carbon and hydrogen atoms was the same, even though the absolute number differed. The differences in the properties of these substances could therefore be attributed to the difference in the absolute number of the constituent atoms.[3]

The Berzelian concept soon came into general use in nineteenth-century chemistry. For example, a well-circulated English chemical dictionary of the 1860s offered the following definition. "Bodies are said to be polymeric when they have the same percentage composition, but different molecular weights; the olefins C^nH^{2n} for example."[4] Encyclopedic writers, such as Leopold Gmelin, compiled lists of many examples of polymeric compounds or "polymers" around mid-century.[5]

Berzelius's polymer concept did not connote large molecules, yet chemists argued for the possible existence of high degrees of polymeric compounds. Among them was the Scottish chemist Thomas Graham. In his landmark paper "Liquid Diffusion Applied to Analysis," published in

1861, Graham reported that certain naturally occurring substances (including starch, dextrin, gums, caramel, tannin, albumen, gelatin, vegetable and animal extractive matters, and some inorganic substances like hydrated silicic acid) in solution showed a peculiar behavior, namely, an extremely slow or even negligible rate of diffusion through membranes such as parchment. He explained, "The comparatively 'fixed' class, as regards diffusion, is represented by a different order of chemical substances, marked out by the absence of the power to crystallize, which are slow in the extreme." He named such substances "colloids" (from the Greek κόλλᾰ, "glue"), as distinguished from "crystalloids," that is, normal substances that could easily crystallize and possessed a high diffusibility.[6] Crystalloids and colloids, he declared, "appear like different worlds of matter."[7]

According to Graham, the unique properties of colloids had much to do with their intimate molecular constitutions. Furthermore, he suggested that such colloidal substances might typically possess a high degree of polymeric constitution. The molecular weight, referred to as "equivalent" of a colloid by Graham,

appears to be always high, although the ratio between the elements of the substance may be simple. Gummic acid, for instance, may be represented by $C_{12}H_{11}O_{11}$, but judging from the small proportions of lime and potash which suffice to neutralize this acid, the true numbers of its formula must be several times greater. It is difficult to avoid associating the inertness of colloids with their high equivalents, particularly where the high number appears to be attained by the repetition of a small number. The inquiry suggests itself whether the colloid molecule may not be constituted by the grouping together of a number of smaller crystalloid molecules, and whether the basis of colloidality may not really be this composite character of the molecule.[8]

Elsewhere in the paper, however, Graham wrote that colloids might also represent a state or condition of matter. He described substances (e.g., some albumen in blood) that could exist in a crystalloidal or colloidal state. In such a case, the colloid would refer to a state of matter, and not to a kind of matter. For he believed, "in nature there are no abrupt transitions, and . . . distinctions of class are never absolute."[9] Hence he stressed the unity of matter, the continuity between "colloid molecules" and "crystalloid molecules."

With the rise of colloid chemistry a half century later, colloid chemists almost unanimously recognized Graham as the father of their science. It was customary for a textbook of colloid chemistry to begin with a discussion of his classic paper of 1861. However, Graham's cautious and guarded language gave rise to varied interpretations and disputes among scientists. Did Graham's emphasis on the unity of matter accord with his view of the colloid as a peculiar molecular species? Was the colloid simply a physical state into which any substance could be brought under certain conditions? Or, on the other hand, was it a fixed class of substances of peculiar molecular constitutions? Was it even possible to reconcile these apparently contradictory aspects? If so, how could "crystalloid molecules" group together to form a "colloid molecule"? Chemists struggled to solve these problems for decades. Indeed, these questions would constitute the central issue during the macromolecular debate in the 1920s.

The Age of Classical Organic Chemistry

Graham, who died in 1869, did not live to see the emergence of colloid chemistry as a scientific discipline or the dramatic expansion of organic chemistry during the second half of the nineteenth century. The major achievement of classical organic chemistry during this period lay in the field of crystalloids according to Graham's classification. The period witnessed the establishment of fundamental theories and methods in the chemistry of organic substances. Synthetic dyes, medical drugs, and numerous organic chemicals—with which chemists successfully applied these theoretical and methodological principles—were all crystalloids. For the time being, colloids remained at odds with the organic chemists' research objectives.

Around the middle of the century, organic chemists shifted their emphasis from elemental composition to molecular structure. After a period of confusion over the use of terminology and discussions at the Karlsruhe Congress of 1860, chemists arrived at a general agreement on the distinction between atoms and molecules.[10] Thereafter, chemistry became primarily the science of the molecule, the smallest portion of a substance capable of existing independently while retaining the properties of the substance. Chemical reactions took place at the molecular level. Molecules were regarded as the entities from which all chemical and physical properties stemmed and with which chemists ought to be primarily concerned. The concept of isomerism, proposed as early as the 1830s by Berzelius, now acquired significance in light of the molecular concept. It indicated that properties of matter depended not only on the kind of constituent atoms, but also on the way in which atoms were arranged within the molecule. The key to understanding distinct properties of organic substances now appeared to lie in the elucidation of the internal geometric structure of the carbon-rich organic molecules.

Kekulé and Archibald Scott Couper were instrumental in providing a theoretical framework of the architecture of organic molecules. Around 1858 they independently arrived at two important principles: one that the carbon atom was tetravalent or of fourfold saturating capacity; the other that atoms of carbon could combine together, forming carbon-carbon links. This so-called valency theory, in which no explanation was given for the cause of the valence forces, provided a new picture of the connection of atoms in a molecule. In addition to the tetravalent carbon atom, each atom possessed a definite limited valency, or capacity for combining with other atoms. The hydrogen atom, for instance, was univalent, oxygen bivalent, and nitrogen trivalent. Thus the individual atom was connected with one or a few neighboring atoms in accordance with its valency. The skeleton of the organic molecule was so constructed that valences of the different atoms satisfied each other. Couper, in particular, introduced the valence line for depicting the chemical bond.[11]

The organic structural theory followed the valency concept. According to this theory, properties of carbon compounds depended on the topological arrangement of atoms in the molecule more than on the kinds and number of component atoms. It now became clear how combinations of carbon and other atoms of only a few different elements, such as hydrogen, oxygen, and nitrogen, could constitute a surprisingly wide variety of organic compounds.

Kekulé later proposed a ring formula of six carbon atoms for the benzene molecule, which he considered a logical extension of his structure theory. He went on to suggest the existence of the benzene ring as a common nucleus in all of what had been called "aromatic compounds," so named because of their characteristic odors. Chemists were thereby able to classify organic compounds into two broad chemical families based on molecular structure — aliphatics, which were represented by the chain structure, and aromatics, which were represented by the ring structure. By the 1870s chemists gave further consideration to spatial arrangement of atoms in the organic molecule, introducing the three-dimensional or stereo-chemical scheme into the molecular structure.

The structural approach yielded a rich harvest in chemical research. Structural representation of molecules replaced the old formulas for known compounds and rapidly expanded its applications to a host of newly discovered or synthesized compounds. The result was a dramatic increase in the number of "known" organic compounds, including both natural compounds and synthesized materials not found in nature. Whereas in 1860 approximately 3,000 organic compounds were reported, by 1883 the number had increased to 20,000. By 1899 they had increased to 74,000, and by 1910 to over 140,000.[12]

Interest in the structural study coincided with the explosive growth of organic-chemical industries in the second half of the century. Particularly notable was the rise of the coal-tar dye industry, which represented a prototype of science-based modern industries. Following the 1856 discovery by the British chemist William Henry Perkin of an aniline purple (the first synthetic dye later named "mauve"), organic chemists synthesized various dyestuffs, such as alizarins, indigos, and other anilines, and their intermediates from a previously almost worthless substance, coal tars. Coal-tar products included not only substitutes for valued natural dyes, but also those producing colors superior to traditional dyes. Large-scale commercial productions were under way in Europe in the late 1860s. By the 1870s German manufacturers began to dominate the field; for example, in the 1890s their annual output reached nearly 90 percent of the world's dyestuff production.[13]

Germany's dye industry greatly benefited from its advanced research and educational programs in organic chemistry. Development of new dyestuffs and improvement of their qualities required adequate knowledge of and a high level of technical skill in organic chemistry. Fortunately, German universities and Technische Hochschulen had adequately trained a large number of organic chemists to do such work in the new industry. The industrial sector turned out to provide them with new career opportunities. By the 1880s major dyestuffs firms, such as the Badische Anilin- und Soda-Fabrik (BASF), Hoechst, and Fabenfabriken vorm. Friedrich Bayer & Co., had opened their own research laboratories.[14] Academic research in organic chemistry in turn was greatly stimulated by the growing industry. University men undertook important studies that directly contributed to the German dyestuffs industry. Among them, Kekulé's student Adolf von Baeyer, at the University of Munich, synthesized the first artificial indigo, and Baeyer's students Emil and Otto Fischer determined the structure of the first aniline dyes.

Of all chemical sub-disciplines, organic chemistry dominated German academia in the latter half of the nineteenth century. Organic chemists occupied many of the important academic positions in that country. The majority of the chemical institutes in German universities, completed between the 1860s and 1890s, were virtually "designed by and for organic chemists."[15] Baeyer's institute in Munich, founded in 1878, stood out as the largest and most expensive in the country. It was designed to accommodate nearly 200 students and advanced researchers in organic chemistry. From this institute emerged a group of the leading organic chemists in Germany, including Heinrich Caro, Ludwig Claisen, Carl Duisberg, Emil Fischer, Otto Fischer, Paul Friedländer, Kurt Heinrich Meyer, Victor Meyer, Rudolf Pummerer, Johannes Thiele, Jacob Volhard, Paul Walden, Heinrich Wieland, Richard Willstätter, and Hermann Staudinger.[16] By incorporating the promising new structural approach, organic chemistry appeared to offer unlimited possibilities for research topics, and its achievements enjoyed predominance both intellectually and institutionally.

"It was the period of crystallizable and distillable materials," as Willstätter later described the heyday of classical organic chemistry. He went on to explain:

Valence theory became the first ordering principle of far-reaching applicability. . . . For decades the problems were simple and had significance for teaching, research, and industry alike. Working methods were simple as well. Liebig's elementary analysis was still the most important quantitative procedure and often the only one. Great discoveries could be based even on qualitative observations.[17]

By the turn of the century, however, some organic chemists had begun to sense that their science might have peaked and that it had reached certain limits. In 1905 Baeyer, now almost seventy years old, told Willstätter, "Chemistry has changed. I would not study organic chemistry again."[18] The reasons organic chemists such as Baeyer felt a cloud hanging over their field were perhaps manifold. On the one hand, chemical laboratories were overcrowded with students specializing in organic chemistry. On the other, the research environment was worsening. Despite his large chemical institute, Baeyer now had only a small number of research assistants to work on his own problems.[19] Beside these institutional problems, there were also methodological problems with classical organic chemistry. Organic chemists were increasingly aware of the existence of complex organic substances with colloidal properties, for example, proteins, enzymes, starch, rubber, cellulose, and synthetic resins, which were not susceptible to conventional methods of organic chemistry. As Baeyer complained, now that studies of sugars and terpenes were nearly completed, "the field of organic chemistry is exhausted . . . and then all that remains is the chemistry of grease (*Schmiere*)."[20] The leading *Organiker* not only showed a distaste for the ill-defined gluey colloidal materials such as proteins and resins, but also he considered the study to be outside the scope of organic chemistry.

Furthermore, organic chemists were confronted by the challenge of two new disciplines, physical chemistry and colloid chemistry. Physical chemistry was founded by individuals who were dissatisfied with the static approach of traditional organic chemistry to molecules and were concerned with the dynamics of matter, or the mechanism of chemical change. The field began to attract a younger generation of chemists during the last decades of the century. Influenced by the concepts and methodology of physical chemistry, colloid chemistry was also beginning to emerge as a discipline, aiming to shed new light on the very problem of *Schmiere*. In order to understand this new climate in chemical science, we must first look at the way organic chemists struggled with the enigma of colloid polymers after Graham's day.

The Early Concept of Large Molecules

Graham's contemporary Kekulé did not fail to see the significance of the former's idea of colloid molecule. If one accepted Kekulé's structural theory, the question of how many atoms could be combined in one molecule was bound to occur. Kekulé himself did not place any upper limit on molecular size. Indeed, there was no reason for him to deny the possibility of large molecules, in which atoms were linked together by the Kekulé valence bond. In his address upon assuming the rectorship of the University of Bonn in 1877, Kekulé envisaged the structure of "colloid molecules" in terms of his valency theory, as follows:

. . . a considerably large number of simple molecules may, through polyvalent atoms, combine to *net-like*, and as we like to say, *sponge-like masses*, in order thus to produce those *molecular masses* which resist diffusion, and which, according to Graham's proposition, are called *colloidal* ones.[21]

Although he did not carry this argument any further, the possibility of the formation of a relatively large, complex chemical structure for Graham's colloid molecule appeared implicit in his structural scheme.

Kekulé's suggestion drew immediate attention from a few contemporaries, including a colleague at Bonn, Eduard Friedrich Wilhelm Pflüger. Seeking a feasible link between colloid molecules and biological processes, the physiologist assumed that these large, complex molecules might compose the "elements of form" of living organisms. He related the growth of organic bodies like plants to the formation of "almost endless carbon chains with their most varied arrangements."[22] Kekulé exhibited a cautious but appreciative attitude toward Pflüger's ideas: "To follow such speculations any further at present would, however, be equivalent to leaving the basis of facts rather too far behind us."[23]

Meanwhile, following the Berzelian definition of polymerism, the notion of polymerization was introduced into the synthesis of polymer compounds during the second half of the century, an epoch noted for laboratory synthesis of organic substances. In 1866 the French master of organic synthesis, Marcellin Berthelot, called "polymeric transformation" (*transformation polymérique*), the conversion reaction of certain compounds, such as styrene, to their polymerides (polystyrene).[24] Likewise, several cases of polymeric transformation were reported towards the end of the nineteenth century. Chemists commonly referred to this process as "polymerization," designating a union of two or more molecules (of the same kind) through chemical reaction to form larger molecules.[25]

Polymerization products, notably those prepared under such drastic conditions as high temperature, exhibited properties corresponding to Graham's natural colloids. They were gelatinous and sluggish to dissolve, and could neither be crystallized from solution nor distilled without decomposition. Often they appeared incidentally as unwanted by-products or residues to be cast aside in the processes of organic synthesis. Dubbed "grease chemistry," the study of these sticky materials did not attract practitioners of organic chemistry, as illustrated by Baeyer's remark noted earlier. These products did not respond to established methods for isolation, purification, and analysis — methods that relied heavily on crystallization or distillation.

That polymerization products discouraged researchers is further exemplified by the attitude of the German chemist Werner Kleeberg. Inspired by Baeyer's experiments on condensation of phenol, Kleeberg obtained a resinous polymerization product from phenol and formaldehyde in the 1890s.[26] Having failed to crystallize the mass, he, like Baeyer, "dismissed the subject and made himself happy with the study of nicely crystalline substances."[27] Ironically, it was Kleeberg's ignored subject that was to be converted into the first commercially successful synthetic plastic, Bakelite.

After the 1880s, the possibility of large molecules was discussed often in the light of molecular weight measurements of naturally occurring polymers. Such techniques were products of physico-chemical studies of solutions. In 1882 François Marie Raoult, of Grenoble, France demonstrated that the depression of the freezing point of a solution was in proportion to the molecular concentration of the dissolved substances.[28] His study led to the establishment of the first quantitative method for determining the molecular weights of substances in solutions. Soon, chemists employed Raoult's method to measure the molecular weights of colloidal substances. They observed negligibly small freezing-point depressions that indicated surprisingly high values of molecular weights. By 1889, the British chemists Horace Tarber Brown and George Harris Morris had arrived at a value of 32,400 for the molecular weight of "soluble starch" from their experiments of freezing point depressions.[29] John Hall Gladstone, Graham's student, and Walter Hibbert applied Raoult's method to raw rubber (caoutchouc), reporting values of 6,500 to "extremely high." They drew the further conclusion that the molecule of a colloidal substance should contain a very large number of atoms.[30] In 1900 A. Nastukoff reported a molecular weight of 5,700 to 12,000 for acetylated cellulose by means of the elevation of boiling point — another version of the Raoult method.[31]

By the years 1887 to 1888, the Dutch physical chemist Jacobus Henricus van't Hoff explained the relationship between the osmotic pressure of a solution and the molecular weight of solute.[32] Following this study, the osmotic pressure method for determining molecular weight was also applied to colloidal substances. Employing this method, in 1900

Table 1.1. High Molecular Weights of Naturally Occurring Colloidal Substances, Reported by 1900

Date	Author	Substance (method)	Mol. weight
1886	Zinoffsky	hemoglobin* (Q.A.)	16,700
1889	Brown & Morris	soluble starch (D.P.)	32,400
1889	Gladstone & Hibbert	rubber (D.P.)	6,500+
1891	Sabanijeff & Alexandrov	egg albumin* (D.P.)	14,000
1893	Linter & Düll	amylodextrin (D.P.)	17,500
1898	Bugarsky & Liebermann	egg albumin* (D.P.)	6,400
1900	Rodewald & Kattein	starch (O.P.)	38,000
1900	Nastukoff	cellulose (E.P.)	5,700–12,000

D.P. = depression of freezing point; E.P. = elevation of boiling point; O.P. = osmotic pressure; Q.A. = quantitative analysis; * = protein. The studies listed here are found in the bibliography at the end of this book.

Hermann Rodewald and A. Kattein obtained a high molecular weight of about 38,000 for starch, similar to Brown and Morris's result.[33] By the turn of the century, then, very high values of molecular weights for such colloidal substances as starch, rubber, cellulose, and proteins were being reported (see Table 1.1), which led a number of chemists to suspect that these substances might indeed be composed of very large molecules.

Revelation of these data on high molecular weights, however, was not followed by wide acceptance in scientific circles of the concept of large molecules. Instead, the concept declined in the early decades of the twentieth century. On the one hand, there was a growing suspicion among colloid chemists that laws of crystalloids might not be applicable to colloids. On the other hand, some physicists and physical chemists expressed doubts about the possibility of large molecules on the grounds that if a molecule was very large and complex, perhaps containing several thousand atoms, then it would be too fragile and unstable to exist. Thus the French physical chemist, Jean Perrin, noted in his famous monograph, *Les Atomes*:

We would expect, moreover, that very complicated molecules would be more fragile than molecules composed of few atoms and that they would therefore have fewer chances of coming under observation. We should also expect that if a molecule were very large (albumins?) the entry or exit of a few atoms would not greatly affect its properties and, moreover, that the separation of a *pure* substance corresponding to such molecules would present no little difficulty, even if its isolation did not become impossible. And this would still further increase the probability that a pure substance easy to prepare would be composed of molecules containing few atoms.[34]

Whereas many researchers were puzzled by the anomalies presented by colloidal polymers, Emil Fischer supplied an answer to this problem from the standpoint of organic chemistry.

Emil Fischer and the Giant Molecule

The physical chemist Wilhelm Ostwald told Emil Fischer at the 1889 Heidelberg meeting of the Society of German Natural Scientists and Physicians (Gesellschaft Deutscher Naturforscher und Ärzte) that organic chemists should thank physical chemists for having developed new methods of molecular weight measurement. But Fischer replied bluntly, "That was entirely unnecessary; I see directly the molecular weight of every new substance, and do not need your methods."[35] Unlike his mentor Baeyer and many other organic chemists, Fischer did not avoid the study of colloids, but took up bio-colloidal substances as his lifework. Yet he was unwilling to rely on physico-chemical methods, choosing instead to approach his subject as an organic structuralist. His pioneering work on the structure of sugars, enzymes, purines, and proteins not only represents a refinement of the organic-structural approach, but also reflects a changing attitude of classical organic chemists toward the large molecular concept.

One of the most successful pupils of the Baeyer school, and the first German to win the Nobel Prize in chemistry, Fischer headed a productive research school of organic chemistry at the newly founded chemical institute of the University of Berlin. He supervised a large body of doctoral students and *Privatdozenten* and published papers at an exceedingly high rate.[36] In order to elucidate the constitution of natural products present in organisms, he developed an innovative synthetic approach. Starting with simple molecules, he synthesized complex ones of known structure closely simulating the natural compound in question. Then he compared properties of the synthetic products with those of the natural products. In this way, Fischer attempted to establish the precise molecular structure of sugars, enzymes, purines, and proteins. Molecular weights of these natural products could be deduced from those of the synthetic models; weights of the models themselves were known from the reactants. Hence there was no need to use physico-chemical methods.

He argued that his synthetic approach sharply contrasted with the tendency in physics to comprehend matter only by dividing, subdividing, and re-subdividing. As he explained:

Molecular physics would do well in the study of high molecular substances to confine itself to the synthetic products of known structure. I will continue the experiments on the building up of giant molecules with the aid of the process described.

Certainly it offers in other respects a great incentive to test the productiveness of our methods. As is well known, modern physics is endeavoring to split matter into smaller and smaller pieces. One is long since past the atom, and how long the electrons will be for us the smallest particles of matter cannot be predicted. It seems to me that organic synthesis is called upon to accomplish the converse, i.e., to accumulate larger and larger masses in the molecule, in order to see how far the compression of matter can go, in the meaning of our present conceptions.[37]

Fischer did consider proteins "high-molecular substances" (*hochmolekularer Stoffe*), namely, substances of a high molecular weight, consisting of large molecules or what he called "giant molecules" (*Riesenmoleküle*). Nonetheless, the reported values of molecular weights for proteins seemed to him too high to accept. He doubted the validity of the conventional method, which attacked the problem of protein structure on the basis of molecular weight measurements. Rejecting some reported physico-chemical estimates of a molecular weight of 12,000 to 15,000 for proteins, in 1907 Fischer claimed, "In my opinion these numbers are based on very uncertain assumptions since we do not have any guarantee that the natural proteins are homogeneous substances."[38] As for the crystalline hemoglobin, which on the empirical formula ($C_{712}H_{1130}N_{214}S_2FeO_{245}$) by Oscar Zinoffsky had a molecular weight of some 16,000, Fischer commented in 1913:

. . . for the beautifully crystalline oxyhemoglobin, as is well known, a molecular weight of 16,000 has been derived from its iron content, but against such calculations the objection can always be made, that the existence of crystals in no way guarantees chemical individuality, particularly since it can be regarded as an isomorphous mixture, such as the mineral kingdom so often presents to us in the silicates. Such objections vanish with synthetic products, whose formation can be controlled by analogous reactions.[39]

Fischer argued that proteins were not composed of polymeric molecules consisting of regularly recurrent atomic groups, but that, unlike many other polymers, the molecules were made up of many different units, that is, different kinds of amino acids. As he put it, "Nature never creates long chains of the same amino acids, but favors the mixed forms in which amino acids change from member to member."[40] In Fischer's view, proteins were composed of "polypeptides," in which many different amino acids were linked together. His study of proteins was therefore directed to the synthesis of various polypeptide chains. In working out the constitution of polypeptides, the structural chemist Fischer made use of the Berzelian concept of isomerism; that is, compounds of the same composition having the same molecular weight could exhibit different properties in accordance with their structural difference. He considered amino acids of eighteen kinds and polypeptides of thirty amino acids. The number of possible isomers of a polypeptide chain consisting of such units was in his estimation 2.653×10^{32}.[41] An extremely large number of possible isomers, he thought, would suffice to explain the wide variation in the properties of natural proteins. Hence there might be no need to assume the existence of very large natural polypeptides! Thus, Fischer's strong emphasis on isomers precluded any further consideration of larger molecular structures.

Fischer admitted a molecular weight of 4,021 for a starch derivative ($C_{220}H_{142}O_{58}N_4I_2$) which he and his student Karl Freudenberg had synthesized. At the 1913 Vienna meeting of the Congress of Natural Scientists (*Naturforscher-Versammlung*), Fischer declared that the value of 4,021 was the highest molecular weight found for any organic substance of known structure derived wholly by synthesis, and further that it was higher than that of any natural protein.[42] His authority among organic chemists made influential the claim that organic compounds of a molecular weight greater than 5,000 might not even exist. Although Fischer's use of synthetic models was later to provide Hermann Staudinger and Wallace Carothers with a powerful tool for the elucidation of the polymer structure, the Fischerian dictum placing an upper limit on the molecular size remained an obstacle that they would have to overcome.

By the 1910s, many organic chemists were likewise in no position to support the concept of very large molecules for colloidal substances, and it therefore became superfluous within the tradition that emphasized structure of organic molecules. Apparently, the concept became even more untenable as the physicalist approach to colloids flourished.

The Rise of the Physicalist Tradition: From Physical Chemistry to Colloid Chemistry

Colloids drew renewed interest from researchers around the turn of the century, four decades after the appearance of Graham's pioneering paper. In part, this new trend owed its success to the expansion of industrial concerns. Chemical industries were manufacturing products of a colloidal nature, including rayon, papers, resins, glues, paints, pigments, rubber, films, ceramics, soaps, and tanning materials, as will be discussed later in more detail. Another stimulus was scientists' growing recognition of the biological and physiological implications of colloidal phenomena in animal and plant cells. A morphologically oriented colloid study of protoplasm, for example, appeared to offer some biologists a more promising guide to the explanation of living processes than the structural organic chemistry of Fischer.

Equally important was the development of new instruments that greatly expanded the scope of understanding colloidal solutions. Among others, the ultramicroscope, which Richard Zsigmondy and Henry F. W. Siedentopf invented in 1903, made a huge impact. By focusing a powerful beam of light within the colloid solution, this instrument enabled researchers to observe bright specks of individual colloid particles dancing and hopping in the solution, particles that were too small to be seen with an ordinary microscope. Ultramicroscopic observations convinced scientists that colloidal solutions were not homogeneous, but heterogeneous, dispersed systems of particles of a certain size, and that dispersion of colloid particles was maintained without precipitation due to their Brownian motions.[43]

Last, but not least important, the new chemistry of colloids emerged as a new branch of physical chemistry. The two fields had common intellectual grounds. Furthermore, the professional identity of early colloid researchers lay largely with the community of physical chemistry. A rapidly growing scientific discipline, physical chemistry — or "general chemistry" (*allgemeine Chemie,*) as its German founder, Wilhelm Friedrich Ostwald, called it — aimed at investigations of the physical nature and behavior of chemical compounds by applying the methods and theories of physics, such as kinetics and thermodynamics, to chemical phenomena. Ostwald disdained structural chemistry and organic syntheses and was ambitious to reform chemical science. The current organic-structural approach was, Ostwald grumbled, too descriptive and static to explain dynamic chemical processes and systems. Using concepts of ions and energy, he theorized about chemical affinity, chemical equilibrium, rates of chemical reactions, and catalysis, for which efforts he was awarded the 1909 Nobel Prize. He was the first to expand upon the relatively unknown published views of the American physical chemist Josiah Willard Gibbs and the values of the phase rule and thermodynamics. Also well known as the champion of "energetics," Ostwald, for a time, went on to argue that scientists should abandon atomic and molecular hypotheses. All in all, Ostwald exerted a powerful influence in redirecting chemists' attention toward a physicalist scheme that interpreted chemical phenomena in terms of matter, force, and energy, rather than molecular structures.[44]

In 1887 Ostwald and van't Hoff founded the *Zeitschrift für physikalische Chemie.* In its inaugural issue, Ostwald declared to his readers (borrowing Emil Dubois-Reymond's words) that "physical chemistry is the chemistry of the future." His strenuous advocacy created a tension with some of the elite classical organic chemists. Richard Willstätter, for example, bitterly recalled that Ostwald unfortunately considered it his mission to battle passionately with the theories and methods of organic chemistry, which made little sense. For years he created discord and anger . . . he criticized, especially in his discussions in the *Zeitschrift für physikalische Chemie*, all of the contemporary literature of organic chemistry, in particular questions of structural chemistry. He had no knowledge or understanding of the development, content, or effect of our organic chemical views which had proved themselves to be uncommonly fruitful. . . . Ostwald's great influence caused us some bad moments.[45]

As Ostwald jokingly though sarcastically remarked, organic chemists did not regard him as a chemist because he had never synthesized a new compound. He proudly added that he had even contributed to *reducing* the number of substances recorded by organic chemists. When an organic chemist claimed to have isolated a new acid, Ostwald proved that it was not a new, but an already known organic acid, by determining its ionization constant.[46]

The increasing popularity of physical chemistry not only brought about an intellectual rivalry but also created an institutional threat against organic chemists who had dominated German academic posts for decades. Occupying one of Germany's first chairs of physical chemistry, Ostwald transformed his old library in the poorly equipped agricultural institute at the University of Leipzig into a world center of physical chemistry, to which eager students flocked from many parts of the globe. The majority of his first pupils were American and British. Germans constituted a small minority, this underrepresentation reflecting the predominance of organic chemistry over other branches of chemistry in the German academy.[47] It was a time, as Ostwald put it, when "what was not organic chemistry was not recognized as chemistry."[48] But within a few decades, the situation changed. In 1897 the Saxon government built a newly independent institute for him at Leipzig, the Physico-chemical Institute. By the 1910s physical chemists managed to hold chairs at Germany's several universities, including Göttingen, Giessen, Freiburg, and Berlin, and at seven of Germany's eleven Technische Hochschulen.[49]

As the prime organizer of this science, Ostwald did not fail to boast about the potential of industrial applications of physical chemistry. Emulating existing applied organizations, which were dominated by organic chemists, in 1894 physical chemists and electrochemists established an industry-oriented academic organ, the German Electrochemical Society (Deutsche Elektrochemische Gesellschaft, later renamed Deutsche Bunsen-Gesellschaft für angewandte physikalische Chemie). As the Society's first president, Ostwald stressed that electrochemistry was "the secret success of the German chemical industry," and called for cooperation between academic and industrial chemists.[50] Ostwald also supported the establishment of the Kaiser Wilhelm Society for the Advancement of the Sciences (Kaiser Wilhelm Gesellschaft zur Förderung der Wissenschaften) that aimed to promote Germany's frontier research by using facilities and equipment superior to those found in existing university institutes. The creation of the Kaiser Wilhelm Institute for Physical Chemistry and Electrochemistry (Kaiser Wilhelm Institut für physikalische Chemie und Elektrochemie) marked the culmination of the institutionalization of physical chemistry. Funded by Leopold Koppel — a Jewish leader of the Berlin gasworks who was concerned with the promotion of this new field — the Institute opened in 1912 in Berlin-Dahlem, under the directorship of Fritz Haber, former professor of the Karlsruhe Technische Hochschule and an authority in both physical chemistry and electrochemistry.

Wilhelm Ostwald himself stressed colloid studies as one of the most important new fields to be explored by physical chemists. In fact, some of his students became notable colloid researchers, among them Frederick George Donnan in England, Wilder Dwight Bancroft and James William McBain in America, and Herbert Freundlich in Germany. These colloid chemists were actually physical chemists by profession. Born in Colombo, Ceylon, Donnan received his doctorate from the University of Leipzig in 1896. After teaching at University College in London, he held the first professorship of physical chemistry in England, at Liverpool in 1914. A native of Canada, McBain studied at Leipzig from 1904 to 1905, taught physical chemistry at Bristol, England, the following year, and later joined the faculty of Stanford University in the United States. Bancroft received his Ph.D. from Leipzig in 1892; four years later he established the *Journal of Physical Chemistry* at Cornell University, also in the United States. Freundlich completed his dissertation on the precipitation of colloid solutions by electrolytes in 1903, under Ostwald's direction, and assisted at the Leipzig physico-chemical institute until he was appointed Associate Professor of Physcial Chemistry and Inorganic Chemistry in 1911 at the Braunschweig Technische Hochschule. Between 1904 and 1906, he helped to do editorial work for Ostwald's *Zeitschrift für physikalische Chemie,* which served as an early vehicle for disseminating their colloid studies.[51]

However, no scientist excelled Wolfgang Ostwald, the son of Wilhelm Ostwald, in organizing and popularizing colloid chemistry as a discipline, and also in exerting influence over academic circles between the 1900s and 1920s. Wolfgang's approach to colloids exhibited the characteristics of the physicalist program promoted by his father. The young Ostwald had studied zoology at Leipzig and spent two years doing postdoctoral research under the German-born physiologist Jacques Loeb at the University of California at Berkeley, where Wolfgang's interest quickly shifted to colloidal phenomena. After returning to Leipzig in 1906, he embarked on a career as researcher, organizer, and proselytizer of this infant field. By 1920, already a prolific and lucid writer, he had published three major textbooks on colloid chemistry: *Grundriss der Kolloidchemie* (1909), *Die Welt der vernachlässigten Dimensionen* (1915), and *Kleines Praktikum der Kolloidchemie* (1920), all of which went through more than seven editions and all of which were translated into English.[52] Founder of the Colloid Society (Kolloid Gesellschaft) in 1922 and editor of the two leading German journals in the field, *Zeitschrift für Chemie und Industrie der Kolloide* (established in 1907 and later renamed *Kolloid-Zeitschrift*) and *Kolloidchemische Beihefte* (established in 1909), Wolfgang Ostwald was responsible for the foundation of colloid chemistry as "an *independent* division of the physico-chemical sciences."[53]

Addressing his textbook readers, Wolfgang Ostwald pronounced that "grease chemistry" should now fall within the purview of the colloid chemist:

All those sticky, mucilaginous, resinous, tarry masses which refuse to crystallize, and which are the abomination of the normal organic chemist; those substances which he carefully sets toward the back of his cupboard and marks "not fit for further use," just these are the substances which are the delight of the colloid chemist.[54]

"I see colloids everywhere," he continued:

It is simply a fact that *colloids constitute the most universal and the commonest of all the things we know.* We need only to look at the sky, at the earth, or at ourselves to discover colloids or substances closely allied to them. We begin the day with a colloid practice — that of washing — and we may end it with one in a bedtime drink of colloid tea or coffee.[55]

The world was full of colloids. But why was it that such an important subject had not been studied systematically? As Ostwald explained,

Physics has until recently busied itself chiefly with the properties of matter in mass; chemistry, on the other hand, has dealt chiefly with the smallest particles of matter such as atoms and molecules. Relatively speaking, we know much of the properties of large masses and we talk much, also, of the properties of molecules and atoms. It is because of this that we have been led to regard everything about us either from the standpoint of physical theory or from that of molecular or atomic theory. We have entirely overlooked the fact that between matter in mass and matter in molecular form there exists a realm in which a whole world of remarkable phenomena occur, governed neither by the laws controlling the behavior of matter in mass nor yet those which govern materials possessed of molecular dimensions. . . . We have only recently come to learn that every structure assumes special properties and a special behavior when its particles are so small that they can no longer be recognized microscopically while they are still too large to be called molecules. Only now has the true significance of this region of the colloid dimensions — THE WORLD OF NEGLECTED DIMENSIONS — become manifest to us.[56]

The ultramicroscope enabled a researcher to count the number of particles in a given volume of colloidal solution by direct observation. With this information, it was then possible to calculate the mean weight of a single particle from the weight concentration of the solution; the information of the weight of a particle and the density of the substance in turn allowed a determination of the mean size of a particle. Expanding on Graham's colloid concept, Ostwald now was able to define colloids as dispersed systems consisting of particles ranging in size from 1/1,000,000 to 1/10,000 millimeter.[57] These dispersed particles, he claimed, were not themselves molecules, but rather their physical aggregates. Rejecting the molecular-structural approach to colloids, he insisted that there existed "no definite connection between chemical constitution and a colloid state."[58] A colloid was a physical state of matter into which any substance might be brought; and under appropriate conditions any substance could form a colloidal solution.

In formulating this bold but ultimately influential tenet, Ostwald was inspired not only by Graham's belief in the unity of matter but also by a recent experimental study of the Russian chemist Peter Petrovitsch von Weimarn. According to Weimarn, over 100 inorganic crystalloidal samples that he examined presented, without exception, colloidal properties when dispersed into fine particles by various means. Weimarn therefore suggested, "colloids and crystalloids are by no means two special worlds."[59] This unionist belief constituted the core of Ostwald's colloid concept. The properties of colloids, whether inorganic or organic, Ostwald claimed, were determined not by the peculiar molecular structure but by the physical state outside the molecules, namely, the degree of dispersion.[60] Colloid chemistry, he believed, "thus presents itself not as a study of colloid substances but as a study of the *colloid state.*"[61]

Ostwald considered colloid chemistry alternatively as "a new branch" or the "sister science" of physical chemistry. The two sciences shared many common concepts and methods. But time and again he pointed out that colloid chemistry was not merely an aspect or side branch of "classical physical chemistry."[62] For example, the ordinary laws of classical physical chemistry, including Raoult's laws of freezing point depression and boiling point elevation and van't Hoff's laws of osmotic pressure, were not directly applicable to colloidal phenomena. For a colloidal solution was not a "true" chemical solution (or a molecularly dispersed system), but a suspension of particles (or a colloidally dispersed system). In other words, a solution of a protein, rubber, or any other colloid in a solvent was regarded not as a single uniform phase but as two phases.

Hence the alleged molecular weights for colloids, measured on the basis of classical theory of molecularly dispersed systems, would not show the real values for their true molecules. Important discrepancies existed between the reported values obtained by different methods, say between the measurement of the osmotic pressure and that of the boiling point of protein solutions. Ostwald claimed that even these facts were sufficient to show that a "careless" application of solution laws should not be made to colloid systems, "for the necessary quantitative relationships are lacking."[63] Thus the chemistry of colloids, he emphasized, had good reasons for forming a special intellectual framework as an autonomous discipline.

There certainly existed variant trends in colloid research. For example, Freundlich, Wilhelm Ostwald's brilliant pupil and Wolfgang's rival, approached colloids more as a career physical chemist. He was regarded as one of the foremost investigators of adsorption phenomena with mathematical treatments. During World War I, when the Kaiser Wilhelm Institute for Physical Chemistry and Electrochemistry in Berlin-Dahlem became the major site for research on chemical warfare, he was invited to the Institute at Haber's request in order to work on agents for gas mask filters. There he was able to demonstrate his ability in colloid chemistry by accomplishing this specialized task. After the war, he held the Kaiser Wilhelm Society's first full-time position in colloid chemistry as Chief of the Division of Colloid Chemistry and Applied Physical Chemistry at Haber's institute. Freundlich pioneered the chemistry of colloid surfaces, or what he called "capillary chemistry" (*Kapillarchemie*). He claimed that capillary chemistry represented "the physico-chemical foundations of colloid chemistry."[64] Since a large surface existed between a colloidal particle and its solvent, he asserted, properties of the dispersed system should be governed by the physico-chemical nature of the surface, with which colloid chemists ought to be primarily concerned. Thus the stability of colloid dispersions and other properties could best be explained in terms of energies, tension, and electrical properties of the surface.

Apart from diverse emphases and approaches, however, most colloid chemists, including Freundlich and Zsigmondy, agreed on the Ostwaldian verdict that described colloids as a physically dispersed state of particles which were too small to be seen and too large to be called molecules.[65] With their emphasis on the physical state of matter as well as their unionist view of matter, colloidalists sought for general principles common to both organic and inorganic colloids, rather than stressing the chasm between the two worlds. As a result, phenomena of inorganic colloids were often taken without hesitation to be identical to those of organic colloids, and vice versa. Investigative efforts were focused on classification and nomenclature of dispersed systems, preparation of colloidal solutions, and morphology and mechanical properties of colloid systems. Colloidalists accumulated a mass of valuable data on diffusion, dialysis, filtration, precipitation, electrophoresis, surface energies, and viscosity.

Martin H. Fischer, a loyal American follower of Ostwald, asserted in his preface to the English translation of Ostwald's *Grundriss der Kolloidchemie:*

> The day is past when the importance of colloid-chemistry to the worker in the abstract or applied branches of science needs emphasis. The endeavor of the "pure" chemist to reduce all substances to crystalloid form and from the knowledge of their behavior to resynthesize the phenomena of nature has been a good one, but the limitations of such a point of view have grown daily more apparent. It happens that nature has chosen the colloid form in which to show her face. Crystalloid behavior is the exception, colloid behavior the rule, in the cosmos. Whether we deal with the regions above the earth, as the color of sky, the formation of fogs, the precipitation of rain and snow, or with the earth itself in its muddied streams, its minerals and its soils, or with the molten materials that lie under the earth, the problems of colloid-chemistry are more to the fore than have ever been the crystalloid ones.[66]

By the mid-1920s, "the world of neglected dimensions" proved no longer neglected. Two Nobel Prizes in chemistry went to colloid researchers, one to Zsigmondy in 1925 for his ultramicroscopic study of colloids, the other to the Swedish chemist Theodor (or The) Svedberg in 1926 for his work on disperse systems. Jean Perrin's work on the discontinuous structure of matter and on equilibrium of sedimentation, which won the 1926 Nobel Prize for physics, also related to colloidal phenomena. These honors clearly indicated the importance of the chemistry of colloids.

Like many other physical chemists, colloid chemists avoided the molecular-structural approach of organic chemistry. Colloid chemists never denied the existence of molecules but, by and large, they were silent about the molecular dimensions inside the colloid particle. The task of pursuing this fundamental problem was taken up by organic chemists who were inspired by the colloidalist view. The aggregate theory was the genuine product of their endeavors.

Formation of the Aggregate Theory

The aggregate theory of polymers arose in the first decades of the twentieth century and soon dominated the study of the nature and structure of organic colloidal substances.[67] According to this theory, colloidal substances, such as cellulose, rubber, starch, proteins, resins, and synthetic polymers, were the physical aggregates of relatively small molecules held together by certain intermolecular forces. The view of colloids as a physical state of matter, as put forth by Ostwald and other colloid chemists, provided organic chemists with a basic framework for the formation of this theory. In addition, the newly emerging concepts of "secondary valence" and "partial valence" helped them elaborate these views.

In the 1890s some chemists reconsidered the nature of Kekulé's valence force (the causes of which had generally been of no concern to classical organic chemists), as they examined structures of inorganic compounds and the nature of some unusual properties of various organic compounds. In 1891, attacking Kekulé's concept of rigidly directed valences, the inorganic chemist Alfred Werner at the University of Zürich described chemical affinity as an attractive force from the center of an atom, acting equally in all directions.[68] In 1902 he introduced the concept of "secondary valence" (*Nebenvalenz*), as distinguished from Kekulé's valence or what Werner called "primary valence" (*Hauptvalenz*). The secondary valence, he proposed, was the residual affinity left in the atom after the formation of the primary-valence bondings in the molecule. In his view, such residual forces were strong enough to hold several molecules together to form "molecular compounds" (*Molekularverbindungen*).[69]

A similar concept can be seen in the notion of "partial valence" (*Partialvalenz*) which Friedrich Karl Johannes Thiele, Baeyer's associate at Munich, introduced in 1899 to explain the unusual reactivity of aliphatic compounds. He stated:

I now assume that in substances to which a double bond is assigned, two affinities of each of the participating atoms are in fact used for bonding the atoms, but that — considering the capacity for addition of double bonds — the power of the affinities is not fully used, and on each atom there still exists a residual affinity, or *partial valence,* an assumption which can be based also on thermal grounds.

In formulas this can be expressed as

$$C=C \quad C=O \quad C=N \quad N=N \quad \text{etc.,}$$

where the marks . . . indicate partial valences. In the partial valences I see the cause of the [double bond's] capacity for addition.[70]

What the theories of Werner and Thiele suggested to their contemporaries was that there were secondary forces other than Kekulé's valence forces, and that these affinities could act as intermolecular forces. Thiele, in particular, attributed the origins of the secondary forces to the double bonds in the molecule. The concept of secondary or partial valence was soon adopted by exponents of the aggregate theory to explain the association of molecules in the colloidal particle.[71]

Between 1900 and the 1920s the aggregate structure was proposed by Carl Dietrich Harries and Rudolf Pummerer for rubber, by Kurt Hess and Paul Karrer for cellulose, by Hans Pringsheim and Max Bergmann and Karrer for starch, and by Emil Abderhalden and Bergmann for proteins.[72] Of these organic chemists, Harries, Hess, Pringsheim, Bergmann, and Abderhalden, not incidentally, were students of Emil Fischer, who had already hinted at a limit in size of organic molecules.[73]

An examination of Harries's view of rubber structure illustrates the fundamentals of the aggregate theory of polymeric compounds. Since the previous century, natural rubber, an elastic solid obtained from a milklike fluid (latex) of certain tropical trees (*Hevea brasiliensis*), had aroused the scientific curiosity of chemists due to its unique properties. A number of investigators, including John Dalton, Michael Faraday, Justus von Liebig, Charles Greville Williams, F. Gustave Bouchardat, and William A. Tilden, had worked on the chemical analysis of this organic substance, often with the aid of destructive distillation, through which rubber was broken down into its fractional parts.[74] By Harries's day, it had been found that rubber was made up of only two elements, carbon and hydrogen, the proportions of which were respectively five to eight (C_5H_8). Williams had coined the name "isoprene" for this unit.[75]

As former assistant to Emil Fischer, head of the First Chemical Institute at Berlin and later professor at the University of Kiel, Harries had led a distinguished career as an organic chemist, especially in the field of rubber.[76] Between 1913 and 1917 he was nominated for the Nobel Prize four times, and in 1920 was elected president of the German Chemical Society (Deutsche Chemische Gesellschaft).[77] The impetus for his study of rubber occurred when he observed that rubber stoppers and rubber tubes were damaged by the vapor of crude nitric acid. His examination of the action of nitric acid on rubber resulted in no new insights into the rubber structure.[78] But he soon developed the so-called "Harries ozonide reaction," a method of adding ozone to unsaturated compounds, such as rubber, to decompose them and detect their structures. By analyzing the ozonized product of rubber, he concluded that the "isoprene" unit (C_5H_8), possibly a constituent of the rubber molecule, could be expressed by the following structural formula:

$$
\begin{array}{c}
CH3 \\
| \\
-CH_2-C=CH-CH_2-
\end{array}
$$

The apparent total absence of end groups in his chemical analysis seemed to preclude the idea of any linear-chain structure and instead, would indicate a ring structure of the rubber molecule. Meanwhile, in 1905 he found that the composition of the ozonized product of rubber was $C_{10}H_{16}O_6$, and its molecular weight seemed in accordance with this formula. This led him to consider the ozonide's structural formula as:

$$
\begin{array}{c}
CH_3 \\
| \\
O-C-\ CH_2-CH_2-CH-O \\
O \diagup\quad |\qquad\qquad\qquad |\qquad \diagdown O \\
\diagdown\ O-CH-CH_2-CH_2-C-\ O\ \diagup \\
| \\
CH_3
\end{array}
$$

Consequently he proposed the formula of an eight-membered cyclic molecule (dimethyl-cyclooctadiene), consisting of two isoprene units, for natural rubber. Colloid particles in a rubber solution were, in his opinion, the aggregates or "physical molecules" of the cyclic "chemical molecules" held together by Thiele's partial valences. The partial valence forces were, he believed, derived from the carbon-carbon double bonds in the "chemical molecule"[79]:

$$
\begin{array}{cc}
CH_3 & CH_3 \\
| & | \\
\cdots\cdots C-\ CH_2-CH_2-CH\ \cdots\cdots C-\ CH_2-CH_2-CH\ \cdots\cdots \\
\| \qquad\qquad \| & \| \qquad\qquad \| \\
\cdots\cdots CH-CH_2-CH_2-C\ \cdots\cdots\ CH-CH_2-CH_2-C\ \cdots\cdots \\
| & | \\
CH_3 & CH_3
\end{array}
$$

Although later altering the size of his ring formula somewhat, Harries maintained throughout the course of these investigations his initial idea of cyclic structure for rubber and the existence of aggregate forces holding together the ring molecules.[80]

Likewise, a cyclic formula was proposed in the 1920s for cellulose by Kurt Hess, an organic chemist at the Kaiser Wilhelm Institute for Chemistry (Kaiser Wilhelm Institut für Chemie) in Berlin-Dahlem. Cellulose was known to be a material that constituted cell walls of all plants. It was the main constituent of dried woods, jute, flax, hemp, and ramie, for example, while cotton was almost pure cellulose. By the 1910s it had been shown that cellulose was a carbohydrate whose empirical formula was $C_6H_{10}O_5$, a formula that corresponded to the anhydride of glucose

($C_6H_{12}O_6$). According to Hess, cellulose was formed by an aggregation of cyclic molecules consisting of five anhydroglucose units ($C_6H_{10}O_5$), and residual valence forces causing aggregation of these molecules:

$$
\left[
\begin{array}{c}
\quad \underset{\diagup}{} \text{OG} \\
\text{HC} \longrightarrow \\
| \\
\text{HCOG} \\
| \quad\quad \text{O} \\
\text{GOCH} \\
| \\
\text{HC} \longrightarrow \\
| \\
\text{HCOG} \\
| \\
\text{H}_2\text{COG}
\end{array}
\right]_n
$$

where G was a glucosyl group.[81]

The popularity of the aggregate theory affected the usage of terminology among chemists of the time. On the synthetic side, the word "polymerization" was used as a synonym for molecular aggregation. For example, Georg Schroeter, professor of chemistry at the Tierärztlichen Hochschule in Berlin, in referring to the polymerization process of ketenes, stated in 1916:

> The concept of molecular aggregates cannot be abandoned, which means that single molecules do not lose their autonomy in a complex. Molecules emit lines of forces as a result of all chemically active forces in their atomic groups. These forces of molecular valences have an independence from the atomic valences. Molecular valences enable single molecules to form a polymer molecule, i.e., a polymolecule.[82]

Molecular aggregates were also called "micelles," a term that had originally been used in the previous century by the Swiss botanist Carl Wilhelm von Nägeli for the crystalline molecular building-blocks of starch.[83] The term was first applied to colloids in the 1900s by Jacques Duclaux and others,[84] but its meaning in the sense of molecular aggregates spread especially after the appearance of impressive studies of soaps in the 1910s and 1920s by the colloid chemist James W. McBain then at the University of Bristol. It was known that soaps were composed of relatively small organic molecules but exhibited typical colloidal characteristics in water.[85] In 1920 McBain demonstrated experimentally that soaps could be dispersed in water to form ionized molecular clusters, or "micelles," which were responsible for the colloidal character of soap solution.[86] The solution of polymers, such as rubber or cellulose, showed characteristics so akin to the soap solution that chemists often argued for the case of an aggregate structure of polymers by analogy to the soap micelle.

The term "molecular weight" referred to the weight of a micelle or a colloidal particle. Thus the apparent high molecular weights of polymers were not generally taken literally as the weights of the real chemical molecules. Max Bergmann, the last chief assistant to Emil Fischer in Berlin, called colloidal substances "pseudo-high molecular substances" (*pseudo-hochmolekulare Stoffe*). Classical structural theory was ill-suited for the study of pseudo-high molecular substances, Bergmann insisted, because the structural theory was based on Avogadro's molecule in gaseous phase. Hence the structural formula would provide little information about variations the molecule underwent in solidification, liquefaction, and solution processes.[87] The cause of colloidal properties, he said, lay largely in the magnitude of the aggregating forces, a factor which classical structural theory could not account for. He concluded:

Therefore, what is especially needed at the present time for chemistry of pseudo-high molecular substances is the development of a structural and spatial chemistry the object of which lies outside the molecule, outside the individual group — a structural chemistry, a spatial chemistry of aggregating forces and of aggregates.[88]

In this respect, the organic chemist Bergmann departed from the traditional structural approach on which Fischer had based his study.

In the 1920s, the aggregate theory was further validated by a physical method, X-ray crystallography, when X-ray diffraction was employed to examine the structure of polymers. Soon after Wilhelm K. Röntgen's discovery of the X-ray in 1895, its study became a popular subject for physicists. In 1912, for example, applying X-rays to crystals, Max von Laue showed that they were diffracted from crystalline atoms, forming a pattern of spots on the photographic plate. Two British physicists, William Henry Bragg and his son William Lawrence, determined how atoms were arranged in crystals, a study which, along with Jean Perrin's work on Brownian motion, demonstrated the reality of the molecule. In 1913 the Japanese physicists Shoji Nishikawa and Suminosuke Ono studied asbestos and silk to obtain fiber diffraction photographs, and were thus able to explain their structures in terms of crystalline orientation. Although this was the first application of X-rays to polymers, the study remained largely unknown to Western scholars.[89] Shortly afterward, Paul Scherrer, using a more powerful, evacuated X-ray tube, independently obtained almost the same result from his study on cellulose.

The intensive use of this type of X-ray research continued at the Kaiser Wilhelm Institut for Fiber Chemistry (Kaiser Wilhelm Institut für Faserstoffchemie, founded in 1920) in Berlin-Dahlem, where Reginald Oliver

Herzog directed a number of capable physicists and physical chemists, including Michael Polanyi, Karl Weissenberg, Erich Schmid, Rudolf Brill, and Herman Francis Mark.[90] It was an exciting time when a new tool shed new light on the riddle of the common structure of various polymers. For example, silk and part of cellulose were known to exhibit a crystalline form to which X-ray analysis could be applied. Then in 1925 Johan Rudolph Katz, a Dutch scientist who had worked with the group in Berlin-Dahlem from 1923 to 1924, applied X-rays to stretched rubber with a surprising result: rubber, when stretched, exhibited a crystalline form with a fiber diagram very similar to those of silk and cellulose.[91] This was a crystallographic demonstration that rubber, silk, and cellulose had a common structure. Why stretching brought about crystallization in rubber remained an open question, however. Meanwhile, X-ray experts observed that unit cells — the recurring atomic groups in the crystalline lattice — of polymers were as small in size as ordinary molecules. During this period, most crystallographers assumed that the molecule could not be larger than the unit cell. From this assumption, some scientists, including Herzog, concluded that the molecular size of the polymer must likewise be small.[92] To proponents of the aggregate theory, this conclusion appeared as clear-cut empirical evidence to support their view of the aggregate structure of polymeric substances.

As we have seen, the aggregate theory was largely formulated within a physicalist program that stressed the physical state of matter and interacting physical forces. The aggregate theory forced changes in the classical concept of polymers and polymerization as well as in the organic-structural approach which had enjoyed its high point in the previous century. Supported by chemists both conceptually and experimentally, the aggregate theory gained overwhelming support in established chemical circles toward the mid-1920s. It served as the first unified theory of structures for diverse polymeric substances until the emergence of the macromolecular theory.

The Industrial Nexus

Despite its unpopularity in academic circles, "grease chemistry" began to win the attention of chemists in the early decades of the twentieth century. Many who worked in the field during this time — including Fischer, Abderhalden, Bergmann, Wolfgang Ostwald, Freundlich, Donnan, Svedberg, and Staudinger — showed an avid interest in biological aspects of colloid polymers. Indeed, such polymers as rubber, cellulose, starch, and proteins are "true organic substances." At the same time, there can be little doubt that a powerful stimulus to the study of polymers came from industry. In fact, the incipient polymer industries had sprung up just before the science of polymers began to take shape. By the early 1910s the production of synthetic plastics such as Bakelite was already under way. Cellulose and its derivatives (such as nitrocellulose, "cellulose nitrate" in modern terminology, and cellulose acetate) were being converted into such useful products as Celluloid, rayon fibers, lacquers, films, cellophane, and explosives. The rubber industry was also growing rapidly, due largely to the rising demand for automobile tires.

The early advances of these new industries relied mostly on experience or on the method of trial and error, rather than on scientific understanding. In short, practice outran theory. But industrial practice in turn stimulated scientific investigation into polymers.

Colloid chemists were eager to boast of the industrial importance of their science. Wolfgang Ostwald, for instance, devoted an entire chapter of his 1917 textbook to the technical applications of colloid chemistry, from which this excerpt is taken:

Colloid chemistry as a systematically studied science is still very young. It cannot be therefore expected that any *conscious* application of colloid chemistry has as yet been made in anything like the degree possible or probable. Many technical experts do not as yet even know that in their every-day practices they are working in colloids and that they should, in consequence, employ the fruits of scientific colloid chemistry in their various endeavors. This fact is often brought home to the colloid chemist who enters into discussion with practical men — something which, by the way, every scientist should do as often as possible.[93]

He provided an extensive list of the "colloid industries," including manufactures of leather, cellulose, rubber, food, photographic materials, and Bakelite. These industries, he insisted, must demand practical applications of the science of colloid chemistry.[94]

Colloid chemists competed with organic chemists for industrial problems of polymers. In 1918 the War Leather Company (Kriegsleder-AG) decided to fund the creation of the Kaiser Wilhelm Institute for Leather Research (Kaiser Wilhelm Institut für Lederforshung) in order to examine scientifically problems of the leather industry. The Kaiser Wilhelm Society was hard put to find a scientist who could direct this special task. Ostwald argued that leather should be considered a colloid chemical object, and lobbied to be appointed director of the Institute. Leather was, as he said, "an animal gel, closely related in its general properties to that prototype of the colloids, gelatin. Leather is tanned with substances of which the majority are colloids, and the whole process of tanning is punctuated with the colloidal phenomena of hydration, dehydration and adsorption."[95] However, the Institute's sponsoring organ, the Central Association of German Leather Industry (Zentralverein der Deutschen Lederindustrie), unanimously recommended for the directorship Edmund Georg Stiasny, a leather expert working on artificial tannins at the Darmstadt Technische Hochschule. But he turned down the offer and recommended as a replacement either Max Bergmann (then at Herzog's Institute for Fiber Chemistry), or Karl Freudenberg. Both Emil Fischer's students, the two were organic chemists. The Kaiser Wilhelm Society, too, came to the conclusion that the basic study of leather — or animal skin — at the new institute should be directed by an expert in biochemistry rather than colloid chemistry. When the institute opened in Dresden in 1922 the directorship went to Bergmann. His mentor, Fischer, had worked on the constitution of natural tannins, and Bergmann now found a suitable opportunity to expand Fischer's study of tannins as well as proteins.[96]

Bergmann's major task was to lay the scientific groundwork of leather research and its manufacture. Within a decade, he and co-workers at the Institute had published over 180 papers. Although most of these papers were of an academic nature, the value of Bergmann's biochemical research was well recognized. Besides academic research, he had to spare much time to consult with leather industrialists and also to serve as president of the International Society of Leather Chemists in 1928–1933. Just as the creation of the Institute for Fiber Chemistry had pointed physical chemists, such as Herzog and Mark, to the study of fibers, the birth of the Institute for Leather Research thus involved Bergmann and other organic chemists in industry-related scientific research.[97]

By the 1910s a good many academic chemists were concerned with the problems of the rubber industry. In the previous century, this industry had started with the advantage of superb craftsmanship, but also existed on shaky scientific ground. Among the pioneers of the early rubber enterprise were the two Britons Thomas Hancock and Charles Macintosh.[98] Without much chemical knowledge, they were able to establish mechanical processes for making waterproof goods, such as garters and "mackintosh" raincoats, out of raw rubber. However, these early products proved impractical, for they became sticky in hot weather and crisp in cold. In an attempt to overcome this deficiency, Charles Goodyear, an uneducated American inventor, accidentally discovered about 1839 a process of heating raw rubber with sulfur to form stable, hard, and elastic rubber. Named "vulcanization" (after the Roman god of fire, Vulcan), this method of "curing" rubber made possible its vast commercial utilization. The discovery was, as Goodyear later stressed, "not the result of scientific chemical investigations," but that of chance observation with a mind prepared by years of labor.[99] Hancock developed the industrial process of vulcanization in the early 1840s. Vulcanized rubber found a host of practical uses, such as tubes, hoses, cushions, and railway buffers, which dramatically raised demand. Annual world production of rubber in 1860, for instance, was 240 times greater than that of 1830.[100]

In the late nineteenth century, the rubber industry partnered the newborn electrical industry, largely because of the wide use of rubber as an insulator for covering electric wires. In the 1880s journals devoted to the interests of the rubber industry were established in Europe and the United States. Not incidentally, these journals also concerned themselves with the electrical industry, as indicated by their initial titles: for example, *India Rubber and Gutta Percha and Electrical Trade Journal* (later renamed *India-Rubber Journal*), established in 1884 in London, and *India Rubber World and Electrical Trade Review* (later renamed *India Rubber World*), established in 1889 in New York.[101]

Until the beginning of the twentieth century, the development of the rubber industry was dominated by the efforts of practical men. The rubber industry had little appreciation for the role that scientific research might play. As mentioned earlier, chemists had occasionally studied rubber in the previous century, but their work remained isolated from the problems of industry. It was not until the turn of the century that Carl Otto Weber launched scientific studies of vulcanization and other practical problems of rubber. A German-born chemist, Weber served as associate editor of the scientific section of the *India-Rubber Journal*. His *Chemistry of India Rubber*, published in 1902, would remain a standard work for more than two decades.[102] Weber investigated the influence of time and temperature on the vulcanization process. Regarding vulcanized rubber as an addition of sulfur and rubber, he determined the degree of vulcanization by calculating the percentage of fixed sulfur. Colloidalist that he was, Weber considered the process of vulcanization to be not only a chemical but also a colloid phenomenon. He believed that rubber — a colloid — was an aggregate of crystalloid molecules ($C_{10}H_{16}$). Addition of the sulfur atom to the rubber molecule made it a chemical process, but vulcanized rubber was thought to be an aggregate of the sulfur-combined crystalloidal molecules, ranging from $C_{100}H_{160}S$ to $C_{100}H_{160}S_{20}$.[103] In the 1910s the mechanism of vulcanization became the subject of lively debate among scholars when colloid chemists like Wolfgang Ostwald asserted that it was a purely physical phenomenon, adsorption.[104]

The growing concern of chemists with rubber problems reflected a new industrial demand for rubber. The demand for vulcanized rubber quickly accelerated after the 1900s, when pneumatic tires began to be produced for the rising automobile industry. World rubber consumption almost trebled between 1900 (49,000 tons) and 1915 (155,000 tons), and the price of rubber soared, reaching a peak in 1910.[105] By the end of the nineteenth century, the British government had transplanted rubber trees from their original habitat, the Amazon basin, to Southeast Asia, including Ceylon, Malaya, and the East Indies, in anticipation of demand. But it would take several more years before rubber plantations came into large scale production. During this period, synthetic rubber therefore became an intriguing and timely topic among chemists. Hence, by the 1900s the era of rubber chemistry had commenced.

At the 1906 meeting of the chemistry section of the British Association for the Advancement of Science held in York, Wyndham R. Dunstan of the Imperial Institute gave an influential speech entitled, "Some Imperial Aspects of Applied Chemistry," in which he anticipated, "it cannot be doubted that chemical science will sooner or later be able to take a definite step towards the production of rubber by artificial means."[106] Not a few chemists foresaw the coming of the age of artificial rubber to replace natural rubber, just as had happened in the dye industry forty years before. Obviously, one way to discover rational methods of making artificial rubber was to begin with a thoroughgoing investigation of the structure and properties of natural rubber. Important studies in the 1900s and 1910s by organic chemists such as Harries, Samuel Pickles, and Staudinger stemmed largely from their interest in synthetic rubber.

Immediately after the Dunstan lecture in 1906, a project for synthetic rubber research was organized by a German firm, the Farbenfabriken vorm. Friedrich Bayer & Company. The Germans had good reason to begin a major effort to make synthetic rubber, for they had no colonies from which to secure a supply of natural rubber. They did have abundant coal to use as raw material for organic synthesis, as well as an impressive tradition in the dye industry that had successfully replaced natural dyes with synthetics by utilizing these coal resources. Companies like Bayer exerted immense power in this field, largely because of their history of organized industrial research.[107]

Fritz Hofmann was made head of the rubber project. Chief chemist in Bayer's pharmaceutical department, he had initially little knowledge about the subject, but had become enthusiastic after reading the published version of the Dunstan lecture. Subsequently, he proposed to Bayer's top management to engage in synthetic rubber research. Friedrich Bayer and Carl Duisberg decided to risk a maximum of 100,000 marks per year for Hofmann's new project.[108]

Many of Hofmann's co-workers were students of Harries, who had himself been associated with the Bayer Company. After learning of Harries's recent formula for rubber, Hofmann first attempted to prepare rubber from dimethyl-cyclooctadiene, the cyclic "chemical molecule" according to Harries's aggregate theory. But this attempt was unsuccessful.[109] He was quick to abandon the initial approach and start over, this time working directly from isoprene. The fact that Hofmann was a neophyte in rubber chemistry was perhaps helpful in this connection.[110]

The idea of preparing artificial rubber from isoprene was not new. By Hofmann's day, a number of chemists from France, Britain, and Germany claimed they had converted isoprene into rubber, although their claims were often cast in doubt by later investigators. Isoprene could spontaneously convert into a rubberlike material. But from an industrial point of view, this process had several disadvantages. For one thing, the yield was exceedingly poor; also the polymerization proceeded too slowly. The British chemist William A. Tilden, for example, pessimistically reported in 1908, "the process occupies several years."[111] Whereas the monomer, isoprene, could be obtained from natural rubber itself by dry distillation, it seemed difficult, though not impossible, to prepare isoprene from other materials.

Despite these shortcomings, Hofmann's group managed to synthesize a small quantity of isoprene from paracresol, a coal product, by using a cumbersome process. By 1909 they had succeeded in obtaining a rubberlike material simply by heating isoprene at a temperature under 200°C in an autoclave for 10–150 hours. Samples were sent to Harries, who concluded from an ozonide analysis that this synthetic product was composed of the same structural unit as that of natural rubber.[112]

Harries was himself interested in the polymerization of isoprene. In 1910, shortly after Hofmann's accomplishment, Harries announced that he had developed a process for preparing synthetic rubber by heating isoprene in the presence of acetic acid. Although aware of Hofmann's success, Harries announced that he was the first to make synthetic rubber, a claim that brought about considerable tension between the two chemists. Meanwhile, a British chemist, Francis E. Matthews, who worked for Strange and Graham, Ltd., in London, discovered a more effective method of polymerizing isoprene by using sodium as a catalyst, and applied for a British patent in 1910. Three days later Harries, perhaps not knowing about Matthews's work, reported his discovery of the same method to the Bayer Company and advised them to submit a patent application.[113]

Russian chemists were successful in finding cheaper homologues of isoprene as monomers for synthetic rubber. As early as 1900, Ivan Kondakov had shown that dimethylbutadiene could be polymerized to a rubberlike product. A decade later his student Serge V. Lebedev published a report that he had polymerized butadiene to a rubbery material. Ivan Ostromislenski polymerized vinyl bromide in 1911 and also discovered a method for synthesizing butadiene from alcohol.[114]

The research rush for synthetic rubber culminated in a heated controversy over priority in the years 1912 and 1913. The issue necessarily centered on the question of who prepared the first synthetic rubber. The debate also brought nationalistic sentiments to the fore. In 1912 William Henry Perkin, Jr., while cooperating with Matthews's group, criticized the German claims for priority and gave credit to antecedent work by the

British chemists Tilden and Williams and the Frenchman Bouchardat. Perkin felt strongly that "much of the credit of the pioneer work belongs to this country and to France."[115] The following year, however, Harries contended that Perkin had aimed to represent all discoveries in the field of rubber as "purely English accomplishments." He openly accused practically all the workers in the field of synthetic rubber, including Bouchardat, Tilden, Perkin, Kondakov, Lebedev, and Ostromislenski, of making false claims that they had made synthetic rubbers.[116]

After all had been said and done, it became clear that the rubberlike materials prepared from isoprene were not identical to natural rubber. Despite their elasticity, they were markedly inferior to natural rubber. In addition, the preparation of isoprene remained arduous and costly. The use of cheap homologues as starting monomers seemed more promising for industrial purposes. Thus, Hofmann's group at Bayer adopted Kondakov's method and prepared dimethylbutadiene from acetylene. By 1912 they had succeeded in the laboratory production of "methyl rubber" by polymerizing dimethylbutadiene. Both Kaiser Wilhelm II's and Duisberg's automobiles were fitted with tires made from a mixture of natural and methyl rubber. But, again, methyl rubber revealed serious shortcomings. Production time was lengthy, taking several weeks, and the product demonstrated poor qualities (e.g., quick aging).

When World War I broke out in August 1914, the Germans were optimistic about their stock of natural rubber. A rubber commission was established in Berlin with Emil Fischer as chair, who urged Duisburg in 1915 that the Raw Material Division of the War Ministry (Rohstoff-Abteilung, Kriegsministeriam) would soon need synthetic rubber.[117] But the rubber industry persistently argued that the production of synthetic rubber would be unnecessary. Within two years, the situation had changed dramatically because the prolonged war and the Allied blockade had led to a serious depletion of rubber supplies. In 1917 the Bayer Company finally started to produce methyl rubber totaling over 2,000 tons by the end of the war. Despite its poor qualities, two types of methyl rubbers (soft and hard rubber) were used for tires, hoses, storage battery boxes for submarines, insulation of electric cables and wires, and gas masks.[118]

Synthetic rubber research stopped after the war, as natural rubber was cheaper to procure due to overproduction of plantation rubber. Nevertheless, recurring shortages and the resulting fluctuating price of natural rubber served as an impetus to sustain the interest of chemists and industrialists in synthetic rubber in the 1920s.

Another major polymer industry at the turn of the century was "artificial silk," a fiber prepared by chemical manipulations of the natural polymer cellulose. The art of manufacturing such natural fibers as silk and cotton had a history of many centuries. But it was not until the mid-nineteenth century that serious attempts were made to prepare artificial ones. In the 1850s, for example, Georges Audemars in Switzerland prepared nitrocellulose from mulberry twigs, the principal diet of the silkworm, in an attempt to make artificial silk (i.e., without the help of the silkworm). The first commercially successful nitrocellulose textile filament was patented thirty years later by the Frenchman Count Hilaire de Chardonnet. While working on silkworm disease at Louis Pasteur's laboratory, he too used mulberry, but soon found other wood more suitable as the raw material for his fiber. In Chardonnet's process, cellulose was first made soluble through nitration, then the nitrocellulose solution was extruded to form filaments, and finally cellulose was regenerated through denitration (treating with acid sulfide solutions) of the filaments. "Chardonnet silk" created a sensation at the Paris Exhibition of 1889; Chardonnet eventually established a plant at Besançon.[119]

Later, cuprammonium hydroxide solution was also used as a solvent of cellulose to prepare artificial silk. However, both processes were soon supplanted by the viscose process, developed in England by Charles Frederick Cross and Edward John Bevan. The two chemical consultants in London had no incentive to make textile fibers; they were studying cellulose to find a carbon filament needed for the recently invented incandescent light bulb. As it turned out, they discovered that alkali cellulose (soda cellulose) reacted with carbon bisulfide to form a water-soluble, viscous material (cellulose xanthate), which they christened "viscose." When treated with acid, viscose decomposed and regenerated into cellulose, an invention patented in 1892.[120] Soon the method became widely used in a new commercial process for producing artificial silk from wood pulp. By the early 1920s, the viscose process produced about 75 percent of the world output of artificial silk.[121] The world production of artificial silk rose from about two million pounds in 1900, to nearly eighteen million pounds in 1910 and thirty-three million pounds in 1920.[122]

Though marketed as "artificial silk" or "fibersilk," it was clear that these cellulose fibers had markedly different properties from silk fiber. Chemically, silk was composed of proteins — not cellulose — a fact of which the pioneer, Chardonnet, had been unaware. In 1924, after lengthy consultation, the industry created a new generic name for the material, "rayon," using a combination of "ray" (meaning to capture the impression of luster and shininess) and "on" (borrowed from cotton).[123]

In the previous century, significant studies of cellulose had been carried out by agriculturists and botanists such as Anselme Payen (who coined the name "cellulose") and Nägeli. As cellulose became increasingly important for technical reasons during the early decades of the twentieth century, a number of chemists began systematic investigations in this area. Wolfgang Ostwald welcomed study of the subject for its application to colloid chemistry: "the whole industry [of viscose silk manufacture] represents an unbroken succession of colloid-chemical processes," since it involved such colloid phenomena as swelling, precipitation, gel formation, and emulsification.[124] As we have seen, physical chemists initiated an X-ray study of cellulose structure. The work of Emil Fischer and other organic chemists had placed the chemistry of simple sugars such as glucose on a solid foundation. In 1920, Emil Heuser, who held a chair of cellulose chemistry at the Technische Hochschule in Darmstadt, established a new German journal, *Cellulosechemie,* which would become a conduit of studies by such chemists as Pringsheim, Hess, Karrer, Mark, and Kurt H. Meyer.[125] Indeed, cellulose was among the most important substances to be studied by polymer researchers in the 1920s and 1930s.

Rayon was not the only successful use of cellulose. The first commercially viable synthetic plastics were also made from cellulose.[126] Their history goes back even decades before the introduction of cellulose fibers. In the 1860s a number of individuals worked on nitrocellulose to find substitutes for expensive natural products such as ivory and tortoiseshell. In 1861 the British inventor Alexander Parkes patented "Parkesine," a product prepared from a mixture of nitrocellulose and wood naphtha. Seven years later his former coworker Daniel Spill came up with "Xylonite" by mixing nitrocellulose, alcohol, camphor, and castor oil. Together, they set up companies to produce items like combs, which were made of their materials.[127]

"Celluloid," discovered in 1869 by the American printer John Wesley Hyatt, created a distinct industry. Celluloid was a hard, moldable substance made by mixing nitrocellulose and camphor under high pressure. Despite problems with its flammability and the high cost of camphor, Hyatt successfully established the industrial production of celluloid in the 1870s. Celluloid at first found limited use in collars and cuffs, for example, but around the turn of the century its application extended to include photographic films, most notably Thomas Edison's motion picture film.[128]

Parkesine, Xylonite, and Celluloid were all results of empirical searches for useful materials. As one story goes, spurred by the prize offer to discover a substitute for the costly ivory used in billiard balls, Hyatt became interested in the field without any theoretical understanding of resins. After all, he was not a chemist. Leo Hendrik Baekeland's 1914 remarks about Hyatt merit quoting:

Hyatt knew no chemistry, but he knew well observed facts intimately connected with the details of the work he had undertaken. His knowledge of nitrocellulose was obtained piece-meal by his own experimenting. Facts found in books he accepted only after he had verified them.[129]

By contrast, the Belgian-born American Baekeland was a Ph.D. chemist who was to serve as president of the American Chemical Society in 1924. Though practical-minded, he certainly knew in some depth the chemistry involved in his invention, Bakelite, the first successful wholly synthetic resin.

This commercial product was not a compound found in nature, nor was it a modified natural product like Parkesine, Xylonite, and Celluloid. It was a purely artificial resin, created from the reaction between phenol and formaldehyde. In 1911 Baekeland founded the General Bakelite Company, forerunner of the Bakelite Corporation, which would later be purchased by Union Carbide in 1939. Bakelite not only replaced Celluloid for billiard balls and other articles, it also achieved great commercial success as an insulating material and as parts of electrical appliances, in response to growing demand from the electrical industry during the early decades of the twentieth century.[130]

Baekeland had begun his search for new materials (including a substitute for shellac, a material made from insects found in India and Burma) by surveying the existing chemical literature. He was aware of Adolf von Baeyer and his students' study of the acetaldehyde-phenol reaction. He knew that Werner Kleeberg restudied that work, using cheaper commercial formaldehyde instead of acetaldehyde in the presence of acid. As mentioned earlier, the result was a non-crystallizable, intractable, greasy mass that dismayed the traditional organic chemist Kleeberg. Aiming at commercially feasible resins, Baekeland repeated experiments by Kleeberg and others who had worked on similar problems. After five years of research, he was able to establish techniques for controlling the reaction by developing the "Bakelizer" autoclave to make Bakelite resins, the first patent for which he applied for in 1907.[131]

Unlike synthetic dyes, Bakelite was hardly the result of a well-designed manipulation of molecules. It was an outcome of Baekeland's bulldog-like empirical study of chemical reactions, rather than a result of the application of scientific theories. To be sure, he was aware that phenol reacted with formaldehyde by a condensation reaction. His rationale — much of which probably came after his invention — loosely reflected the aggregate theory. In 1913 he reasoned that

after the so-called condensation has taken place, polymerization sets in, with the result that the molecules of condensation product form by aggregation or regrouping, so-called polymerized molecules of much higher molecular weight; and the chemical and physical dynamics of these enlarged polymerized molecules are proportionately lessened.[132]

This would explain the distinctive inertness and resistivity of the Bakelite resins. "Until we have anything better," he proposed a ring molecular formula for his product that appeared as follows[133]:

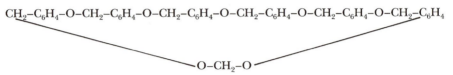

$$CH_2-C_6H_4-O-CH_2-C_6H_4-O-CH_2-C_6H_4-O-CH_2-C_6H_4-O-CH_2-C_6H_4-O-CH_2-C_6H_4$$
$$O-CH_2-O$$

He was, however, candid about his uneasiness at such theoretical speculations. Likening the chemistry of resins to the "zones where readable charts, governors and compass are unavailable," he admitted:

Unfortunately, the study of the chemical reactions involved in these processes is certainly not an easy one. Any organic chemist knows that bodies of a resinous or colloidal nature, which neither crystallize, nor melt, nor distill, nor dissolve, do not belong to the class of substances easy to purify nor to be studied with some hope of accuracy, approaching even in the most remote way what can be accomplished with clearly crystalline or volatile substances.[134]

Experience, he said, "has taught me enough sober sense not to be dominated exclusively by any purely theoretical considerations, when referring to such complicated substances as these condensation products."[135] Furthermore, he believed, "As soon as we try to propose any theory about their molecular structure, we leave the terra firma of established facts, and enter the nebulous realm of gratuitous assumptions and hypotheses."[136]

Bakelite ignited research on phenolic resins, as indicated by subsequent papers and patents brought by other chemists such as the Germans H. Lebach and Friedlich Raschig and the Canadian Lawrence Vincent Redman. Baekeland fought with his rivals for patent priority, a dispute that invoked further detailed study of the chemistry of these resins. But he liked to quote Raschig: "Concerning the chemistry of Bakelite, we still grope entirely in the dark."[137] Baekeland appreciated the power of science but maintained that the scientific rationalization of synthetic plastics was yet to come.

With Bakelite on the market, chemists at the General Electric Company embarked on investigations of another condensation reaction, the glycerol-phthalic anhydride reaction. The result was the discovery of new resins that proved useful for coating electrical appliances, among other applications.[138] Called "Glyptal," these alkyd resins were to catch particular attention from a young Harvard chemist, Wallace H. Carothers. His encounter with the industrial product triggered inspiration for a new theory of polymerization that would lead to his landmark study at the DuPont Company, which will be discussed in Chapter 3.

As we have seen, the polymer industries flourished as key industries, achieving both economic and national significance during the early decades of the twentieth century. They advanced by interacting with other fledgling industries, such as the photographic, electric, and automobile industries. As with other technologies, many early inventor-entrepreneurs had developed their craft and attempted to solve technical problems without depending much on existing academic studies. Whether personally or through institutional channels, industrial practices spurred academic chemists into systematic investigations of polymers, a subject formerly regarded as intractable. Their studies further stimulated other scientists. Industrial management also began to concern itself with research on polymers. The growth of organized industrial research encouraged chemists in the study of rubber, cellulose, plastics, and other polymers. Concurrently, various scholarly organizations and institutions associated with the polymer industries had formed by the 1920s, including Germany's two Kaiser Wilhelm institutes, the Colloid Society, the German Rubber Society (Deutsche Kautschuk-Gesellschaft, founded 1926), Britain's Institution of the Rubber Industry (1921), and four divisions of the American Chemical Society (Rubber Chemistry, 1919; Cellulose Chemistry, 1922; Colloid Chemistry, 1926; and Paint and Varnish, 1927).[139] These organizations, one way or another, served to promote exchange of knowledge and information between academia and industry. The science-based polymer industry was yet to be born. But, whether in the form of colloid theories, the aggregate theory, or the macromolecular theory, these scientific explanations of polymers were all developed in the context of a remarkable upsurge of the polymer industry between the 1900s and the 1920s.

Chapter 2
Staudinger and the Macromolecule

It was in the early 20's when the X-ray physicists overstepped the boundaries of their science and invaded the territory of organic chemistry. Today it seems amazing that many organic chemists [too] were gullible enough to believe that such high molecular substances as cellulose were composed of single small molecules.
— Hans Z. Lecher (a former associate of Hermann Staudinger) to Staudinger, 1953.

Despite the large number of organic substances which we already know today, we are only standing at the beginning of the chemistry of true organic compounds and have not reached anywhere near a conclusion.
— Hermann Staudinger, "Die Chemie der hochmolekularen organischen Stoffe im Sinne der Kekuléschen Strukturlehre," 1926.

It is no secret that for a long time many colleagues rejected your views which some of them even regarded as laughable. Perhaps this was understandable. In the world of high polymers almost everything was new and untested. Long-standing, established concepts had to be revised or new ones created. The development of macromolecular science does not present a picture of a peaceful idyll.
— Arne Fredga, Presentation Speech to Staudinger at the Nobel Celebration in Stockholm, 1953.

. . . I was most impressed by his [Staudinger's] tremendous enthusiasm and energy and his powerful personality, which, I imagine, would make him a most formidable opponent in any argument.
— Clement H. Bamford, quoted in V. E. Yarsley, "Hermann Staudinger," 1967.

The macromolecular theory was first proposed during World War I by the German chemist Hermann Staudinger, who at the great Technische Hochschule in Zürich raised the banner of revolt against the current colloidalist interpretation of polymers. Assuming the existence of organic molecules of an almost arbitrarily large size, he advanced the idea that colloidal substances, such as rubber and cellulose, were formed by these very large chain molecules. In his approach to polymers, Staudinger proved himself to be a staunch disciple of the eminent nineteenth-century chemist Kekulé. Trained along traditional lines of organic chemistry, Staudinger belonged to the organic-structural tradition, which interpreted properties of matter in terms of the structure of organic molecules. In his view, colloidal particles were in many cases "macromolecules" that were composed of between 10^3 and 10^9 atoms linked together by the "Kekulé" bonds. Macromolecular compounds were therefore structured, he claimed, according to the same principles as those of classical organic chemistry, Kekulé's structural theory. In this respect, Staudinger was strongly opposed to the predominant physicalist approach to colloids.

However, Staudinger was not a single-minded defender of the structural tradition of his period. By then, as we have already seen, the Fischer school and Kekulé's German successors had intentionally eliminated the classical concept of big molecules from their structural scheme. Although rooted in the organic-structural tradition, Staudinger's macromolecular concept departed from their path in some crucial ways.

Staudinger's theory soon met with vehement opposition from contemporaries on both sides, physicalists and structuralists alike. The stormy controversy over the macromolecule was to continue for the next fifteen years until his theory began to receive significant approval in the mid-1930s. As a result, Staudinger devoted the rest of his scientific career to settling this debate and to establishing his science as a new branch of organic chemistry.

As teacher and proselytizer, Staudinger used all possible means at his disposal — students, associates, lectures, and publications — to spread his views and spur those interested in the field into action. His contemporaries and even some of his own students regarded the methods used by the charismatic professor to push his agenda forth as somewhat dogmatic and high-handed. Yet, as one of Staudinger's students stated, it was just this same determined and strong characteristic in Staudinger that enabled him to challenge accepted views, to survive a decade-long controversy against leading scientists of his time, and eventually to reign over scientific circles.[1] The emergence of the macromolecular theory caused an intellectual upheaval in the German scientific community during the Weimar period.

The Making of a Classical Organic Chemist

Hermann Staudinger was born in Worms on March 23, 1881, into a well-educated Protestant family of Hessian origin. His mother, Auguste Wenck, was the daughter of a physician from Darmstadt; his father, Franz, was a Gymnasium professor in philosophy and also a prolific writer. Hermann was their second son.[2] Franz Staudinger instilled in his younger son a broad interest in culture, history, art, literature, and science. A neo-Kantian and Marxist philosopher associated with the Marburg school of philosophy, Franz had taken pains to unify Kantian moral philosophy and socialism. In his masterpiece, *Wirtschaftliche Grundlagen der Moral*, which was published in 1907, he claimed that the moral goal of socialism was to actualize Immanuel Kant's "Kingdom of Ends," a metaphorical realm in which rational beings would act or be acted upon according to moral law.[3] In the wake of German social democracy at the turn of the century, Franz Staudinger played a key role as a leading thinker in both the Social Democratic Party (SDP) and the cooperative movement. Hermann's younger brother, Hans Wilhelm, followed his father's footsteps. After receiving a doctorate in political economy and sociology from the University of Heidelberg in 1913, Hans served as secretary for the Association of the Southwest German Cooperatives (Südwestdeutscher Verband der Konsumvereine) and became a Social Democratic representative in the national parliament in Berlin during the Weimar period.[4] Unlike Hans, Hermann never pursued a political career but showed an early concern with German politics. Hermann's pacifist outlook, evident in antiwar statements made by him during World War I, was no doubt shaped by Franz's and Hans's ideology.

When Hermann graduated from his father's Gymnasium in Worms, the eighteen-year-old schoolboy decided to study botany, a subject that had captured his interest from an early age. On his own initiative, he had studied the flora of the area and taken great pleasure in microscopic observations. In 1899 he enrolled at the University of Halle in order to specialize in this field with the botanist Georg Albrecht Klebs. Simultaneously, he began studying chemistry in the Halle laboratory of Jacob Volhard (a student of Justus von Liebig and a former associate of Adolf von Baeyer), as a result of advice given to Franz by a botanist that thorough training in chemistry would prepare Hermann for a future career in botany. Ironically, this preliminary study in chemistry turned out to be the making of his professional career, although his love of biology continued throughout his life. He would go to any length to identify and pursue the connections between biology and chemical investigations, as can be seen in his later study of macromolecular chemistry. Sixty years later, when he wrote his scientific memoir, *Arbeitserinnerungen*, at the age of eighty, he noted that he had not yet concluded his "preliminary" studies and that his sixty-year investigations in chemistry were but the prelude to his long cherished study of botany.[5]

After a brief period at Halle, Staudinger completed training in analytical chemistry at the Technische Hochschule in Darmstadt in 1900. He spent the following two semesters in Adolf von Baeyer's Institute at the University of Munich, a mecca for classical organic chemistry. There Staudinger took Baeyer's one-year course "Chemisches Praktikum," as well as other courses taught by his disciples, including "Stereochemie" by Johannes Thiele, "Electrolytische Methoden" by Oskar Piloty (Baeyer's son-in-law), and "Geschichte der Chemie" by Walter Dieckmann.[6]

His Munich experience marked a great step toward the making of Staudinger as a classical organic chemist. It is likely that Dieckmann's lectures also helped arouse Staudinger's keen interest in the history of chemistry, especially of organic chemistry. A great admirer of Baeyer, Staudinger considered him the legitimate successor of Kekulé. As Staudinger later stressed in a lecture on the life of Baeyer, "The encounter of von Baeyer with Kekulé was not only significant for both men, but also of decisive importance for the development of organic chemistry in Germany."[7]

Back in Halle, in 1903, Staudinger completed his doctoral dissertation on the malonic esters of unsaturated compounds, under Daniel Vorländer, the organic chemist who at the time was studying the synthesis and structure of esters.[8] For the next four years, Staudinger served as assistant to Johannes Thiele, while the latter was professor at the University of Straßburg. Between 1901 and 1907, both Vorländer and Thiele inspired Staudinger further in the field of theoretical organic chemistry. As noted in the previous chapter, in 1899 Thiele had proposed his theory of partial valences explaining the unusual nature of unsaturated organic compounds—a theory which inspired a number of chemists to formulate the aggregate theory of polymers. Under Vorländer's direction, Staudinger attempted to test Thiele's theory with respect to addition products of unsaturated malonic esters.[9] The young organic chemist was much concerned with the classical structural theory during this period. As early as 1905, for example, Staudinger thought critically about the development of the benzene theory, beginning with Kekulé's ideas and leading up to the views of such contemporaries as Baeyer, Thiele, and Vorländer. This review, which was recorded in his notebooks, exhibits Staudinger's cautious and critical attitude toward then-current theories in organic chemistry, an attitude which later manifested itself more fully in his study of high-molecular organic substances.[10]

In 1905 Staudinger unexpectedly discovered a new chemical family of highly reactive unsaturated compounds, ketenes. This discovery at once promised a productive research program, and was to form the subject of his *Habilitation*, a qualifying examination to become university lecturer. After its completion in the spring of 1907, he was immediately promoted to the position of Associate Professor of Organic Chemistry at the Technische Hochschule in Karlsruhe. There, at the Chemical Institute of Carl Engler, Staudinger continued his investigation of ketenes by studying the preparation of various ketenes and their reactivity. All told, working alone and together with his doctoral students, he produced 57 papers on this subject over the next two decades.[11] His first book, *Die Ketene*, published in 1912, eventually won a wide readership as the standard textbook on ketene chemistry.[12] During this period, he also carried out research in Karlsruhe on aliphatic diazo compounds and oxalyl chlorides[13] — studies that sufficient to earn him early recognition as an organic chemist of some stature.

In 1912 Staudinger, at the age of thirty-one, was appointed professor at the Eidgenössische Technische Hochschule (ETH) in Zürich, succeeding Richard Willstätter, a future Nobel laureate who had moved to the newly established Kaiser Wilhelm Institute for Chemistry in Berlin-Dahlem. One of the most prestigious technical schools in Europe, the ETH offered outstanding faculty and facilities. Albert Einstein — a 1900 ETH graduate — accepted a professorship of physics there, which coincided with Staudinger's appointment.[14] Other distinguished ETH colleagues included Hans Eduard Fierz-David, William D. Treadwell, Paul Niggli, Paul Scherrer, Eugen Wiegner, Emil Ott, and Carl Schröter. Before leaving Karlsruhe, Staudinger had received warm congratulations from his Karlsruhe colleagues on his appointment to this distinguished professorship. Only Fritz Haber, a close Karlsruhe colleague who had done postdoctoral work at the ETH, prophesied that "the happy days of undisturbed research were now over."[15] Indeed, what awaited Staudinger at the ETH was a heavy teaching load (more than eight hours of lectures a week) as well as numerous administrative chores. This new situation forced him to abandon some of the research problems he brought with him from Karlsruhe.

ETH's chemistry department consisted of two sections: a technological chemistry section that focused on applied and engineering chemistry, and a general chemistry section that covered theoretical and general aspects of the chemical sciences. The dyestuffs chemist Fierz-David headed the former section and Staudinger the latter. The two sections did not always maintain a good relationship, each insisting on its primary importance over the other. Furthermore, the two heads, both with strong personalities, were prone to clash.[16]

In Zürich, Staudinger's lectures attained wide popularity among students. According to Willem Quarles, "Staudinger's students came under

Figure 1. Hermann Staudinger in his laboratory at the Eidgenössische Technische Hochschule, Zürich, 1912. Courtesy Magda Staudinger.

the spell of this tall German [about 6 feet 4 inches] who, with his curt nervous gestures and whispering voice, penetrated into the mysteries of the carbon atom and kindled a love for the strange organic world."[17] Another pupil, V. E. Yarsley, recalled, "Staudinger as I knew him was what is popularly described as a fine figure of a man. Robust, genial and kindly, I looked on him very much as a father figure." Yet also strict, punctual, and intolerant of inefficiency, Staudinger often scared young students. According to Yarsley, "many was the time that the laboratories rapidly cleared of the more junior students who may have been conscious of deficiencies in their work, when they heard that Staudinger was coming along the corridor."[18] As a researcher, he conducted his work with tremendous enthusiasm, paid meticulous attention to detail, and carefully planned his experimental work. At the same time, as one of his contemporaries put it, Staudinger was "not exactly a gentle and compromising protagonist of his work"[19] — a scientific personality that especially manifested itself in the macromolecular debate of the 1920s and 1930s.

Busy with teaching and administration, Staudinger had no time to do experiments himself.[20] But he did have a great number of students working on their theses as well as some outstanding co-workers, including two future Nobel laureates, Leopold Ružička and Tadeus Reichstein.[21] Ruzicka most likely had mixed feelings about his mentor. After earning his doctorate at Karlsruhe in 1910 for work on ketenes, directed by Staudinger, Ruzicka accompanied his teacher to Zürich, where he was assigned work on insecticides. When Ruzicka told Staudinger about his desire to pursue his own interests in the terpenes, he lost his assistantship and "found his research facilities to be severely curtailed," an action that disappointed Ruzicka.[22] Staudinger was an inspiring teacher, but also a harsh taskmaster for the young chemist.

World War I and Staudinger as a Pacifist

In the summer of 1914, two years after Staudinger's move to Zürich, the World War I broke out. Germany's food shortage led Staudinger to work on the synthesis of a pepper substitute and also on the aroma of roasted coffee. As a structural chemist, he took this opportunity to investigate the relationship between their properties, that is, the taste of pepper and coffee aroma, and their molecular structures.

Watching the war escalate from a neutral zone, Staudinger felt a growing concern about the nature of warfare in which science and technology were playing a decisive role for the first time. In October 1914, ninety-three German intellectuals signed the notorious "Manifesto to the Civilized World" to justify necessity for Germany of the Great War. The signers included his mentors and former colleagues, such as Adolf von Baeyer, Carl Engler, Fritz Haber, and Richard Willstätter.[23] The extreme nationalism, the introduction of chemical warfare, and industrial and scientific mobilization — all these anomalies aroused Staudinger's suspicion of and repugnance to this war.

Politically tolerant Switzerland harbored many pacifists and socialists who shared Staudinger's antiwar sentiments. Dorothea Förster (the daughter of an Evangelical Church minister from Halle), to whom Staudinger had been married since 1906, was an active socialist.[24] Together, they joined a pacifist group organized by Leonhard Ragaz, an evangelical socialist and professor of theology at the University of Zürich.[25] There are also indications that Staudinger was in contact with Georg Friedrich Nicolai, the German physiologist and pacifist, most noted for his book, *Die Biologie des Krieges* (1917).[26]

In a pedantic but less ideological tone, Staudinger made public statements implicitly criticizing Germany for continuing this desperate, technology-based war. In July 1917, shortly after the United States entered the war, he published an article, "Technik und Krieg," in *Die*

Friedens-Warte, a Zürich journal promoting the international peace movement, which was founded and edited by Alfred Hermann Fried, the Austrian pacifist and co-winner of the 1911 Nobel Peace Prize. Technology and industry, Staudinger stated in the article, were playing the decisive role in the ongoing total war, even though the Central Powers were inferior to the Allies in industrial resources such as coal and iron. While the "industrial power" of the Allies had been about equal to that of the Central Powers, in Staudinger's estimation, American entry into the war increased the strength of the former to three times that of the latter.[27]

In November, with the hope of encouraging peace negotiations, Staudinger wrote Erich Ludendorff, head of the German military, setting down his opinions. The German chancellor, Georg von Hertling, sent an official representative to Zürich to meet with Staudinger. The reply Staudinger received disregarded his concerns and instead informed him of Germany's plan to gain victory by widespread use of poison gas at the Front. Moreover, the official sounded out Staudinger's willingness to join the German research effort to develop poison gas, an offer he immediately refused.[28]

Chemical warfare had been introduced by the German Army in 1915, when they released chlorine gas at Ypres on the Western Front. Within half a year, French and British troops began to retaliate by projecting their own poison gases. Both the Central Powers and the Allies mobilized a considerable number of chemists to develop chemical weapons and gas protectors. At the head of the German gas research project was the patriotic Fritz Haber, Staudinger's former colleague at Karlsruhe, best known for his epoch-making invention of ammonia synthesis, and now director of the Kaiser Wilhelm Institute for Physical Chemistry and Electrochemistry in Berlin-Dahlem. Reportedly, about 2000 German scientists and assistants were engaged in research and development for chemical warfare. Such reputable chemists as Richard Willstätter, Heinrich Wieland, Reginald O. Herzog, Herbert Freundlich, Kurt H. Meyer, and Rudolf Pummerer were joining the project. The United States too embarked on poison gas research at its Chemical Warfare Service, created in 1918, of which James B. Conant of Harvard University was a leader.[29]

Realizing Germany's intention to continue the war, and fearing expansion of chemical warfare, Staudinger conceived of a broader appeal to the warring powers to stop the use of poison gas. This time, he chose the International Red Cross in Geneva as the best agent to take the initiative. In January 1918 he reported this plan to his friend Herbert Haviland Field, an American zoologist who resided in Switzerland. Field in turn sent the message to Frédéric Ferrière, a Vice President of the International Committee of the Red Cross. Much of the inside story was recorded in the diary of Ferrière's peer, Romain Rolland, the French novelist, who at the time lived in Switzerland and was active in the anti-war movement.[30] According to Rolland, Field had communicated to Ferrière:

I want to inform you of a very important matter which I heard last night from my close friend who is a German chemist [Staudinger]. . . . It concerns a revelation of the intention with which his fellow-countrymen will introduce into their methods of war a means more terrible than anything the world has ever known. The professor in question considers himself a patriotic German, and his only motive to report to me about the matter is to unburden his conscience. He did not himself want to go to the Red Cross directly, but he is prepared to make specific for a neutral representative of the Red Cross what he knows. He estimates that the Red Cross, perhaps as well as the Pope, is the only one that possesses the power to exert some influence. It [Staudinger's report] is about attack by means of poisons on a scale unknown until now. He himself has been consulted, and knows that many of his colleagues [in Germany] are working on this prostitution of science in their laboratories. . . . I think that the precise details, provided today by a professional who was inspired by his sense of duty for human beings, can be useful.[31]

Field added:

I thoroughly guarantee the informant's sincerity. It is possible that he might be an unconscious tool of the intrigue of those [Germans] who apprehended his inner psychology, and who thought, "If we inform him of this plan, it is evident that he will not keep it secret; then we will succeed in terrifying the opposite camp." I wish I could believe this interpretation, and would like to dismiss the other repugnant, monstrous hypothesis from my mind.[32]

Ferrière visited Staudinger to obtain further information, and subsequently the latter's suggestion was accepted. In February 1918 the International Red Cross issued an appeal calling for the renunciation of all chemical warfare. The appeal was wired to all governments of the Central Powers and the Allies, as well as to the Pope. It was also immediately published in Swiss newspapers.[33]

Meanwhile, Staudinger submitted an article on poison gas to Ferrière, and granted permission to use it at his discretion. In his diary, Rolland summarized the article:

[Staudinger] says that he passionately salutes this [Red Cross] appeal, because he clearly knows what monstrous possibilities of destroying human lives modern technology, especially modern chemistry, has created. Thus, he explains that prussic acid was discovered by Scheele in 1782. In 1811, it was shown as a pure form by Gay-Lussac. Today, a century later, it is possible to produce such a large quantity that it would exterminate at least nine billion people. As Prof. Standinger [sic] pursues, the poisons, which were used only in exceptional cases to poison human beings because they were rare and difficult to procure, are today quite widely spread in a much shorter period due to the progress of technology.

The annual [world] production of cyanide was 22,000–24,000 tons in 1914. A little less than half of it was produced in Germany, and the greater part of the rest in America and England. The possibility of employing poisonous arms is therefore obviously the same on both sides. One would think that the human race is blind in front of the danger of threat.[34]

Staudinger did not know which kinds of gas were actually being used on the Front. In practice, gases were frequently changed to outwit opponents. Whatever the kind — each country was capable of quickly producing large quantities of poison gas — the war appeared to Staudinger to reach an unprecedentedly horrifying stage.

The Red Cross appeal produced no positive results, and, as Staudinger predicted, the war shifted in favor of the Allies. Staudinger's German manuscript was translated into French, and published in the *Revue internationale de la Croix-Rouge* in May 1919. The article, entitled "La technique moderne et la guerre," as we have seen in Rolland's summary, provided an historical and chemical synopsis of poison gases and, in general, denounced their war use. A passage read, "In all previous European wars . . . it would have never been possible to employ them as a means of destruction, even if that criminal idea had occurred to the mind of an army commander."[35] However, the war was coming to an end by the time the article finally appeared.

In October 1919 Staudinger sent a reprint of the Red Cross article to Fritz Haber.[36] Perhaps he did so intentionally, knowing that Haber had played a critical role in German research on, and development of, poison gas. In reply, Haber wrote bitterly to Staudinger, "What bothers me is that you do not realize the effect of your French paper at all. . . . You betrayed Germany at the time of her greatest crisis and helplessness." Staudinger's Red Cross article had appeared on May 15, 1919, just a week after the Allies ratified provisions of the Versailles Treaty that made heavy demands for German war reparations. Germany reluctantly signed the Treaty on June 28. Haber believed that Germany's use of poison gas did not violate international law; that during the war poison gas was first used by the French army — not by the Germans — despite Allied claims to the contrary; and that poison gas was, after all, not especially inhumane compared with other weapons. Hence Germany's use of poison gas was not a "criminal" act. To enforce their "Punic peace," he continued, the enemy needed "punishable mean tricks" to justify their vindictiveness. Haber insisted that this French article by a German professor, which had appeared in a timely manner and was written with an authoritative tone, must have helped support the enemy's demands, thereby causing irreparable damage to the Reich. In conclusion, Haber announced an end to his friendship with Staudinger.[37]

Figure 2. Staudinger and his son Hansjürgen in Zürich, ca. 1920. Courtesy Deutsches Museum.

Staudinger found Haber's accusations intolerable. A month later, he wrote Haber that the Red Cross article had been intended to bring public attention to the ominous nature of modern technology. Technology had changed the structure of modern warfare, and its abuse could destroy the world. He added, "I only want to express my astonishment that you, the man who in essential ways contributed to this change in the world situation with your ammonia synthesis and the resulting possibility of endless production of explosives, do not understand the fundamental significance of this change."[38] With this letter, their correspondence ended. As the exchange reveals, Staudinger's wartime remarks had aroused enough animosity to ruin their old friendship. And the anger of this influential German scientist in turn was to affect Staudinger's postwar career, to which we shall return later.[39]

Origins and Development of the Macromolecular Theory

It was at the Eidgenössische Technische Hochschule during World War I that Staudinger first conceived his new theory of polymers. Because of a growing interest in polymeric substances during the last years of his stay in Zürich, Staudinger set aside his investigations of ketenes and other classic organic compounds. By the mid-1920s, he had virtually discontinued work on earlier research topics, and instead devoted his time to the study of polymers. It is not surprising, given the success of his early work in the chemistry of ordinary compounds, that his colleagues were skeptical about his shift to *Schmierenchemie* or the study of "very unpleasant and poorly defined compounds, like rubber and synthetic polymers."[40] However, it was in this "unpleasant" field that Staudinger was able to bring about an upheaval in modern chemistry and eventually to establish his reputation as the founder of a new science.

Staudinger had had frequent occasion to consider phenomena of polymers and polymerization before his Zürich days. Daniel Vorländer, his thesis adviser at Halle, had discussed "polymolecular ester" in his study of a polyester prepared from ethylene glycol and succinic acid. Vorländer assigned it the formula of 16-membered cyclic dimer, the formula later to be disproved by Wallace H. Carothers.[41] While at Straßburg, Staudinger worked for Johannes Thiele, whose partial valence concept was being widely applied to the aggregate structure of polymers. Staudinger's mentor at Karlsruhe's Chemical Institute, Carl Engler, also had some concern with the general problem of polymerization.[42] Engler's assistant A. Kronstein had published papers on polymerization years before Staudinger joined Engler's Institute.[43] Another student of Engler's, Carl Ludwig Lautenschläger, did thesis work in 1913 on the autoxidation and polymerization of unsaturated hydrocarbons, to which Staudinger lent assistance. However affected by the current aggregate theory, this work barely hints at Staudinger's later concept of macromolecules. At any rate, Staudinger's association with Vorländer, Thiele, and Engler probably helped lay the groundwork for his interest in the subject of polymerization.[44]

Staudinger's investigation of ketene compounds provided an incentive to become directly involved in polymerization research. He observed that highly reactive ketenes readily polymerized to four-membered cyclic dimers, that is, derivatives of cyclobutane. Such four-membered ring molecules were easily split by pyrolysis. This observation induced him to examine the pyrolytic decomposition of other ring systems, particularly those of such six-membered rings as terpenes $(C_{10}H_{16})$. In 1911 Staudinger was able to report that, among terpenes, "limonene" and its isomer "dipentene" decomposed with very good yield (up to about 70 percent) into isoprene (C_5H_8), a substance known as the basic unit of natural rubber.[45] Consequently, he attempted to polymerize isoprene into synthetic rubber shortly after this study, probably around 1911. However, he found that, while the synthetic product resembled natural rubber, it differed in a number of properties, such as its colloidal nature. "These differences," he recalled, "in particular, stimulated my research in this field."[46] In summary, his work on ketenes had led Staudinger to the study of isoprene and its polymerization, which in turn brought his attention in the early 1910s to the structure of rubber.

There can be little doubt that Staudinger's shift in interest to rubber synthesis was in some part a result of a boom in the synthetic rubber market at that time. As we have seen in the previous chapter, because of the increased demand for rubber as well as the rising price of raw rubber, synthetic rubber had been by the 1910s a captivating topic among chemists. There was intense rivalry among various groups engaged in the study of synthetic rubber, with bitter controversies often erupting over the issue of priority. As Staudinger later recalled, when he initially became involved into isoprene research the topic of rubber "had gained much importance" as a consequence of the priority debate between Carl D. Harries at the University of Kiel and Fritz Hofmann at the Bayer Company over the synthesis of rubber from isoprene.[47]

Perhaps Staudinger's move from Germany to Switzerland in 1912 provided him with an opportunity to reconsider more impartially the state of the art in his field. A memorandum written around 1912 reveals that he surveyed some thirty-two pieces of historical literature reflecting the state of the art in the chemical sciences, from F. Gustave Bouchardat's 1875 work to F. E. Barrow's 1911 paper. There he credited the British chemist William A. Tilden for the first preparation of isoprene from sources other than rubber (terpenes) and for the conversion of isoprene into a rubbery material well before the Germans Hofmann and Harries.[48] Skeptical of the synthetic rubber craze in Germany, Staudinger lamented that the immense amount of research devoted to synthetic rubber made him feel that organic chemistry had "gone wrong."[49] Going against the tide, Staudinger's own attention was to turn increasingly to the purely scientific problem of molecular structure of polymers.

Staudinger's 1911 paper on isoprene contained a single brief, uncritical reference to Harries's interpretation of the aggregate structure for rubber.[50] Six years later, however, he rejected this interpretation, and for the first time expressed a view favoring the long-chain structure for rubber. In a lecture, "Über Kautschuksynthese," given on October 7, 1917 before the 36th general meeting of the Swiss Society for Chemical Industry (Schweizerische Gesellschaft für chemische Industrie), Staudinger remarked:

> In his important work on the influence of ozone on organic substances, Harries in 1904 and 1905 was able to pave the way for understanding how the isoprene molecules form rubber. I will not go into details here on his initial wrong assumption that rubber is composed of a hydrocarbon of cyclooctadiene series. Afterwards I adopt Pickles's work which states that in the rubber formation the isoprene molecules get together at the ends, namely, the 1,4-position, and that hundreds of isoprene molecules form the large rubber molecules, which can be observed through the ultramicroscope, and which determine the colloidal properties of rubber.[51]

It is clear from this statement that he endorsed views proposed by the English rubber chemist Samuel Shrowder Pickles.

A relatively unknown young researcher, Pickles received his chemical training at Owens College, Manchester, under Harold B. Dixon and William H. Perkin, Jr. Between the years 1905 and 1908, Pickles carried out doctoral research on rubber under the direction of Wyndham R. Dunstan at the Imperial Institute in London.[52] As noted in the last chapter, Dunstan declared at the 1906 annual meeting of the British Association for the Advancement of Science that the laboratory preparation of artificial rubber from isoprene was virtually complete — an announcement that inspired Bayer's management to initiate a research program for the industrial production of synthetic rubber. During this same meeting Dunstan's assistant, Pickles, also presented a paper in which he expressed views at variance with Harries's theory of rubber structure.

The objections raised by Pickles at the 1906 meeting were not widely known until his paper, "The Constitution and Synthesis of Caoutchouc," appeared in 1910 in the *Journal of the Chemical Society.*[53] The paper was primarily targeted at Harries's interpretation of rubber structure as the physical aggregate of dimethylcyclooctadiene molecules ($C_{10}H_{16}$), consisting of two isoprene units (C_5H_8). According to Pickles, Harries's formula was unsatisfactory, "as its arrangement demands vague and unnecessary conceptions of polymerisation." Further, he wrote, "The single molecule [dimethylcyclooctadiene] is regarded [by Harries] as the 'chemical' molecule, and the polymerised aggregate as the 'physical' molecule. The extent of polymerisation . . . is undefined." If rubber were made up of aggregates of small molecules held together by loose physical forces, one could obtain these simple chemical molecules by heating or through chemical treatments of rubber, such as saturation of the double bonds in the molecules by means of bromination. Pickles argued, however, that this was contradictory to chemical experience: for example, bromination of rubber did not result in simple molecules ($C_{10}H_{16}Br_4$), but in a rubber-like material. Pointing out that polymerization of isoprene into rubber was not a physical process, but "strictly chemical," he suggested an alternative view in which isoprene units were united in long chains of the structure:

$$\text{Me} \qquad \text{Me} \qquad \text{Me} \qquad \text{Me}$$
$$\cdots CH_2 \cdot C{:}CH \cdot [CH_2]_2 \cdot C{:}CH \cdot [CH_2]_2 \cdot C{:}CH \cdot [CH_2]_2 \cdot C{:}CH \cdot CH_2 \cdots$$

Pickles's formula, though, was not a linear chain structure. As he noted, "the two ends of the chain should be linked together, which, of course, leads to the formation of a ring. . . . Rubber probably contains at least eight C_5H_8 complexes."[54] In this respect, Pickles shared with Harries the idea of a ring structure for the rubber molecule; their difference of opinion lay in the ring size.

Pickles's paper — short and sketchy, especially in the experimental part — did not have an immediate impact on his contemporaries. From the outset, Harries would not concede the validity of Pickles's views.[55] In response to Pickles's criticism, however, Harries eventually conducted experiments on the addition of hydrogen chloride to rubber, and subsequently on the removal of hydrogen chloride from this addition product by treatment with pyridine. Ozonolysis of the hydrocarbon thus regenerated yielded decomposition products containing 11 to 15 carbon atoms. Despite his initial intention to counterattack Pickles, these results forced Harries to alter his own formula. In a 1914 article Harries conceded that the "chemical" molecule of rubber might be larger than he had initially thought, comprising at least five isoprene units.[56] It is apparent that Harries had to withdraw his early formula of two isoprene units in the face of Pickles's criticism.

Staudinger probably favored Pickles's view after recognizing the consequences of Harries's failure to stand by his initial explanation of rubber structure during the course of the controversy.[57] More important, what appealed most to Staudinger was Pickles's structuralist claim that properties of matter could be explained solely in terms of molecular structure. While he never accepted Pickles's assumption regarding the closed chain formula, Staudinger adopted one of the most important arguments in Pickles's paper, namely that polymerization of isoprene was a purely chemical process and that during it the isoprene molecules united to form very long chains only by the Kekulé valence bond.

Whereas Pickles confined his study to rubber, Staudinger aimed at general principles of polymers and polymerization by means of a coherent research program of organic structural chemistry. Comparing their divergent approaches, G. Stafford Whitby, an early supporter of Staudinger's theory, commented that Pickles's views "hardly possessed the clear cut, unequivocal character of Staudinger's later ideas, and they were not, as Staudinger's ideas were, pursued and supported by a serious experimental program."[58] About to enter the rubber manufacturing industry as a practical chemist, Pickles chose not to pursue further discussion on the issue of rubber structure.[59]

In his 1917 lecture, Staudinger suggested a long chain formula in which hundreds of isoprene units build up the large molecules of natural rubber:

$$CH_3 \qquad\qquad CH_3 \qquad\qquad CH_3$$
$$| \qquad\qquad\qquad | \qquad\qquad\qquad |$$
$$-CH_2-|-CH_2-CH=C-CH_2-|-CH_2-CH=C-CH_2-|-CH_2-CH=C-CH_2-|-CH_2-$$

$$\text{isoprene} \qquad\qquad \text{isoprene}$$

He proposed a similar but slightly different structure for the synthetic rubber prepared from isoprene[60]:

$$CH_3 \qquad\qquad CH_3 \qquad\qquad CH_3$$
$$| \qquad\qquad\qquad | \qquad\qquad\qquad |$$
$$-CH_2-|-CH_2-CH=C-CH_2-|-CH_2-C=CH-CH_2-|-CH_2-CH=C-CH_2-|-CH_2-$$

$$\text{isoprene} \qquad\qquad \text{isoprene}$$

During the formation of synthetic rubber, unlike natural rubber, the isoprene units were therefore not arranged side by side in a regular pattern. He thought this was the reason synthetic rubber exhibited properties differing from those of natural rubber. In this regard, Staudinger's observations about rubber were based wholly on Kekulé's classical structural theory, which stated that the properties of compounds depended on the arrangement of atoms in the molecule. By using the analogy of the large molecular structure of rubber, he proposed a long chain formula for paraformaldehyde (polyoxymethylene), a polymerization product of formaldehyde.[61]

A published version of Staudinger's 1917 lecture appeared in a Swiss journal two years later. But it attracted little attention, no doubt due in part to the war situation.[62] Shortly after the war, he published a refined work, "Über Polymerisation," which appeared in 1920 in the renowned German journal *Berichte der deutschen chemischen Gesellschaft*.[63] In this article — now considered a classic of macromolecular chemistry — he repeated that polymerization products could be explained satisfactorily by normal or Kekulé valences. He thought it unnecessary to assume that molecular compounds were held together by secondary-valence forces:

And therefore I believe that from the available observational material, such assumptions [of molecular compounds which are held together by secondary valences] as to the origin of polymerization products do not have to be made; it is much more likely that different types of polymerization products . . . can be explained satisfactorily by normal valences, and as far as possible the properties of the compounds can continue to be expressed straightforwardly in organic chemistry by normal valence formulas.[64]

Maintaining Berzelius's classical definition of polymerism, Staudinger restated that polymerization was a chemical reaction in which two or more molecules combined into a product of the same composition, but with a higher molecular weight.[65] In this way, Staudinger was able to provide long-chain formulas for such polymerization products as paraformaldehyde and metastyrol (polystyrene) as well as for rubber[66]:

$$\text{paraformaldehyde} \quad \cdots \text{C} \cdot \overset{H_2}{O} \cdot \text{C} \cdot O \cdot \overset{H_2}{C} \cdot O \cdot \overset{H_2}{C} \cdot O \cdots \cdots \text{C} \cdot \overset{H_2}{O} \cdot \text{C} \cdot O \cdot \overset{H_2}{C} \cdot O \cdot \overset{H_2}{C} \cdot O \cdots$$

$$\text{metastyrol} \quad \cdots \overset{C_6H_5}{\text{CH}} \cdot CH_2 \cdot \overset{C_6H_5}{\text{CH}} \cdot CH_2 \cdot \overset{C_6H_5}{\text{CH}} \cdot CH_2 \cdot \overset{C_6H_5}{\text{CH}} \cdot CH_2 \cdot \overset{C_6H_5}{\text{CH}} \cdot CH_2 \cdots$$

The colloidal properties of these substances were attributed entirely to the sizes of their primary valence molecules, which he speculated might contain on the order of a hundred units. In contrast to the ring structure, Staudinger's chain formulas suggested the existence of unsaturated, free atomic groups at the ends of the long chain. He argued (as it happens, incorrectly) that those unsaturated free valences could remain nonreactive owing to the enormous size of the molecule:

Take, for example, hundreds of formaldehyde molecules, and we have twice the molecules in the unpolymerized condition. If we accept that these hundreds of molecules themselves polymerize to form a paraformaldehyde molecule, then we have there only two unsaturated positions; the reactivity is thus many hundred times less. This more or less agrees with the observation that high-molecular compounds are far less reactive than monomolecular end-products, and that they however still to some extent exhibit the reactions of monomolecular substances.[67]

It is striking that Staudinger's 1920 paper did not carry any substantial experimental support for his concept of long chain molecules of polymeric compounds. Unlike earlier exponents of the idea of large molecules (such as Horace T. Brown and John H. Gladstone), he provided no data for the molecular weights of colloids to support his argument. He simply pointed out the technical difficulties in measuring the molecular weight of these substances.[68] It was not until two years later that he proposed what he called "the first evidence for the existence of macromolecules."[69] Perhaps, as his contemporary Herman Mark suspected, Staudinger "had postulated intuitively" the idea of long chain molecules before developing his evidence.[70] In other words, the idea may have occurred to Staudinger first, and its experimental evidence or justification came later. But it is clear that his "intuitive postulate" was conceptually well-grounded in his firm belief in the classical structural approach, as opposed to the prevailing physicalist views of polymers.

Staudinger's "first evidence" in 1922 was produced from an experiment on the hydrogenation of rubber which he had carried out with his student Jakob Fritschi. As mentioned earlier, Carl D. Harries had concluded that colloidal particles in a rubber solution were aggregates of relatively small molecules of a ring structure. The partial valences holding these molecules together, Harries thought, were generated from the unsaturated double bonds of the molecules. Harries's theory predicted that hydrogenation of rubber (like the bromination method Pickles had employed) would yield a normal low molecular weight substance, because saturation would occur and destroy the partial valences between the molecules. In their experiment, however, Staudinger and Fritschi obtained a contradictory result. They hydrogenated rubber at 250°C and 100 atmospheres without using any solvent. The properties of the saturated hydro-rubber, they found, were quite similar to those of natural rubber; the hydro-rubber did not crystallize but produced a colloidal solution like rubber. They could therefore only conclude that colloidal particles of rubber were not aggregates of small molecules held together by partial valences, but rather giant molecules. For the first time, they now applied the term *Makromoleküle* to designate these molecules.[71] In 1924 Staudinger presented the following definition of "macromolecule" in organic-structural terms:

For those colloidal particles in which the molecule is identical with the primary particle and in which the individual atoms of the colloidal molecule are linked together by normal valences, we propose the term *macromolecules*. Such colloidal particles form true colloidal materials, which, in accordance to the bonding power of carbon, occur particularly in organic chemistry and in organic natural substances. Here the colloidal properties are determined by the structure and size of the molecule.[72]

Staudinger's coinage of the term "macromolecule" was a manifestation of his conscious departure from tradition. With this new word, he sought to avoid terminological confusion surrounding the name "polymer" or "high polymer," which had been used by supporters of the aggregate theory to designate the small molecular substance.

Whereas the Ostwaldian colloid concept had reinforced an aspect of Thomas Graham's views, that is, those pertaining to the unity of matter, Staudinger interpreted Graham's concept differently. He stated that Graham had rightly suspected that the nature of colloids was determined by the peculiar chemical structure of the substance. In most cases, Staudinger claimed, colloidal particles were themselves organic macromolecules.[73] Hence he reasoned that the interpretation of colloidal phenomena should be based on the principles of organic chemistry, not those of colloid chemistry. Staudinger thus reinterpreted Ostwald's realm of colloidal dimensions by considering it to be a new field within organic chemistry.

In an attempt to overcome experimental difficulties surrounding the use of natural substances, Staudinger eventually took the bold step of investigating synthetic polymers as model substances. Synthetics, he thought, would serve to explain the structure and behavior of the more complex natural polymers. For example, polyoxymethylene was used as a model for cellulose, and polystyrene for rubber. He prepared various degrees of these synthetic polymers from simple compounds and showed that physical properties, such as the viscosity of their solutions, correlated with the degree of polymerization — a characteristic shared with the homologous series of ordinary paraffin hydrocarbons. The result appeared to strongly support his view of a macromolecular structure for colloidal substances.[74] He also subjected these polymerization products to chemical reactions, such as hydrogenation, methylation, nitration, and saponification, only to find that the degree of polymerization was not affected. In other words, polymers were converted into their derivatives without changing their sizes. The method whereby such conversions were made, later called "polymer analogous reaction," acquired growing importance for Staudinger as a demonstration of macromolecularity, as reflected in the following passage:

The essential proof for the existence of macromolecules was adduced by classical organic chemical methods via polymer analogous reactions; thus, polymers were converted into their derivatives without their degree of polymerization being changed. This is proved further when such polymer analogous reactions are carried out on the high and low molecular parts of a polymer homologous series, as in many cases was done. The argument for the existence of macromolecules is based on the same consideration as that for the existence of small molecules in organic chemistry one century earlier. Thus Wöhler and Liebig in their research on the radicals ethyl and benzoyl in 1832 . . . were able to convert organic compounds into derivatives with other properties, whereas a large part of the molecule — the radical — remained unchanged in size. This discovery was very surprising at that time. In the same manner it can be demonstrated that under suitable conditions macromolecules leave their "macroradicals" unchanged with respect to size when they are converted into their derivatives.[75]

Thus, by using classical techniques of organic chemistry for ordinary compounds, Staudinger attempted to demonstrate that colloidal substances were composed of macromolecules in which atoms were linked together by powerful main valence forces. Undoubtedly Staudinger, the traditional organic chemist, was quite content with his approach, for he believed that the problems of organic compounds could only be solved by means of organic chemistry. Despite his confidence, this theory provoked antipathy among contemporaries, including X-ray crystallographers, colloid chemists, and physical chemists, as well as organic chemists. So began a decade-long controversy over the existence of the macromolecule.

Moving to Freiburg

Increasingly devoted to defending and demonstrating his macromolecular theory, Staudinger had apparently put aside his earlier political concerns. Patriotic German scientists, however, did not forget his "political past."

In July 1925 the University of Freiburg in Breisgau, Germany informed Staudinger that he had been selected as a candidate for a position formerly held by the organic chemist Heinrich Otto Wieland.[76] Wieland had been appointed professor at the University of Munich to take over the position of Richard Willstätter, who had resigned, in part, because of anti-Semitic remarks made by the dean of the Munich faculty.[77] It was Wieland who recommended Staudinger as his successor at Freiburg. As a peer in the Baeyer school, he knew Staudinger well and held his early work on low molecular organic compounds, such as ketenes, in high regard. Certainly, Wieland did not nominate Staudinger because of his recent work on macromolecules. This is evident from "friendly advice" Wieland gave Staudinger a year later:

Dear colleague, drop the idea of large molecules; organic molecules with a molecular weight higher than 5000 do not exist. Purify your products, such as rubber, then they will crystallize and prove to be low molecular compounds![78]

A protégé of Staudinger's, Hans Z. Lecher, in his recollection of a meeting of 1926 also confirms this interpretation:

I remember a meeting of the Freiburg Chemical Society [Chemische Gesellschaft Freiburg] which I regard as a historical one, a meeting in which you [Staudinger] reported your work on the polymers of formaldehyde. This was a great revelation and the chairman, [Walter Otto] Madelung, praised your work as such. It was significant that in this meeting Wieland disagreed with Madelung and also with your conclusions, which showed to what extent the . . . wrong concept has confused even prominent organic chemists.[79]

Influenced by Emil Fischer's tacit dictum on the maximum size of organic molecules and by Carl Harries's aggregate theory (he had assisted Harries from 1901 to 1902), the 1927 Nobel laureate Wieland likewise considered Staudinger's macromolecular theory untenable.

Staudinger's current scientific theories were of less concern, apparently. The appointment committee of the University of Freiburg wrote to Staudinger that they would consider his employment only on the condition that he satisfactorily explain his political past.[80] The Freiburg faculty had no doubt been informed about that past, whether directly or indirectly, by Haber. This situation posed a dilemma for Staudinger, since he now desired to return to the German scientific community. There were also burdensome tensions within the ETH's chemistry department; his general chemistry section and Fierz-David's technological chemistry section had long suffered a poor relationship, often involving personal conflict. Freiburg's offer promised that he would head the entire department of chemistry, which was, as his student Rudolf Signer witnessed, "in accordance with his [Staudinger's] intentions."[81]

After due consideration, Staudinger went to Freiburg in September 1925 to inform Wieland and the committee members that his political statements in the 1910s had created some "misunderstandings." Two days after the interview, he wrote a letter to the head of the appointment committee, stating in part:

I learned that my position during and after the war was misunderstood in many ways, and I gladly take this opportunity to report to you again in writing as follows.

My views on gas war, as it appears, provoked several colleagues. This is understandable to me, since an error, which was not my fault, crept in the French translation of my article for the "Revue internationale de la Croix-Rouge," which caused a misconception. It translated "der unheilvolle Gedanke [that disastrous idea]" into "cette pensée criminelle [that criminal idea]." At that very time renewed studies on this problem guided me to the view that it was upside down to put down the poison gas war as especially inhumane.[82]

The "mistranslation" in his Red Cross article was critical to Staudinger. But now we might question whether it distorted in any essential way his wartime position. We do not know whether he proofread the French translation before publication.[83] We do know, however, that he did not object to the mistranslation when Haber quoted the same passage in his letter to Staudinger. As the previously quoted entry in Rolland's diary indicates, it would not come as a surprise if the pacifist thought in general that the use of poison gas was criminal.

In the letter mentioned above, Staudinger also pointed out that his application for Swiss citizenship might be another source of misunderstanding. He said that he had no political reasons for taking out foreign citizenship; it would be the same situation as if a Swiss professor at a German university wanted to obtain German citizenship. "Since then I have always maintained an upright relation to Germany," he added, "and I would feel very sorry if there was the opinion that my naturalization was motivated by an antipathy against Germany."[84]

Shortly before his visit to Freiburg, Staudinger had completed an article for the *Revue internationale de la Croix-Rouge*, in which he corrected the mistranslation and remarked that gas weapons, as compared with other weapons, might not be any less inhumane. Then, to clear himself, he sent the manuscript to Wieland before publication.[85]

This apology and attitude "completely satisfied" Wieland and the Freiburg faculty.[86] Although one member—we know him only by the initial "H"—continued to object to his recruitment for political reasons, Staudinger, having received Wieland's strong recommendation, won enough votes to beat the other three finalists. It should be noted that among the finalists was Rudolf Pummerer, who championed the aggregate theory of rubber in the 1920s.[87]

Divorcing Dorothea and leaving his four children in Zürich, Staudinger moved to Freiburg in April 1926 to assume the chair of chemistry. An active socialist, Dorothea later remarried a communist worker, a marriage which did not last long.[88] In Germany, Staudinger would meet Magda Woit, the daughter of a Latvian ambassador to Germany and a young, bright plant physiologist. Their common scientific interests—chemistry, biology, and particularly the living cell and its molecular construction—led them to marriage in 1928. Magda, more than anyone else, would inspire Staudinger to explore the biological implications of macromolecules and would further this work with her microscopic study of the morphology of macromolecules. As his collaborator and close partner, Magda would take part in the remainder of Staudinger's scientific life—in his research, debates, academic meetings, and recognition.[89]

The Macromolecular Debate, 1920–1928

The controversy over Staudinger's macromolecular theory had begun years before his move to Freiburg. The line of evidence he marshaled was by no means enough to convince most contemporaries, from whom he encountered vigorous opposition. A month after the appearance of Staudinger's 1922 article on hydrogenation of rubber, Pummerer published a paper on the same topic, but with conflicting conclusions. A former student of Willstätter's and later an associate professor at Munich, Pummerer had become interested in the chemistry of rubber during World War I while working at Haber's institute on methyl rubbers intended to improve gas masks.[90] In the 1922 article, Pummerer claimed that he hydrogenated rubber in a solvent at a relatively low temperature, using platinum black as a catalyst, and obtained a normal low-molecular compound (C_5H_8). He added that this substance was so susceptible to autoxidation that it soon transformed to an isorubber $(C_5H_8)_x$.[91] This was taken to be new evidence that rubber consisted of small molecules.

Harries immediately extolled Pummerer's success. In a paper published in 1923 — the year of his death — Harries reported that he had obtained similar results using the same experimental conditions as Pummerer. Harries now criticized Staudinger's conclusions as "premature." Since Staudinger's drastic experimental conditions, such as high temperature and disuse of solvent, should involve pyrolytic scission of rubber, they appeared inappropriate for the analysis of rubber structure.[92] Even though Staudinger was convinced that Pummerer's experiment was not reproducible, he heeded Harries's warning.[93] Heating was necessary for hydrogenation, but it would inevitably be accompanied by thermal decomposition of natural rubber. It was this realization which caused Staudinger to pay more attention to other ways of demonstrating his theory, such as the use of synthetic models.

From the perspective of X-ray crystallographers, such as Johan R. Katz, Staudinger's arguments contradicted their own interpretations. Katz recalled,

> At the meeting of the Naturforscherversammlung in Innsbruck in September, 1924, I first heard him [Staudinger] defend this theory [of macromolecules], especially for the case of polyoxymethylenes, but also some other cases. Neither I myself nor some others to whom I spoke were convinced by his very interesting exposition. His conception seemed possible, but, many of us thought, not proved. And the whole subject did not yet look attractive to many of us.[94]

And again the following year a meeting of the Zürich Chemical Society (Chemische Gesellschaft Zürich) elicited a similar reaction. Staudinger recalled:

> During a lecture given in 1925 I thought I had given good evidence for the existence of macromolecular structures, using as examples rubber, polystyrene, and polyoxymethelene; then, however, the well-known mineralogist Paul Niggli rose and his only discussion remark was, "Such a thing does not exist!"[95]

The principal objection to Staudinger's view that emerged during the meeting again came from data of X-ray analysis, appearing to suggest that molecules of polymers were small. In contrast, Staudinger's arguments for the macromolecule were based solely on the principles and methods of classical organic chemistry, with few references to X-ray analysis, the new physical technique. Indeed, to supporters of the aggregate theory, such as Niggli, Staudinger's theory flew in the face of results that ostensibly had a better established scientific basis. The issue came to no resolution at the meeting. As a witness reported:

All the great men present: the organic chemist Karrer, the mineralogist Niggli, the colloid chemist Wiegner, the physicist Scherrer, and the X-ray crystallographer (later cellulose chemist) Ott tried in vain to convince Staudinger of the impossibility of his concept because it conflicted with exact scientific data. The stormy meeting ended with Staudinger's shout [quoting Martin Luther's words], "Here I stand, I cannot do otherwise!"[96]

Karrer, Alfred Werner's successor at the University of Zürich and a future Nobel laureate, was president of the Swiss Chemical Society (Schweizerische Chemische Gesellschaft) from 1924 to 1926 and a well-known proponent of the aggregate theory of starch, cellulose, and other polymeric carbohydrates. Staudinger observed that most of his colleagues at the ETH, including Niggli, Wiegner, Scherrer, and Ott, were unwilling to accept his macromolecular theory, especially after Karrer's work on polymeric carbohydrates won the Marcel Benoist Prize in 1923.[97] Even his colleague Hans E. Fierz-David, a luminary in dyestuffs, had much interest in the aggregate concept. In 1925, for instance, he assigned a post-doctoral student to base his study of cellulose acetate on the aggregate theory.[98]

By the mid-1920s, Staudinger found himself isolated in the face of heavy criticism. It is worth remembering, however, that by this time he was already an organic chemist of considerable repute, who had successfully worked in different areas (such as ketenes) and had adhered to conventional lines of research. Because of this reputation, his opinions demanded careful consideration at any scientific gathering. While in Zürich between the 1920 and 1926, he had worked on macromolecules with no fewer than seventeen doctoral students, who, under his direction, compiled a mass of data on polymerization and the structures of rubber, cellulose, polysaccharides, polyoxymethylene, polystyrene, and polyindene (see Appendix Table 1).[99]

Almost immediately after Staudinger's return to the German scientific community, he was confronted by his German critics. In 1926, several months after Staudinger had moved to Freiburg, the Society of German

Natural Scientists and Physicians decided to hold a special symposium on polymers at its annual meeting in Düsseldorf; it eventually would go down in history for its famous debate. The symposium was chaired by Richard Willstätter, but, behind the scenes, Fritz Haber helped arrange the meeting. Haber pitted leading advocates of the aggregate theory, such as Max Bergmann and Hans Pringsheim, against Staudinger. Herman Francis Mark of the Kaiser Wilhelm Institute for Fiber Chemistry was also chosen to speak. As Mark reminisced about the event:

> In the summer of 1926, Haber called me into his office. Herzog was present, and after friendly greetings, Haber told me . . . the following: " . . . [The Society of German Natural Scientists and Physicians] has its annual meeting in Duesseldorf in September 1926, and the Chemical Society intends at this opportunity to organize a general discussion about the constitution of the so-called *high polymeric* organic substances. We would like you to give a lecture on the general topic of X-ray structure examinations of organic substances and especially to explain whether a small crystallographic elementary cell excludes the presence of a very large molecule."
>
> Haber was, as usual, right to the point and had understood the situation very clear[ly]. Most of the arguments — pro and con — were qualitative, even subjective. On the other hand, the small volume of the elementary cells of cellulose and silk had been clearly established; each of them could only contain four units. Now, if it were true that a molecule could not be larger than the elementary cell established by crystallographic investigation there would be a clear-cut and objective argument against the existence of very large molecules.[100]

Although polymers were not Haber's own research specialty, it was no accident that he showed unusual interest in the macromolecular debate, and that he was at this point in favor of the aggregate theory.[101] The faculty of the Kaiser Wilhelm institutes, in which he was a leading figure, included a number of specialists in this field, most of them passionate supporters of the aggregate theory. Among them were Reginald O. Herzog, director of the Institute for Fiber Chemistry in Berlin-Dahlem; Bergmann, director of the Institute for Leather Research in Dresden; Kurt Hess, head of the Organic Chemical Division of the Institute for Chemistry in Berlin-Dahlem; and Herbert Freundlich, head of the Division of Colloid Chemistry and Applied Physical Chemistry in Haber's Institute for Physical Chemistry and Electrochemistry in Berlin-Dahlem. In Berlin-Dahlem, Herzog (who had headed a section on respirators for chemical warfare during World War I) conducted an X-ray crystallographic study of polymers; he supported the view that substances like cellulose and silk fibroin were physical aggregates of molecules smaller than the crystallographic unit cell. Under Herzog, Mark had been working for the past four years on the X-ray analysis of organic substances. Haber informed

Mark that at the symposium "Staudinger will tell his story and the others will tell their story. They are all organic chemists, and I am afraid that this controversy goes beyond classical organic chemistry." Apparently, Haber expected this young physical chemist to make a strong case against Staudinger from the physico-chemical standpoint, especially from his X-ray study. Mark accepted the invitation to be a speaker at the symposium, and, as Haber had promised, he shortly received an official letter of invitation from Willstätter.[102]

At the Düsseldorf symposium, held in late September 1926, three organic chemists — Bergmann, Ernst Waldschmidt-Leitz at Munich, and Pringsheim at Berlin — presented arguments favoring the aggregate structure of proteins, inulin, and other polysaccharides, before an audience of several hundred chemists. They stressed the significance of aggregation forces in determining the properties of these colloidal substances.[103]

Alone on the opposing side, Staudinger, in a paper titled, "Die Chemie der hochmolekularen organischen Stoffe im Sinne der Kekuléschen Strukturlehre," defended his theory by presenting data of his experiments on homologous series of various polymers and the hydrogenation of rubber, polyoxymethylene, and polyindene. Emphasizing the classical interpretation of the molecule, he insisted that these kinds of polymeric compounds were built up according to Kekulé's structural theory. "The possibility of the existence of macromolecules," he asserted, "can be deduced from wholly universal considerations."[104] To his list of macromolecular compounds, Staudinger, for the first time, added proteins as true organic compounds. He declared that organic chemistry was far from exhausted, as many classical chemists, including Baeyer, thought:

The world of organic compounds lies between the simplest carbon compounds such as methane, carbon monoxide, cyanogen, and the largest molecules, the high polymeric carbon compounds. . . . Despite the large number of organic substances which we already know today, we are only standing at the beginning of the chemistry of true organic compounds and have not reached anywhere near a conclusion.[105]

Although the final speaker, Mark, belonged to the aggregate theory school, his position at the meeting turned out to be noncommittal. By referring to the work by Alfred J. Reis and Karl Weissenberg, he suggested that there were some cases in which the molecule could be larger than the crystallographic unit cell.[106] He explained that these cases would happen when primary valences penetrated the entire crystal. He therefore stated that the fact that unit cells were small did not preclude the existence of macromolecules, nor did it prove their existence either.

According to Staudinger, however, Mark's position was by no means neutral; Mark was clearly in league with the other three lecturers by failing to endorse Staudinger's theory.[107]

A heated, free-form discussion followed the formal talks. Staudinger once again found himself deserted by all and fought alone. Shaking his forefinger, the vehement Staudinger approached the other speakers several times to refute opposing arguments; Willstätter, who presided over the symposium, had to take him back to his chair each time.[108]

As had happened in 1925, this scientific gathering also concluded with no agreement on the issue of macromolecules. Only Willstätter was moved to favor Staudinger's view. According to Mark,

Willstätter thanked all lecturers and discussion speakers in friendly words and said: "For me, as an organic chemist, the concept that a molecule can have a molecular weight of 100,000 is somewhat terrifying, but, on the basis of what we have heard today, it seems that I shall have to slowly adjust to this thought."[109]

Rudolf Signer, concurring with Mark's recollection, added that Willstätter, still somewhat dubiously though, said that if Staudinger was right, "he opens to chemistry a field which is much bigger than all organic chemistry."[110] In Staudinger's memory, Willstätter was particularly impressed by Staudinger's argument on the hydrogenation of polystyrene.[111]

Most of the audience, however, felt more personal distaste than Willstätter for the notion of such enormous organic molecules — molecules that were a thousand times larger than those which they had studied in their own laboratories. One of the chemists reportedly said, "We are shocked like zoologists would be if they were told that somewhere in Africa an elephant was found who was 1500 feet long and 300 feet high."[112]

In their defense, Staudinger's Düsseldorf lecture did not contain any decisive argument in regard to the problem posed by X-ray crystallographers, although he believed that he had given sufficient evidence for his theory on the basis of organic chemical methods. The X-ray crystallographer Katz, who also attended the symposium, reported that "Staudinger's conceptions did not seem to many of us really convincing, nor was the decisive value which X-ray spectrography could have for the subject yet understood at this meeting."[113]

During the years 1926 and 1927, activity in the field steadily expanded. Like Mark, some scientists began to regard with suspicion the assumption that the molecules of polymers were smaller than the X-ray elementary cell of crystals. As early as 1926, for example, the American botanist Olenus Lee Sponsler of the University of California at Berkeley, together with Walter Harrington Dore, independently suggested that the results of X-ray diffraction from cellulose were compatible with a long chain formula composed of glucose residues.[114]

Staudinger was slow to address the physico-chemical issue, but finally decided to ask physicists for assistance. In 1927 fellow faculty members Gustav Mie and Josef Hermann Hengstenberg, working in collaboration with Staudinger at Freiburg, published papers on the X-ray analysis of polyoxymethylene. According to Staudinger, polyoxymethylene was made up of long chain molecules of $-CH_2-O-CH_2-$ with end groups H and OH (a part of water). With the action of acetic acid anhydride, these long chains could be broken into smaller pieces of different length. Staudinger's student Rudolf Signer broke up polyoxymethylene and separated the complex mixture by chemical means into twenty-two groups of $-CH_2O-$ chains, or oligomers with identifiable molecular structures. Then Mie and Hengstenberg took X-ray pictures of the oligomers. They had a characteristic interference pattern that varied with the number of $-CH_2O-$ groups in the polymer molecule, which indicated a long-chain structure of polyoxymethylene, along which the unit $-CH_2O-$ repeated itself. The elementary cell contained only four $-CH_2O-$ groups which corresponded to only a small portion of the entire molecule. Hence they concluded that "The molecular size of high molecular compounds cannot be determined by X-ray measurements."[115] To X-ray investigators like Katz, this work seemed to be "the first direct experimental evidence of Staudinger's hypothesis, and a decisive one."[116] Niggli also changed his mind around this time.[117] After 1927, the issue of the small elementary cell no longer played as important a part in the debate as it once had.

Although one of the arguments in favor of the aggregate theory thereby apparently lost ground, the opinions of many organic chemists remained unchanged during the late 1920s. Staudinger's rival Pummerer, now a chemistry professor at the University of Erlangen, continued to compete with him. In 1927, on the basis of the freezing-point depression of camphor solution of rubber, Pummerer and his collaborators obtained a low molecular weight of approximately 600 for rubber, which convinced them that the rubber ring molecule consisted of eight isoprene units $(C_5H_8)_8$.[118] Staudinger recalled the comments of a colleague made sometime around 1928, which referred to the work of Pummerer — once Staudinger's contender for the Freiburg chair:

About two years after my call to the chemistry chair at the University of Freiburg, one of my older colleagues from the Freiburg faculty told me that he had heard at Erlangen that through Pummerer's work the question of the structure of rubber was solved, and that my concept therefore was untenable. I asked him if he and the faculty were now of the opinion that they had made a mistake in offering me the chair![119]

Pummerer was not the only formidable anti-macromolecularist of the late 1920s. The former assistant at the Chemical Institute of the University of Freiburg from 1914 to 1918, Kurt Hess also continued to attack Staudinger's theory. His magnum opus, *Die Chemie der Zellulose und ihrer Begleiter,* published in 1928, became the manifesto of a still active camp of aggregate theorists.[120]

The macromolecular debate was not painless to the fervent disputant Staudinger himself. During the 1920s, when he returned to Freiburg from scientific meetings where debates were conducted, Staudinger often looked in deep meditation, walking up and down in the corridor before his laboratory. It was not rare that his assistant Signer tried to encourage Staudinger. Staudinger's strife created a feeling of solidarity among his loyal students and assistants. One of his doctoral students said that he had never experienced such a touching esprit de corps in other laboratories.[121] Eiji Ochiai, who in 1930 came from Japan to study with Staudinger, recalled: "As I eventually noticed, there was an unusually tense atmosphere in the whole laboratory, and I was even under the impression that people were like a cult of believers who would sacrifice themselves for a religion."[122] "We all clung very much together. Our laboratory was very much like a dedicated brotherhood," related Magda Staudinger.[123]

Mark, Meyer, and the New Micelle Theory

In January 1927, Herman Mark left Berlin-Dahlem to join I. G. Farben (Interessengemeinschaft Farbenindustrie Aktiengesellschaft), a giant conglomerate formed in 1925 by merging such large German chemical firms as BASF, Bayer, Hoechst, AGFA, and Cassella. Sometime earlier, on Haber's suggestion, Mark had met Kurt Heinrich Meyer, a member of I. G. Farben's board of directors and also manager of their Ludwigshafen plant. I. G. Farben then operated five large research laboratories (specializing in various areas, including inorganic chemistry, dyestuffs, pharmaceuticals, intermediates, and fertilizers) in Bitterfeld, Leverkusen, Hoechst, and Ludwigshafen. Its management had just decided that the lab in Ludwigshafen should concentrate on fiber research. Because Mark had ample experience, stemming from his X-ray study of the structure of cellulose and silk, as well as a particular interest in the physico-chemical approach to polymers, Mark was offered the job by Meyer to head the laboratory. Thus also began a long-lasting friendship between the two.

Mark was primarily a physical chemist, but one who had been trained in organic chemistry. Meyer was primarily an organic chemist, but with a strong inclination to apply physical chemistry to organic chemistry. At Farben, their approach to polymers was to reflect their common background, eventually resulting in a novel combination of the two chemistries.

Mark was a native of Vienna, Austria. After serving in the Austrian Army during World War I, he studied organic chemistry at the University of Vienna under Wilhelm Schlenk, who had been a student of the Baeyer school and later succeeded Emil Fischer at Berlin. Mark's 1921 doctoral thesis was in organic chemistry, dealing with the synthesis of a new compound containing a highly reactive free radical; however, his primary interest always lay in physical chemistry. While working for Michael Polanyi at the Kaiser Wilhelm Institute for Fiber Chemistry, he firmly established himself in physical chemistry, in the process gaining a strong background in organic chemistry.[124]

Born in Dorpat, Estonia, Meyer studied chemistry under Arthur Hantzsch — an authority in the subject of stereochemistry of organic nitrogen compounds — at the University of Leipzig. Under the influence of Wilhelm Ostwald, also on the faculty there, research in organic chemistry was strongly influenced by the field of physical chemistry. In this environment, Meyer became attracted to the application of concepts and techniques of physical chemistry to the problems of organic chemistry. After being awarded a doctorate in 1907, Meyer spent a few months in Ernest Rutherford's laboratory at Cambridge University in England. Then, drawn by Adolf von Baeyer's personality and his work in classical structural chemistry, he joined Baeyer's laboratory at the University of Munich and in 1913 became assistant director of its Laboratory of Physical Chemistry. During World War I he was invited by Haber to the Kaiser Wilhelm Institute for Physical Chemistry and Electrochemistry to work on poison gas. There Meyer started a friendship with Haber which was to last for the rest of their lives. After the war he returned to Munich, where he served as director of the Laboratory of Organic Chemistry. In 1920 he accepted an offer by the Badische Anilin- und Sodafabrik (BASF) to direct its central research laboratory in Ludwigshafen, which would later merge with I. G. Farben.[125]

By the mid-1920s, I. G. Farben controlled nearly half the entire capital of the German chemical industry. Accordingly, Farben officials had a strong voice in the German government concerning chemical education. In 1926, Meyer, together with Carl Bosch, a Farben representative, cautioned the Prussian Ministry of Religious and Educational Affairs (Ministerium der geistlich und Unterrichts-Angelegenheiten) that there was a serious problem with young chemists. In Germany, there were too many university-trained chemists, especially in organic chemistry. Yet in the year 1925, for example, approximately five hundred chemists were turned down for positions at I. G. Farben not because of a lack of posts but because their academic training did not meet current industrial research demands. The golden age of synthetic organic chemistry was now over. The huge strides made in chemical research in recent years were, according to Meyer and Bosch, based on the study of reaction processes, a situation calling not only for organic chemistry but also for inorganic and physical chemistry. Therefore Meyer and Bosch proposed that a major emphasis be placed on training young chemists in inorganic and physical chemistry.[126] They also wrote Richard Willstätter, then head of the Association of Laboratory Directors (Verband der Laboratoriumsvorstände) at German universities, that I. G. Farben did not intend to hire chemistry graduates who had neither passed a practicum in physical chemistry nor been tested satisfactorily in physics, since "the organic chemist is today assigned physico-chemical works far more often than before," as is seen in the investigation of reaction processes, and colloid chemical, photochemical, electrochemical and catalytic problems.[127] Mark — whom Meyer recruited to serve in the capacity of research manager — had an impressive combined expertise in organic as well as physical chemistry, which met with official approval at Farben.

To join Mark's research group, I. G. Farben hired a number of able physical chemists and physicists, including G. von Susich from Berlin and Hengstenberg from Freiburg. Within a few months, Mark and his coworkers had set up an X-ray apparatus far better than what he had had in Berlin. With this tool, they were able to begin systematically investigating the crystal structure of cellulose and other fibers. Management encouraged Mark's group to conduct research that would meet academic standards. They published papers and attended conferences; Mark maintained his academic contacts by giving lectures at the Technische Hochschule at Karlsruhe.[128] But he also had to spend considerable time in applied research and in resolving technical problems:

I was manager of a relatively large industrial research and development laboratory — almost 50 scientific collaborators — in which existing manufacturing processes were supervised and new commercially feasible products developed. Professor Meyer and I spent very little time on fundamental research usually only in occasional discussions, over weekends or during a joint visit to Frankfurt. My laboratory was essentially occupied with the examination and testing of viscose and acetate fibers which were produced by the company, and with the synthesis and evaluation of new polymers.[129]

In the late 1920s, Mark and his collaborators successfully developed the catalytic production of styrene from benzene and ethylene, a process which lowered the cost of styrene and made possible I. G. Farben's first commercial manufacture of polystyrene in 1930.

Besides their applied research, Mark and Meyer carried out a theoretical study of natural polymers. Mark had come over to Staudinger's view of long-chain molecules shortly after the Düsseldorf meeting.[130] Now that the issue of the small elementary cell in X-ray crystallography no longer seemed essential, Mark, with Meyer's guidance, conducted sophisticated X-ray analysis to establish the exact arrangement of individual atoms in the long chains. Based on their experiments, Meyer and Mark, between 1928 and 1930, developed a new theory which appeared to compromise between Staudinger's macromolecular theory and the aggregate theory. According to them, colloidal particles in solution were not themselves macromolecules. Rather, they were "micelles" (*Micellen*), a term often used by proponents of the aggregate theory for clusters of molecules. Because Meyer and Mark considered micelles to be aggregates of primary valence chains (*Hauptvalenzketten*), or long-chain molecules held together by "special micellar forces" (*besondere Micellarkrafte*), they claimed that the molecular concept could not be applied to the micelle, that is, the colloidal particle. The weights of colloid particles determined by physico-chemical methods (e.g., osmotic pressure measurements) represented not molecular weights but "micellar weights" (*Micellargewichte*). Micelles were stable in solution due to the considerable cohesive power of the Van der Waals-type micellar forces. Their colloidal properties depended on the micellar structure of the colloidal particles, rather than on the structure of the long-chain molecules. Meyer and Mark estimated the size of micelles by measuring the widths of the X-ray diffraction spots. For example, they suggested that a cellulose micelle was formed by 40–60 primary valence chains, each of them composed of 30–50 glucose units ($C_6H_{10}O_5$). The micellar size of 30–50 glucose units was taken to be the molecular length of cellulose. A micelle of 40–60 chain molecules was like a matchbox, the matchbox being the same length as each match (see Figure 3).[131] Apart from the size of primary valence chains, the basic structure they proposed would prove to be a substantial contribution to modern cellulose chemistry.

Ever the conservative organic chemist, Staudinger suspected that his much loathed physicalist approach was inherent in what he called the "new micelle theory" (*neuer Micellarlehre*) of Meyer and Mark, because the two chemists stressed molecular cohesion and physical molecules in their explanation of colloidal properties of polymers. In both published papers and personal correspondence, Staudinger opposed the work of Meyer and Mark. He believed their estimate of the main chains to be too short. To Staudinger, the 40–60 chains would be, in the matchbox model, connected through primary valences to form a very long but thin chain-

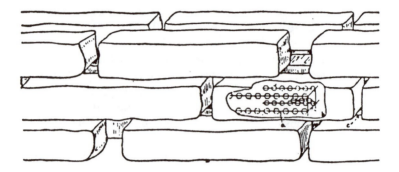

Figure 3. Meyer's sketch of the micellar structure of cellulose. Each box indicates a micelle in which primary valence molecular chains are arranged regularly. From Meyer (1930), p. 13.

molecule. This view led Staudinger to deny the existence of the micelles. The colloidal particle was identical to the macromolecule, which determined colloidal properties of the polymer. Challenging claims based on X-ray crystallography, Staudinger maintained that neither the size of the elementary cell nor that of the crystallite would have any bearing on the length of the polymer molecule.[132]

Although they fundamentally agreed on the long-chain structures of polymers, the difference in methodology led Staudinger to a more serious attack on Meyer and Mark. In retrospect, Mark has explained:

We [Staudinger and Mark] both favored the concept of long-chained molecules. He [Staudinger] did, on the basis of organic chemistry; and I did, on the basis of X-ray diffraction. He only trusted organic chemistry. I said trust both (techniques); we have two methods which do not contradict. My God, they could have contradicted![133]

This statement epitomizes Staudinger's anti-physicalist position that the organic chemist maintained throughout the debate.

Meyer in particular responded strongly to Staudinger's attack, which undoubtedly left considerable bitterness between the two chemists.[134] Time and again they made their standpoints clear, emphasizing their difference in perspectives, rather than seeking to find common ground. On certain occasions, Staudinger and Meyer argued with each other more vigorously than Staudinger had with defenders of the aggregate theory.

One aspect of the debate was that it developed into an argument of priority. Agreeing with the existence of long-chain molecules, Meyer preferred to use the term "primary valence chains" instead of "macromolecules." Extremely sensitive about priority, Staudinger claimed that Meyer employed Staudinger's concept as if it were his own idea. In a letter to Mark dated October 31, 1928, Staudinger wrote:

I cannot agree with him [Meyer] on two points. First, in my opinion, the statements of K. H. Meyer are not new, but coincide with the views that I have held for years and established experimentally; secondly, I do not believe that the introduction of "primary valence chains," instead of macromolecules, serves to clarify the problem.[135]

In response, Mark suggested that the terms "primary valence chains" and "macromolecules" were essentially identical. And those who agreed upon the concept of large molecules should join forces against the school attacking their common beliefs, as Mark proposed:

I believe that we should put forward the viewpoint of this inquiry [the concept of long-chain molecules] and not emphasize the differences in our own conceptions. Otherwise, the high polymer camp could easily fall into error, which is well known merely from politics. Because of slight differences in neighboring opinions, a larger point of view does not get sufficient attention, nor does it find any vigorous expression.[136]

This suggestion, however, was ignored. While Staudinger continued to reproach Meyer for his use of the terms "primary valence chains" and "micelles," Meyer objected that Staudinger presented Meyer's opinions as his own.[137] This situation can be explained by the fact that Staudinger did not exclude the possibility of inter-macromolecular forces (to the order of the magnitude of Van der Waals forces), at the same time that he rejected the possibility of micellar forces. Furthermore, Staudinger held to the idea of crystallite — that there was a crystalline part of polymeric substances which consisted of a bundle of macromolecules — an idea which appeared somewhat similar to the concept of micelles.

Meyer and Mark were winners in the race to publish the first textbook encompassing the entire field of polymers based on the long-chain molecular concept. In 1930 a comprehensive book, *Aufbau der hochpolymeren Substanzen,* appeared which summarized their previous papers and won wide acclaim.[138] As Mark recalled, in the main this book was intended to "demonstrate to the chemical community that macromolecules actually exist, without using Staudinger's arguments but using our own." Predictably, the publication further antagonized Staudinger. Mark later explained:

[T]he Reviews [of the book] were very, very good. Only Staudinger didn't like it. . . . The real reason Staudinger didn't like it was that we anticipated his book. He was working on a larger, bigger book, but because it was larger he was a little bit slower. It [*Die hochmolekularen organischen Verbindungen*] came out only in 1932.

I think if our book had come out in 1934, he would have said, "Beautiful book. Just a copy of mine." But we came out first. I can understand. It disappointed him; somebody pulled the carpet out from under him.[139]

Staudinger faulted Meyer and Mark not only for ignoring his own work in this book, but also for plagiarism. Attacking Meyer in particular, he inscribed an embittered comment in a later edition of their book: "This book is not a scientific work but propaganda. . . . Meyer takes essentially the results of the Freiburg Laboratory without citing them."[140] By the same token, Meyer continued telling his colleagues that what Staudinger was doing was not science at all.[141] Their conflict, which encompassed personal animosity as well as professional rivalry, continued until the next decade.

The Colloid Society, 1930

The emergence of the Meyer-Mark micelle theory, which embraced aspects of both the macromolecular concept and the aggregate theory, clearly gave rise to a renewed spirit in German scientific circles. The theory drew favorable attention even from the advocates of the aggregate theory. In the long run, the work of Meyer and Mark played a significant role in widely disseminating the concept of long-chain molecules. Much like Tycho Brahe's eclectic system of the universe that compromised Copernicus's heliocentric system and Ptolemy's geostatic universe, the new Meyer-Mark micelle theory acted as a bridge between the two opposing views.[142]

The interpretation of polymeric compounds diverged around 1930 between Staudinger's school at Freiburg, Meyer and Mark at I. G. Farben, and supporters of the aggregate view, who included a large number of colloid chemists. In light of the situation, the Colloid Society arranged a discussion on "Organische Chemie und Kolloid Chemie" at its annual meeting, held in Frankfurt in September 1930. Meyer was nominated as president of the large and influential society, which at that time had about a thousand members.[143] Wolfgang Ostwald, the German leader of colloid chemistry, presided over the meeting. Staudinger, Meyer, Mark, Herzog, Hess, and Pummerer were invited to be the principal speakers.

Before an audience of colloid scientists, Staudinger stressed that colloidal substances should be examined on the basis of organic chemistry rather than relying on current colloidal doctrines, and that the study of these substances represented a new field within organic chemistry.[144] For this occasion, he presented the "Staudinger Law of Viscosity," as it came to be known, which expressed a relationship between the viscosity and the molecular weight of polymers in dilute solutions. He had developed this concept by 1929, breaking with the tradition of such colloid scientists as Wolfgang Ostwald, who claimed that there was no linear relationship between viscosity and the molecular size of colloids.[145]

Einstein's equation, which related concentration of solute particles to viscosity, was known to be applicable only to spherical molecules. Staudinger put two of his students, Ryuzaburo Nodzu and Eiji Ochiai, both from Japan, to work on the problem. They experimentally confirmed that the Einstein equation did not apply to low-molecular linear chain compounds, but also found that there was a proportionality between viscosity and chain length.[146] The implication was clear to Staudinger: viscosity depended on the shape and length of the molecule. He then applied this notion to macromolecules. By the autumn of 1929, Staudinger considered a simple formula expressing the proportional relationship between molecular weight and specific viscosity. The formula was in accordance with his assumption that the linear macromolecule behaved like a rigid rod spinning around an axis perpendicular to its long rod.

To test his formula, Staudinger needed an independent means of molecular weight estimation. Well aware of the limits of conventional techniques such as end-group analysis and osmometry, Staudinger badly wanted an ultracentrifuge, an instrument that had been introduced by The Svedberg at the University of Uppsala, Sweden. The ultracentrifuge was a device that could bring about sedimentation of colloid particles by subjecting them to high-speed centrifugal forces of some 100,000 times that of gravity. With this tool, Svedberg and his associates were able to estimate the molecular weight of some proteins to be in the several millions.[147] Not that this work led Svedberg to adopt the macromolecular concept. He was concerned little with recent developments in organic chemistry, much less Staudinger's theory of macromolecules, as indicated by no citations of it in his papers in the 1920s and 1930s. Clinging to the colloidalist views with which he had started his colloid study, Svedberg maintained that high-molecular weight substances like proteins were made up of aggregates of smaller molecular units, or what he called "submolecules," though he considered these submolecules to be far larger than ordinary molecules. To Svedberg's way of thinking, that proteins had very high molecular weights was one thing, but that proteins were true giant organic molecules was another matter.[148]

For his part, Staudinger considered Svedberg's ultracentrifuge to be almost decisive in his demonstration of macromolecules. Hopeful of purchasing this expensive piece of equipment, Staudinger applied for a grant to the Emergency Community of German Science (Notgemeinschaft der Deutschen Wissenschaft), which Fritz Haber had played a leading role in organizing. Typically, the community administered grants for needy researchers to cope with the postwar inflation; about 700 scientists enjoyed the benefits of this fund during the 1920s. However, to Staudinger's bitter disappointment, his request was refused.[149]

Consequently, Staudinger had to choose the so-called "hemicolloid" for his experiment. With molecular weight of only 2000–5000, it could be measured relatively easily by conventional means. The data he produced confirmed the validity of his formula, and he was convinced that this cheap though handy method of viscometry could be applied to compounds of larger molecular weight. After this discovery, Staudinger mobilized many students to work intensively on viscosity measurement of various polymers. A visitor to Freiburg was amazed to observe Staudinger's laboratory full of viscometers standing "like a forest."[150]

At the Frankfurt meeting, Staudinger employed various degrees of polymers (polystyrene, polyoxymethylene, and decomposed cellulose and rubber), to demonstrate that their molecular weights, measured by end-group analysis and the depressions of freezing point, were in proportion to the viscosity of solution.[151] On the basis of this relation, he was then able to obtain the molecular weights of high polymers (which could not have been previously determined by available methods) from viscometry. According to his estimate, natural rubber was composed of macromolecules of about 1000 isoprene units (C_5H_8), and a cellulose molecule consisted of 500 to 1000 glucose units ($C_6H_{10}O_5$). Confident of these findings, Staudinger concluded by contrasting his macromolecular viewpoint with the new micelle theory and the aggregate theory: for example, the chemical formula of rubber was $(C_5H_8)_{1000}$ by Staudinger's estimate; $(C_5H_8)_{100}$ according to Mark; and $(C_5H_8)_8$ in Pummerer's view.[152]

Meyer confronted the audience with an impressive array of X-ray data on cellulose, starch, silk fibroin, and rubber, showing slide pictures.[153] Mark, for his part, spoke about theoretical treatment of polymers, such as osmotic pressures and temperature dependence of polymer solutions.[154] Although their lectures were based on the new micelle theory, they joined Staudinger in supporting the concept of long-chain structure for these colloidal substances.

After presentations from the school of large molecules, Herzog, Hess, and Pummerer defended the aggregate theory of cellulose and rubber.[155] Attending the meeting, Ichiro Sakurada, Hess's loyal student from Japan, could not help but sense that the aggregate theory was on the verge of falling apart. It was a hot autumn afternoon. When Hess saw Sakurada on his way to the lecture theater, he asked him "Leben Sie noch [Are you still alive]?" Hess probably was referring to the hot weather, but Sakurada understood him to question whether their theory was still alive.[156]

However, Hess's dignified lecture was an encouragement to Sakurada.

Figure 4. Staudinger in his Freiburg laboratory, ca. 1932. Courtesy Deutsches Museum.

Realizing that the aggregate view was facing a formidable challenge, the cellulose chemist closed with the following words:

My discussion does not intend to refute the view that the number of the C_6-group in the cellulose molecule can be very large. The intention in my presentation is to illuminate only experimental facts a little more sharply than before. One can hardly avoid the impression that the phenomena are more complicated than they appeared at the beginning, and that in order to elucidate them further we need new experimental data.

In this sense, if it provokes new investigations, I am a friend of the old view of Emil Fischer and Bernard Tollen, revived by the work of Meyer and Mark in a new form. Even if this view must be abandoned or revised some day, it will not lose its great value; because the expression, "the truth is what is fruitful" would hold good in this difficult field more than for many others.[157]

Hess evaluated the work of Meyer and Mark highly; the new micelle theory, if not necessarily Staudinger's macromolecular theory, appeared to him to be acceptable. He had implicitly reminded the audience that any scientific theory was destined someday to be rewritten, which might hold true for the aggregate theory as well; yet, the contributions of the aggregate school to the development of this new field must not be overlooked. Impressed by Hess's appeal, Wolfgang Ostwald—champion of

Figure 5. Staudinger and collaborators of *Die hochmolekularen organischen Verbindungen*, 1932. Top row (standing, left to right), Ernst Trommsdorf, Heinz Haas, Werner Heuer; second row (seated), Hellmut Scholz, Ernst Leupold, Hermann Staudinger, Heinrich Freudenberger; bottom row, Werner Kern, Heinrich Lohmann. Courtesy Magda Staudinger.

the colloid doctrine and chair of the session—could not help but exclaim, "Bravo!"[158]

While favoring the Meyer-Mark theory, Hess—one of the most powerful defenders of the aggregate view—would seek to maintain his own standpoint for years following the Frankfurt meeting of the Colloid Society. However, the climate of opinion in German academia was steadily shifting to the side of the macromolecular theory. As we have seen, an early sign had been the receptivity of X-ray crystallographers back in 1927. Others were impressed by Staudinger's new viscosity law. More important, the impact of Meyer-Mark's new micelle theory, a bridge that had existed since 1928 between the macromolecular view and the aggregate theory, opened the way toward this shift. Proponents of the aggregate view, such as Pummerer, Bergmann, Pringsheim, and even Hess, began to cite Meyer and Mark's work. Haber too came to discredit the aggregate theory after learning about Meyer-Mark's work at a colloquium in his institute.[159] Willstätter considered their findings to be additional powerful evidence for Staudinger's macromolecular theory.[160]

Meanwhile, Staudinger's school of macromolecules had grown; by

1930 he had trained eight doctoral students at Freiburg, bringing the total number to twenty-five including those graduated from the ETH (see Appendix Table 1).[161] In 1932, two years after the Meyer-Mark textbook appeared, Staudinger published his masterpiece, *Die hochmolekularen organischen Verbindungen,* in which he summarized the results of his and co-workers' decade-long research on macromolecular compounds.[162]

Staudinger's theory now found supporters outside Germany as well. One of the most crucial movements in the world of organic chemistry was inspired by a young American researcher, Wallace Hume Carothers, whose work restated and expanded the ideas of Staudinger, Meyer, and Mark, in the process providing indispensable evidence for macromolecules. We shall return to this subject in the next chapter.

From Organic Chemistry to Macromolecules: Staudinger's Intellectual Stance

Macromolecular debates in the 1920s and the early 1930s demonstrate to a large extent the conceptual and methodological clash between Staudinger and his opponents, that is, a clash between the organic-structural tradition and the physicalist tradition. As Mark testified, at the time of the controversy,

Many of his colleagues in high academic positions remained for a long time skeptical and overcautious. They did not approve of the strong terms with which Staudinger elevated his own working field to a "new branch of organic chemistry," and they displayed mistrust in a number of his methods and results.[163]

Because of his strong allegiance to traditional organic chemistry, Staudinger has been called "one of the last great organic chemists of the old school."[164] The triumph of Staudinger's approach in the field of polymeric substances would mean a victory for the old school of organic chemistry. His former colleague and opponent, the X-ray physicist Emil Ott, stressed that, "Staudinger succeeded where others failed because he knew and believed in organic chemistry."[165]

As we have seen, Staudinger brought back the organic-structural approach to the study of colloidal substances, where previously this field had been dominated by the physicalist view of colloids as a state of matter. Unlike colloidalists, he distinguished organic colloids from inorganic ones and concerned himself with the former. Regarding organic colloidal particles themselves as giant molecules consisting of 10^3 to 10^9 atoms (with a few exceptions such as soap, which, despite the small molecule, exhibited colloidal behavior), he claimed that these macromolecules were built on the same principles as those of small molecules, that is, in accordance with Kekulé's classical organic structural theory. In this respect, it may be argued that Staudinger established his concept of macromolecules by returning to classical organic chemistry. Given this context, one may raise the question: was Staudinger's macromolecular theory merely a revival, or an extension, of the classical concept of molecule? In other words, to what degree was it based on classical structural chemistry? To answer this question, it is important to examine more closely Staudinger's view of the properties of polymers.

As an organic chemist, Staudinger regarded the molecule as the entity from which all physical and chemical properties of the substance stemmed. He shared with traditional chemists such as Emil Fischer a belief in the structural concept, holding that the properties of compounds largely depended on their molecular structure, especially the arrangement of constituent atoms in the molecule. This point of view was in stark contrast to Max Bergmann's claim that the properties of polymers could best be understood by studying physical conditions outside rather than inside the molecule. The structural concept also differed from the Meyer-Mark micelle concept, which stressed intermolecular-micellar forces rather than the structure of primary valence chains. While chemists traditionally had maintained that a pure substance must consist exclusively of a single and definite molecular component, Staudinger broke with them on this issue. He wrote in 1926:

> There is an essential difference between a simple and uniform material and a high molecular substance, the neglect of which prevented application of the molecule concept. All molecules have the same size in simple uniform compounds. On the contrary, high molecular compounds are mixtures of molecules of similar structure but different size. A separation into uniform products is not possible due to the small differences in their physical and chemical iproperties. If a molecular weight for high polymers is given, it can only be an average value.[166]

He stressed that polymers were composed of molecular chains of different lengths owing to their great size; that molecules in the compound were not identical in size. In other words, macromolecular substances were "polymolecular" (*polymolekulare*), or composed of molecules of various sizes. Their molecular weights therefore had certain ranges and could only be expressed by average values, not by precise numbers. In this respect the macromolecular conception differed from the traditional notion of chemical compounds, which was held both by structural chemists and by exponents of the aggregate theory.

The structural chemist Fischer, while unaffected by the aggregate theory, doubted the existence of exceedingly large molecules. He believed, as explained earlier, that since isomerism would suffice to explain the complexity of natural organic compounds such as proteins, the assumption of giant molecules ought to be superfluous. Staudinger went even further. He insisted that because of the great size of macromolecules (1,000–100,000 times the size of low molecules), there were an almost infinite number of structural possibilities for molecules. This was especially true, he thought, in the case of protein molecules, in which even slight differences in the structure would yield different biological properties.[167] Furthermore, according to Staudinger, the great variety of shapes of macromolecules caused numerous variations in the properties of polymers, such as fibrousness, elasticity, tensile strength, viscosity, and swelling phenomena. He therefore classified macromolecules into two large groups, linear and spherical, according to their shape. On one hand, substances with spherical macromolecules were usually powders in the solid state. They dissolved in water without swelling and formed solutions of low viscosity. Glycogen was a typical polymer of this group. Linear macromolecular substances, on the other hand, were fibrous and tough, and dissolved with considerable swelling to give gel solutions of high viscosity. Cellulose and many other macromolecular compounds, he claimed, belonged to this group.[168] Thus, he was able to conclude, "The shape of macromolecules affects the physical and chemical properties of the substances considerably more strongly than is the case with the low molecular compounds."[169] With his emphasis on molecular size and shape in explaining the properties of the polymer, Staudinger departed from Fischer, who confined his study to the arrangement of atoms within molecules.

Staudinger claimed that the macromolecule — when viewed as a whole, like a building — exhibited its own unique properties that could not be simply deduced from those of the low molecular units.

Molecules as well as macromolecules can be compared with buildings which are built essentially from a few types of building stones: carbon, hydrogen, oxygen, and nitrogen atoms. If only 12 or 100 building units are available, then only small molecules or relatively primitive buildings can be constructed. With 10,000 or 100,000 building units an infinite variety of buildings can be made: apartment houses, factories, skyscrapers, palaces, and so on. Constructions, the possibilities of which cannot even be imagined, can be realized. The same holds for macromolecules. It is understandable that new properties will therefore be found which are not possible in low molecular materials.[170]

In short, the whole was much more than a mere sum of its parts. For this reason, he focused his attention on the properties of the giant molecule itself, for example, size and shape, factors with which Kekuléan structural chemistry was little concerned. In this instance, Staudinger no doubt revealed a holistic viewpoint grounded in biology that distinguished him from many of his contemporaries.[171]

An appreciation of the whole, rather than its isolated parts, also set Staudinger's thinking apart from the mechanistic trend, which reduced colloidal phenomena to smaller units and forces. A friend, Bernhard Welte, addressed this point in a speech to Staudinger, from which I quote in part:

> Amidst the quantitative methods of modern chemistry, your attention was attracted by a zone in the chemical structure of our material world—and this seems very notable to me—in which the quantum, the number of building units and elements of the material substances, are not anymore only a quantum but go over to quality, to cite a thought from Hegel's logic. You became aware of configurations and formations in which law, order, and shape—e.g., the shape of chains and rings of various types—are more important than the mere number of building units or the elementary active forces. Perhaps I may illustrate the point, where the quantum changes to quality, by an allegory. Let us go with you and your science in the way you first went, thinking and exploring the way from classical quantitative chemistry to macromolecular chemistry. On this way, it seems that we first walked over a big construction area, filled with a lot of valuable raw material, stones, iron bars, woods, and more things of this sort, and then we stepped in front of a building, a big house, perhaps a cathedral. A cathedral which certainly is more than the sum of the many single working units and building materials from which it is built up. It is more because the multitude of single building units are combined in it to a formed unity of higher level and of higher range. The quantum became a quality.[172]

Quality came from the whole rather than the parts. In this regard, Staudinger stated that macromolecular compounds exhibited properties which "cannot be predicted even by a thorough study of the low molecular [weight] substances."[173] It was this conviction that led Staudinger to consider macromolecular chemistry as "a new field of organic chemistry," rather than a part of classical organic chemistry which dealt with low molecular substances.[174]

In summary, the following points might be made about the conceptual background in which Staudinger's macromolecular theory emerged. His theory was rooted in the organic-structural tradition, both theoretically and experimentally. In explaining the properties of compounds, however, Staudinger departed from the classical concept of chemical structures. He maintained that the physical and chemical properties of polymers were determined not only by the internal structure of the molecule, but more significantly by its external structure, such as size and shape. Perhaps it was this very departure from the traditional structural approach that enabled him so firmly to establish the macromolecular view—in contrast to earlier nineteenth-century attempts to establish the concept of the large molecule, which was subsequently rejected by the organic-structuralists. Staudinger's holistic conception of matter, stressing the whole rather than its individual parts, played a crucial role in making this transition. His strong rejection of mechanical reductionism in the physicalist approach to polymers can also be understood in the context of his holistic conception. Thus the development of Staudinger's macromolecular theory cannot be seen merely as the replacement of one theory by another; nor can his theory be interpreted simply as a revival of classical chemistry. Rather, the emergence of the macromolecular concept was accompanied by an epistemological shift in the way that scientists observed matter.

Staudinger and Industry

Staudinger's primary interest remained in the pure science of polymers; he was unwilling to be directly involved with its industrial applications. This was a facet of his scientific personality. As Magda Staudinger relates,

he never was interested in industrial processes of production and applications; he preferred to work on the whole field of macromolecules . . . he was very interested to see what was going on in industry and to hear what his pupils were doing there — but not to work himself for such applications.[175]

During his lengthy academic career Staudinger published over 500 papers on the subject of macromolecules. By contrast, one is struck by the small number of his patents in macromolecular chemistry, a field of great potential for technology and industry. He secured but eight patents in this field, three of which dealt merely with the preparation of monomers.[176] By and large, applied research was not in the scope of his own research program at Freiburg. Within the walls of the university, the German professor could enjoy the traditional values of academic scientists and pursue the ideal of German *Wissenschaft*.

But this is not to imply that Staudinger had no contact with industry. On the contrary, he maintained close connections with German industry, especially I. G. Farbenindustrie. While the macromolecular debate of the 1920s and early 1930s was going on, Staudinger often observed that "people working in this field in industry accepted new ideas much faster than scientists at the universities." University men, he reasoned, tended to spend enormous amounts of time scrutinizing newly proposed concepts and comparing them with other extant ideas they had clung to. Men of industry, on the other hand, would readily adopt theories that best solved their practical problems. Staudinger believed that his theory was readily accepted by industrial chemists, because to them "the macromolecular theory was more useful than others."[177] It is doubtful that such industrial researchers as Meyer, Mark, and Carothers evaluated Staudinger's theory on a criterion of utility. But certainly there were a number of industrial leaders who showed interest in Staudinger's work from its early stages and who tried to incorporate his ideas into their research and development. Their interest stemmed largely from the synthetic polymers Staudinger studied for his demonstration of the macromolecular theory.

Among the important industrial chemists who had early access to the Staudinger school was Georg Kränzlein, director of dyestuffs work and later head of the plastics research group at Hoechst in the I. G. Farben combine. Recalling Staudinger's lecture at the 1926 Düsseldorf symposium, Kränzlein said that he "agreed with the views of high polymers with an innermost conviction."[178] In his lecture Staudinger presented an extensive list of synthetic models: polystyrene, polyphenylbutadiene, polyisobutylene, polycyclopentadiene, polyindene, polyanethole, polyvinylacetate, polyvinylalcohol, polyvinylbromide, poly(acrylic acid), poly(acrylic ester), and polyoxymethylene. He subjected these polymers to analogous reactions in order to show that they converted into their derivatives without changing the long-chain structures. For example, he demonstrated that saponification of polyvinylacetate would yield polyvinylalcohol and esterification of polyvinylalcohol would give back the original polyvinylacetate.[179]

Staudinger had requested the sample of polyvinylacetate from Arthur Voss of I. G. Farben Hoechst. This polymer was first prepared in 1912 by Fritz Klatte at Griessheim-Elektron, a company which merged with I. G. Farben in 1925. Kränzlein's group at Hoechst, including Voss and Klatte, initiated a project of industrial development of the polymer for potential use in adhesive and film. Thus Staudinger's structural demonstration of polyvinylacetate at Düsseldorf caught Kränzlein's immediate attention. Soon after the symposium, Staudinger and Kränzlein's group began a frequent exchange of information and materials. Staudinger received funds and chemicals from I. G. Farben, and in return he offered Hoechst occasional consultation and lectures on the subject of this polymer. Under Kränzlein's direction, Farben chemists successfully worked on the manufacturing process of polyvinylacetate, leading in 1928 to its small-scale production at the Hoechst plant. Although Staudinger hardly played a leading role in this industrial undertaking, the Freiburg school continued to receive enthusiastic support from the Hoechst group.[180]

Besides, industry provided an important job market. All but two of Staudinger's forty doctoral students trained in the polymer field by 1935 went into industry and carried his science into industrial research and practice (see Appendix Table 1). Thus industry became an important partner in the growth of his school and the dissemination of his views. During the Nazi regime, this industrial nexus would prove to be critical to the survival of Staudinger's school, a subject to which we return in Chapter 4.

Chapter 3
Carothers and the Art of Macromolecular Synthesis

I went into another room [at the Grand Academy of Lagado], where the walls and ceiling were all hung round with cobwebs, except a narrow passage for the artist to go in and out. At my entrance he called aloud to me not to disturb his webs. He lamented the fatal mistake the world had been so long in of using silkworms, while we had such plenty of domestic insects, who infinitely excelled the former, because they understood how to weave as well as spin. And he proposed farther, that by employing spiders, the charge of dyeing silks would be wholly saved, whereof I was fully convinced when he showed me a vast number of flies most beautifully coloured, wherewith he fed his spiders, assuring us, that the webs would take a tincture from them; and as he had them of all hues, he hoped to fit everybody's fancy, as soon as he could find proper food for the flies, of certain gums, oils, and other glutinous matter to give a strength and consistence to the threads.
— Jonathan Swift, *Gulliver's Travels*, 1725

Fischer synthesized a compound having a molecular weight of 4021. . . . There is no reason why it might not be possible to prepare substances of known structure and having still higher molecular weights, and since at present no substances either natural or synthetic, having known structures and molecular weight greater than 4021 exist, the preparation of such substances would evidently constitute a significant contribution.
— Wallace H. Carothers, "Proposed Research on Condensed or Polymerized Substances," 1928

Our studies of polymerization were first initiated at a time when a great deal of scepticism prevailed concerning the possibility of applying the usually accepted ideas of structural organic chemistry to such naturally occurring materials as cellulose; and its primary object was to synthesize giant molecules of known structure by strictly rational methods.
— Wallace H. Carothers, "Artificial Fibers from Synthetic Linear Condensation Superpolymers," 1932

A half century later Mark said two developments prevented his "more active" involvement. One was his and Meyer's belief that Staudinger had completely established his priority in proposing long chains. The second was the work of W. H. Carothers, which convinced him in 1929 that the long chain connection of natural and synthetic polymers would soon be irrevocably resolved.
— G. Allan Stahl, "Herman F. Mark," 1981

During the 1920s the macromolecular debate remained largely a German phenomenon. By the opening of the next decade, however, the contestants on both sides of the debate had become increasingly conscious of a new development on the other side of the Atlantic. Herman F. Mark was struck by the contributions to the field of a young American named Wallace Hume Carothers. Mark immediately recognized the importance of Carothers's study on polymerization, and was convinced that "the long chain connection of natural and synthetic polymers would soon be irrevocably resolved."[1] While his bitter rivalry with Staudinger persisted, Mark turned into an ardent admirer of this American industrial counterpart, who was one year his junior. "No investigator has excelled," Mark would appraise a decade later, "Wallace Hume Carothers in advancing our knowledge of high polymeric chemistry and at the same time providing a basis for the development of technically useful synthetic polymeric materials."[2]

Carothers was largely responsible both for the emergence of macromolecular chemistry in America and for the birth of a science-based polymer industry. Shortly before Mark and Kurt H. Meyer began publishing their views on long-chain molecules, Carothers had entered into the world of polymers. That Carothers had no personal ties with German schools was perhaps advantageous to the stance he took in his study. Trained as an organic chemist and without a background of German chemical training from which earlier generations of Americans had often benefited, he was the product of American higher education during and immediately after the First World War. A latecomer to the field, he was in a position to be able to scrutinize impartially but critically the pros and cons of German work on polymers. Belonging to the organic-structural tradition of Staudinger at heart, he had no difficulty in envisioning polymers as true giant organic molecules.

While admiring Staudinger, Carothers looked on his German counterpart as a scholarly competitor rather than as a mentor to follow. Both excellent theorists and experimentalists, the two represented the best among organic chemists in their respective countries. In addition, they had broad knowledge of, and interest in, philosophy, history, literature, arts, and politics. Yet in many ways Carothers's personality and scientific style were in contrast to that of his German counterpart. An introvert who suffered from periodic depression, Carothers did not possess Staudinger's charisma and arrogance. He lacked the administrative force and strong leadership qualities Staudinger used to boost his field into a higher academic standing. Carothers, whose logical and clear lucid style impressed and attracted many scientists, was disdainful of Staudinger's somewhat redundant expository style of scientific writing.[3] He discredited Staudinger's dogmatism and challenged his method of demonstration. After all, Carothers was a research chemist in industry, unlike the academician Staudinger, who was imbued with the ideal of German *Wissenschaft*. The whole of Carothers's work on polymers and polymerization between 1928 and 1937 was carried out with a small number of co-workers in the fundamental research program at the DuPont Company. His approach to polymers was characteristically synthetic. He selected condensation polymers as his research problem — a subject which Staudinger's school had neglected. With his theory of condensation polymerization, Carothers sought to make artificial macromolecules, the existence of which was then still questioned in scientific circles. Within the industrial framework, Carothers's theoretical study soon found practical applications, resulting in the commercial production of the synthetic fiber nylon and the synthetic rubber neoprene. Such industrial equivalents had not emerged from Staudinger's research program.

In industry, Carothers best acted the part of pure scientist, the kind of role industrial management came to find indispensable for the new business strategy. While Carothers was attracted to pure science at an early age, the formative period of his intellectual development coincided with an era of dramatic expansion for the American chemical industry. As a closer examination of his educational background will reveal, he not only favored a purely theoretical pursuit of chemistry, but was also affected in many ways by the utilitarian ethos of the chemical science. This dual outlook helps explain the making of Carothers's industrial career, his initiation into polymer research, his scientific style, and above all, the dilemma of working in a precarious industrial environment. The way Carothers pursued his research on macromolecular synthesis exhibits a remarkable character, perhaps inherent in the context of American science and industry.

The Making of a Midwestern Organic Chemist

Wallace Hume Carothers was a product of America's Midwest; born and raised in Iowa, he graduated from a Missouri college, and acquired a doctorate in Illinois. In many respects, the shaping of Carothers's later scientific career was integral to his midwestern context. In 1896 — the same year that August Kekulé died and Staudinger was a Gymnasium student in Worms — Carothers was born into a High Presbyterian family of Scottish descent and of modest means.[4] Whereas all his ancestors were either farmers or artisans, Wallace's father, Ira Hume Carothers, chose an unusual career for one of such a background. Like Staudinger's father, Ira was an educator, though his focus was more practical. Born on a farm near Burlington, Iowa, Ira taught at a country school in his teens and subsequently went into commercial education. He married Mary Evalina

McMullin, who was of Scottish-Irish heritage and a native of Burlington. There Wallace was born, the eldest of four children, two sons and two daughters. In 1901 the family moved to Des Moines, Iowa, where Ira taught at Capital City Commercial College. He was to remain there for the next four decades until his retirement in 1941, at which time he held the position of a vice president. All four children began their higher education at Capital City.

As a child, Wallace regarded his religious and straitlaced father as he would a preacher — an "incredibly remote wise and exalted" being to be feared. As Wallace grew older, however, his emotional reaction to his father turned to one of contempt. While away at college and relieved of the watchful eye of his father, he began to feel that Ira "was lacking in many desirable qualities," that he was vapid, vain, vulgar, and even lacking in intelligence.[5] But certainly, Ira's gloomy looks as well as his physique were transmitted to his son.

Wallace's deeply emotional, shy, modest, and romantic character was perhaps inherited from his sensitive mother, Mary. "It is quite impossible to put into words," Ira later admitted, "the value and power of the mother's influence and guidance in the early years of his life."[6] Above all, Mary was a great lover of music. Wallace's intense interest in and appreciation of music, especially classical music — which Mary instilled in her son — continued throughout his life. He had a melodious singing voice that "might have developed, under training, into something worthwhile."[7] Carothers was later known to remark that were he to start over he would devote his life to music.[8] The artistic talent of his beloved younger sister Isobel was also imparted from Mary. Isobel studied speech, and during the mid-1930s performed professionally in Boston as Lu in a popular radio trio singing show, "Clara, Lu, and Em," which Wallace loved to listen to.

As a young boy, Wallace indulged in reading books, finding special pleasure in Jonathan Swift's *Gulliver's Travels,* the works of Mark Twain, and a biography of Thomas Edison. He also possessed a marked mechanical aptitude and spent much of his time with boyhood friends in experiments using batteries and coils. A moody perfectionist, the hard-driving Carothers, even from an early age, had acquired a habit of leaving no work unfinished; to begin a task meant to accomplish it.[9] His relentless devotion to work, evident in later scientific study and research, mirrored a characteristic of the stern Protestant work ethic of his midwestern upbringing.[10]

A great admirer of Edison, the young Carothers once seriously considered electrical engineering for a profession.[11] But his interest soon shifted to chemistry. This interest began with Robert Kennedy Duncan's popular books, which he read while he attended North High School in

Des Moines.[12] An enthusiastic science writer and at that time professor of industrial chemistry at the University of Kansas, Duncan had published two major books for lay readers. His first, *The New Knowledge: A Popular Account of the New Physics and the New Chemistry in Their Relation to the New Theory of Matter*, published in 1905, was a succinct and attractive exposition of modern scientific ideas, ranging from atoms, molecules, elements, ions, various rays, and radioactivity to inorganic evolution and cosmic problems.[13] By contrast, his second book, *The Chemistry of Commerce: A Simple Interpretation of Some New Chemistry in Its Relation to Modern Industry*, which appeared in 1907, dealt with practical problems, such as the manufacture of sulfuric acid, the nitrogen fixation needed to produce fertilizers, glass-making, the synthesis of perfumes and medicines, and the industrial production of cellulose fiber. Through these examples, the book, in part, aimed to convince men of commerce, "how absolutely applicable is modern science to the economy and progress of manufacturing operations." Criticizing the "pure science" ideal of most American academics, which tended to regard applied science as impure or degraded, Duncan also hoped that his book would induce young people who were considering entering the chemical field to understand that "the application of science to the material needs of men is just as much science and just as much research as that which is pursued solely for its own ends." While not intending to belittle pure science, he stressed that "the world needs both" pure science and its application. Citing favorable impressions of the German chemical industry gained from his own European tour, Duncan appealed in this book for cooperation between academic science and manufacturing technology, which, he lamented, the American chemical industry severely lacked.[14] These two books, both enjoying tremendous success in the early decades of the twentieth century, opened Carothers's eyes to the world of chemistry and the chemical industry.

Carothers spent his first academic year completing the accounting and secretarial curriculum at his father's commercial college. On Ira's suggestion, in the fall of 1915 Carothers enrolled in Tarkio College, a Presbyterian liberal arts college in northwestern Missouri. His prior experience immediately qualified him as an assistant in Tarkio's Commercial Department. Two years later, he assumed an assistantship in English, perhaps because of his unusual skill in writing and interest in English literature. He also served as private secretary to the president of Tarkio, a position of which his father was proud.[15] On account of the need to earn his own way, it took him five years to acquire a bachelor's degree.

It was during his Tarkio days, at the encouragement of a chemistry teacher, Arthur McCay Pardee, that Carothers definitely decided on chemistry as his life's work. At that time, Pardee observed a discrepancy between Carothers and his father: "Although Carothers seemed to come from very nice responsible people, his family had really very little educational background. . . . [H]is father throughout Wallace's college career didn't really sense the possibilities of the field to which his son wanted to devote himself."[16] Failing to conceive of the possibility of his son as the first scientist stemming from an ordinary midwestern family, Ira probably expected Wallace to enter the commercial world. Surely, that was why Ira encouraged him to study at his college and then sent him on to Tarkio's Commercial Department. Whatever his father's expectations, Wallace set his mind on chemistry. Pardee recognized that Carothers's "interest in chemistry and the physical sciences was immediate and lasting, and he rapidly outdistanced his classmates in accomplishment." He had a number of talks with Carothers in which he persuaded him that "the sky was the limit in what he could accomplish."[17]

The inspiring young instructor Pardee had received his bachelor of arts degree in 1907 from Washington and Jefferson College in Washington, Pennsylvania, where Duncan held a position as professor of chemistry and mineralogy from 1901 to 1906. While appointed to Tarkio, Pardee spent two academic years (1910–11 and 1915–16) at Johns Hopkins University. At Hopkins, he was profoundly influenced by Ira Remsen, professor of chemistry and later president of Johns Hopkins, who successfully introduced into his teaching a Germanic style dedication to pure research. In 1916 Pardee completed his doctoral thesis on the conductivity of organic salts under the physical chemist Harry Clay Jones.[18] Pardee's interest then shifted toward organic chemistry, in which he pursued research with Hopkins's organic chemist, Ebenezer Emmet Reid.

At Tarkio, Pardee taught Carothers organic chemistry and physical chemistry during his sophomore and junior years. During World War I, when Pardee was called to his alma mater, Washington and Jefferson, Carothers was appointed to assume responsibility for teaching chemistry. Exempted from military service on account of "a slight physical defect" — reportedly a goiter — the student thus served in the joint capacity of undergraduate as well as instructor until his graduation.[19]

Completing every chemistry course offered at Tarkio with straight As, Carothers was especially fascinated by organic chemistry. He recalled later,

Organic chemistry appealed to me from the first more than any other branch because its realm includes among a great many other things the chemistry of living things and because, as I still think, its fundamental theory is at once much simpler and much more adequate than any other science.[20]

Leaving Tarkio College in 1920 with a bachelor of science degree, Carothers began graduate work at the Department of Chemistry of the

University of Illinois in Urbana, an emerging center of organic chemistry in the Midwest. William Albert Noyes, another student of Ira Remsen, was instrumental in the phenomenal growth of Illinois's chemistry program. Editor of the *Journal of the American Chemical Society* and future founder of *Chemical Abstracts,* as well as department head for two decades, from 1907 to 1926, he brought to Illinois a Remsen-inspired ideal of pure scientific research, and transformed what had been a second-rate applied program in a land-grant college into a nationally renowned center for the training of professional chemists.[21] The theoretically-minded Noyes was among the earliest American organic chemists to encourage application of physical chemistry to organic chemistry, with the electronic theory of valence, the correlation of ionization and structure, and physico-chemical measurements in determining reactivity of organic compounds.[22] Carothers probably took Noyes's graduate courses during his first year at Urbana.

Earning a master of arts degree from Illinois in 1921, Carothers spent the following academic year as an instructor at the University of South Dakota in Vermillion, where his former teacher Pardee had moved to head their Chemistry Department. During this year, Carothers developed his idea of applying the electronic theory of valence to organic compounds, an idea that echoed Noyes's thinking. Inspired by recently proposed theories in physical chemistry, particularly Gilbert Newton Lewis's theory of the shared-pair electron bond and also Irving Langmuir's octet theory of valency, Carothers investigated their implications in his chosen field, organic chemistry.[23] By measuring the density, viscosity, and vapor pressure of two organic compounds, phenyl isocyanate and diazobenzene-imide, at various temperatures, he demonstrated that they had identical atomic and electron arrangement, though containing different atoms. This experiment resulted in his first independent paper, "The Isosterism of Phenyl Isocyanate and Diazobenzene-Imide," which appeared in 1923 in the *Journal of the American Chemical Society.*[24] Carothers's theoretical paper of the following year, "The Double Bond," expanded his views into a general argument for reaction mechanisms (e.g., addition processes, reduction, and hydrogenation) of the double bond on the basis of the electronic theory.[25]

Despite his ambitious effort to combine the Lewis-Langmuir theory and organic reactions, Carothers's early work remained an isolated effort for the period. Whereas physical chemistry grew rapidly in the United States in the early decades of the twentieth century, American organic chemists tended to resent the intrusion of physical concepts of subatomic structure into the realm of organic chemistry.[26] Even as late as 1928, one organic chemist ridiculed Lewis's cubic atom: "the structural formula of the organic chemist is not the canvas on which the cubist artist

Figure 6. Wallace Hume Carothers (center) and Arthur M. Pardee (right) at the University of South Dakota, ca. 1921. Courtesy A. Truman Schwartz and the University of South Dakota.

should impose his drawings which he alone can interpret."[27] As his close friend John Raven Johnson testified, Carothers's article on the double bond was considered "too fanciful" by many of his contemporaries. Indeed, "it barely escaped the editor's wastebasket in 1924."[28] Apparently the electronic interpretation of organic reactions captured Carothers's attention for several years thereafter. However, after a later paper on this subject met with a referee's strong criticism that "the author seems to attach an unwarranted measure of sanctity to the 'octet rule,' " Carothers no longer pressed the issue publicly.[29]

Returning to the University of Illinois in 1922, Carothers turned more and more to a traditional mode of research in organic chemistry. The shift was significant in light of his eventual reshaping as a classical organic chemist. Under the direction of a new mentor, Roger Adams, Carothers began working on his doctoral thesis. Adams had been educated at Harvard University, receiving his doctor of philosophy degree in 1912 under Charles Loring Jackson.[30] For the year following, 1912–13, he studied in Germany with Otto Diels, then *Privatdozent* in Emil Fischer's laboratory at the University of Berlin, and later with Richard Willstätter, a rising star in German classical organic chemistry at the Kaiser Wilhelm Institute for Chemistry.[31] In Berlin, Adams also attended lectures by Fischer, who at age sixty was still actively studying proteins. After he returned to the United States, Adams's interest in organic structural chemistry and organic syntheses continued along traditional lines, little influenced by emerging trends of physico-chemical applications to organic chemistry. With this background, Adams joined the faculty of chemistry at the University of Illinois in 1916.

An especially significant factor in Adams's research and education was the emphasis on synthetic organic chemistry. After German sources of organic chemicals were curtailed during World War I, Adams turned a summer project to prepare organic chemicals for classroom and research use — an enterprise called "Organic Chemical Manufactures" — into a pilot plant to produce over a hundred chemicals for war and industrial use. In this way Adams not only created a financial success but also demonstrated to the nation the practical power of an Illinois education in chemistry. Adams's venture at Illinois continued after the war, in order to introduce students to industrial operations. Bulletins were soon issued to describe the synthetic methods developed there; in 1921 they were transformed into the annual monograph series, *Organic Syntheses*.[32] During the next forty-nine years, Adams served as a chief member of the Advisory Board for the series. Dubbed the "Adams Annual," the monographs were, and still are, widely consulted by chemists around the world.

Adams's adherence to synthetic chemistry reflected his pragmatic approach to the field. When he took over Noyes's position as department head in 1926, the "Yankee scientist" vigorously challenged the tradition of academic chemistry "for its own sake," an ideal preferred by Noyes. Not that Adams disdained pure science. Schooled in the German concept of *Wissenschaft*, he, like Noyes, did stress the importance of pure scientific research. But he simply refused to let chemistry just stay there. In many respects Adams's views afforded a parallel to what Duncan had advocated. At a time of explosive growth of industrial research in the United States, immediately following World War I, Adams held the view that one of the primary responsibilities of universities ought to be to train high quality professional scientists for industry. Chemistry must exist for and serve industrial society, he firmly believed.[33] Through close contacts with the chemical industry, he found positions for many of his students. Of 184 Ph.D.s trained by Adams over a forty-year period beginning in 1918, 120 pursued industrial careers upon graduation. Ties with the DuPont Company were especially close. While he himself consulted for DuPont, twenty-five of his doctoral students including Carothers, became DuPont chemists.[34] The synthetic approach and industrial orientation of the Adams school undoubtedly exerted a profound influence upon Carothers's style in chemistry and career, culminating in distinctive scientific activities at DuPont.

While serving as a research assistant in the Chemistry Department, Carothers proceeded with doctoral research on the catalytic reduction (hydrogenation) of aldehydes over the "Adams catalyst," a platinum oxide. A year before graduation, he wrote a friend about a "reverie" of his future, which suggests his concern with the practical aspects of chemical research:

. . . looking beyond to the time after graduation I meditate complete escape — New York and a laboratory where I could test out some ideas of vast commercial importance which occasionally occupy my reveries — or even beyond that to Paris, Vienna, Berlin. But the means are not obvious, no money to start a laboratory of my own.

He added, "Paris might represent a possibility," since holders of the Carr Fellowship at Illinois had customarily gone to Paris to continue their work.[35]

Whatever his ideas of commercial importance, Carothers — a 1923–24 Carr Fellow — would not translate his dream into action in Paris. After earning his doctorate in 1924, he did tour Paris, but shortly he returned to Illinois. His dissertation consisted of a series of three papers titled "Platinum Oxide as a Catalyst in the Reduction of Organic Compounds," which was co-authored with Adams in the *Journal of the American Chemical Society*.[36] According to Adams, at graduation Carothers "was considered by the staff as one of the most brilliant students who had ever been awarded the doctor's degree" in the Department of Chemistry at the University of Illinois.[37] Impressed by Carothers's talent, Adams promptly arranged for his student's appointment as an instructor in organic chemistry at Illinois. Carothers remained for two years until he accepted an invitation from Adams's alma mater, Harvard University, to teach there, in the fall of 1926.

In Cambridge, Massachusetts the new instructor taught three courses, "Experimental Organic Chemistry," "Structural Organic Chemistry," and "Organic Chemistry."[38] "So far as teaching goes," Carothers wrote to a friend, "Harvard is the academic paradise."[39] But in reality he was not a successful teacher in the classroom or in public. His superior, James Bryant Conant, never estimated his lecturing capacity very highly: Carothers at that time "was extremely shy and nearly had nervous prostration every time he was asked to speak in public."[40] Carothers's close friend John Raven Johnson related that the former was "modest and unassuming in manner, shunned publicity, and, shy and sensitive by nature, was ill at ease in a large group, although within his small circle of close friends he was a witty conversationalist."[41] He was not the type of academician who would become a high ranking professor with administrative power; his dominant quality was that of a researcher rather than

Figure 7. Carothers and Carl S. Marvel fishing in Squaw Lake, Wisconsin, ca. 1925. Courtesy Chemical Heritage Foundation.

teacher. Thus DuPont's newly established program for fundamental research would appear to suit Carothers's scientific personality well, since it strictly demanded research conducted in a small group, with chemists to assist him in facilities that "would be difficult or impossible to duplicate in most university laboratories."[42]

Carothers and DuPont's Fundamental Research Program

The phenomenal growth of American industrial research in the early decades of the twentieth century coincided with the birth of a number of giant corporations, such as American Telephone and Telegraph, U.S. Rubber, General Electric, and Eastman Kodak. The depression of the 1890s had helped consolidate a large number of companies; in 1899, for example, 340 companies had been merged into large corporations, and in 1900 the number of amalgamated companies rose to over 1,200.[43]

These newly established companies began creating research organizations in order to obtain technological hegemony in the face of growing patent competition. They equipped their own laboratories and employed freshly minted Ph.D.s to conduct industrial research. The number of American industrial laboratories rose dramatically from 4 in 1890 to about 50 in 1900, 180 in 1910, 520 in 1920, and 1030 in 1930.[44] As in Germany, the early industrial laboratories in the United States centered on fields relating to chemistry and electricity, those showing keen international technological competitiveness and also implementing the visible effects of scientific research on their technologies. Among the first chemical companies to establish such laboratories were General Chemical (1899), Dow (1901), DuPont (1902), Standard Oil of Indiana (1906), Goodyear (1909), Eastman Kodak (1912), and American Cyanamid (1912).[45] Most of these early laboratories dealt with applied research aimed at improvement of existing manufacturing processes and products. But from this outgrowth there emerged a new perspective among a few of the companies. Basic research programs at such firms as General Electric in 1920 and DuPont in 1927 were among the earliest that were created as part of an incipient movement in American industry to do basic science research.[46]

E. I. du Pont de Nemours & Company originated as a gunpowder manufacturer. The company was founded in 1802 on the banks of the Brandywine River near Wilmington, Delaware, by a French émigré, Eleuthère Irénée du Pont de Nemours, a protégé of Antoine Laurent Lavoisier. During the nineteenth century, DuPont dominated the nation in the manufacture of explosives. Shortly after World War I, the company had been transformed into a maker of a wide variety of chemical products. Owing to its prior experience in the explosives industry, DuPont's new business included manufacture of products of nitrocellulose and other cellulose derivatives, such as "Fabrikoid" (artificial leather), "Pyralin" (a celluloid-type plastic), "Duco" (lacquers), cellophane, and viscose rayon. By this time, DuPont had research facilities such as the Eastern Laboratory in Repauno, New Jersey, which was begun in 1902, and the Experimental Station in Wilmington, established a year later. With few exceptions, however, the company accomplished this transformation by purchasing other promising companies rather than through its own research and development effort.[47]

In 1927 the DuPont management agreed on a "radical departure from previous policy," by setting up a fundamental research program in its Chemical Department at Wilmington. Charles Milton Altland Stine, the Chemical Director, was its founder. Although modeled on such precedents as the German chemical firms and the research laboratory at the General Electric Company (where Irving Langmuir was successfully working on the incandescent light bulb), the DuPont program went even further. The purpose of the program was to discover and establish new scientific facts without regard to practical problems or immediate practical use. Stine intended to separate fundamental research from ongoing applied research, which was concerned with existing processes and products. In this way, he aimed to create an improved research environment that would closely approximate conditions of the university laboratory yet offer better facilities and more funds.[48]

A Johns Hopkins Ph.D., Stine was a great admirer of Ira Remsen, whom he considered a "true scientist" rare to his generation. Stine was later to draw a parallel between his own situation and the way in which Remsen, after returning from Germany in the 1870s, had struggled to discover a "real opportunity for fundamental research" in the American university. Remsen eventually found the opportunity at Johns Hopkins University, when President Daniel Coit Gilman offered him a professorship in chemistry. Stine liked to quote Remsen's words:

President Gilman's injunction was simply this: "Do your best work in your own way." What could be finer? I bought all the apparatus I wanted and all the books I wanted. A simple laboratory was built. I had but three or four students and we went to work. Now I am well aware of the fact that chemistry was not revolutionized as a result of our efforts, but we made a start in a new direction.[49]

Thus Remsen was able to implant pure scientific research in the soil of American chemical education.

By the same token, Stine set out to implant pure scientific research in industrial soil. With this step, he did not intend to revolutionize chemistry, but certainly wanted to chart a new direction for industrial research. To succeed, he believed it essential to secure the highest grade of scientists "and then turn them loose." Or, as he liked to say, "Go as far as you can see, and then see how far you can go!"[50] At first glance, Stine's program appears to be a paradoxical attempt to induce a Remsenian spirit of anti-pragmatism into the profit-oriented corporate world. Stine was, however, an astute businessman who did not fail to caution that fundamental research in industry "is not conceived to be a labor of love; it is conceived to be sound business policy, a policy that should assure the payment of future dividends."[51] He anticipated that this type of research would eventually yield new products without depending on university science or outside technology, thereby benefiting the company in the long run. Yet he also stressed other important results that the new program should have: scientific prestige or advertising value, improvement of staff morale, and establishing good contacts with scientists in universities and other research institutes.[52]

Stine's somewhat idealistic proposal to DuPont's Executive Committee, given to them in December 1926, evoked considerable argument within the company. In particular, Elmer Keiser Bolton, then director of research in the Organic Chemicals Department, was strongly opposed to the new program. He responded to Stine that "he did not consider that fundamental research could be administered logically as part of an industrial organization."[53] Ironically, it was Bolton who in the 1930s would head this very program and would bring to fruition the impressive results that would change DuPont's future.

To make his proposal more concrete and persuasive, Stine gathered opinions from experts, both DuPont managers and outside consultants, on a course of research to be pursued in this program. Through this process, colloids and polymerization caught his most attentive interest. He immediately received suggestions pertaining to colloid research from many quarters. Cole Coolidge, assistant director of the Experimental Station, was of the opinion that research subjects "should tie in with the present duPont lines of manufacture, or with new lines of manufacture on which it seems logical for the Company to embark." In connection with this, he advised the colloidal study of cellulose derivatives, because nitrocellulose was the basic intermediate in "Duco" paint, "Fabrikoid" fabric, and "Pyralin" materials industries of the company, about which "the prior art reveals little real information." In fact, DuPont had already begun a small amount of research work in this area at the Experimental Station. Coolidge's idea was an extension of this infant project.[54]

A DuPont consultant and young assistant professor at the University of Wisconsin, Elmer O. Kraemer, also encouraged the company to pursue colloid chemistry, his own field. A former student of The Svedberg, Kraemer recommended the construction of an ultracentrifuge for determining the degree of dispersion of colloids, such as cellulose derivatives, rubber, and pigments. As he explained:

The ultracentrifuge has finally been developed to the point where it may be accepted safely as a powerful instrument for studying solutions containing large molecules. No other method would be so effective in determining the real degree of dispersion of cellulose derivatives and other such solutes in various solvents and under various conditions. Researches along this line, properly executed, would command the attention and respect of "pure" scientists over the country, and justify the claim of fundamental research in an industrial laboratory. I dare say, it will be some time before such studies could be duplicated in this country or elsewhere, outside of Svedberg's laboratory.[55]

Later, Stine would also consult with Wilder D. Bancroft, the leading colloid chemist in the United States, about the best way to conduct fundamental research on colloids at DuPont.[56] High opinions of the colloid approach clearly reflected the strong interest in the discipline in America, as will be discussed below.

Meanwhile, E. Emmet Reid, a Johns Hopkins University professor and DuPont consultant, included among his recommendations the study of polymerization. In a letter to Stine in late December 1926, he wrote, "The causes and mechanism of polymerization need investigating. The results might be of great commercial interest. . . . Much study and experimenting will be required to develop this problem."[57] Stine regarded Reid as "an old-fashioned, organic chemist" who had "very wide knowledge in organic chemistry" but unfortunately had "never taken one field and cultivated it intensively."[58] Indeed, Reid had personally done little work on polymerization, but, quite predictably, expressed a desire to study it. For DuPont's ultimate benefit, Stine took Reid's advice. By the mid-1920s, DuPont's chemical products included artificial leather, nitrocellulose plastics, rayon, and cellophane — products already recognized not only as colloidal products but also as polymers, even though their molecular structure and formation process remained largely unsolved. Also, a small project on synthetic rubber had just been initiated by Bolton. In this context, polymerization, like colloids, matched DuPont's interest perfectly.

Taking all this advice into consideration, Stine rewrote his proposal and submitted a new version to the Executive Committee in late March 1927. Therein he outlined five lines of investigation to make up the fundamental research program: catalysis, colloids, polymerization, physical and physico-chemical data (or chemical engineering), and synthetic organic work. Of these, colloids and polymerization would be the most vital areas of research. Stine explained that colloid chemistry was "one of the fields in which fundamental information is badly needed, and in which very little is as yet being done." Since the field would cover DuPont products such as paints, varnishes, spray and brush lacquers, celluloid plastics, photographic film, Fabrikoid, rubber top materials, and dynamite, Stine pointed out that "its importance to our Company can be readily appreciated." Furthermore, the field was "so large that it constitutes an almost unlimited opportunity for profitable fundamental research." As for polymerization, Stine stated:

The phenomenon of polymerization . . . is illustrated by natural resins and in the production of synthetic resins and treated drying oils. . . . In most cases, very little is known about the actual mechanism of the change which takes place, so that the methods in use are based almost solely upon experience. It is believed that a more thorough understanding of polymerization would permit us to develop oil and resin products and similar materials which would possess properties markedly superior to the properties of materials of this sort at present available.

Stine considered polymerization not simply a subject within organic chemistry. "To a considerable extent," he felt, "studies in this field would tie in with studies in the field of colloid chemistry, because the materials we should be dealing with are of a colloidal nature." In general, he stressed that such subjects as catalysis, colloids, and polymerization should involve "a combination of physical and organic chemistry." "Properly carried on," Stine now restated, "fundamental research is bound to result in the discovery of new, highly useful, and in many cases, indispensable knowledge."[59] Thus Stine skillfully suggested feasible links between scientifc investigations and the company's products.

In April 1927 the Executive Committee approved, in principle, Stine's proposed program, deciding to risk a few hundred thousand dollars for the first year of this venture.[60] On a practical level, the Chemical Department's fundamental research program was organized into four sections based on disciplines (rather than specific research topics): organic chemistry, colloid chemistry, physical chemistry, and physics.[61] The organic chemistry group was considered to be pivotal to the program, since it would encompass the two lines of work Stine proposed: polymerization and organic synthesis. Arthur P. Tanberg, director of the Experimental Station, stated in the fall of 1927:

Our Company's interest in organic chemistry is, of course, very great and it has been my general feeling that of the total amount we spend perhaps 40 or 50% should be spent on organic subjects. I hope that eventually we shall have fifteen or more men working along organic lines.[62]

Stine's first attempt to hire men of established reputation for this section met with little success. He then targeted "young men of exceptional scientific promise but no established reputation."[63] Harvard's thirty-one-year-old instructor Wallace Carothers belonged in this category. Relatively unknown, he had thus far published only eight papers in organic chemistry. DuPont approached Carothers, however, based on recommendations by Adams and Conant, both of whom found exceptional promise in him as a researcher rather than a teacher.[64] Stine and his two assistants, Tanberg and Hamilton Bradshaw, assistant director of the Chemical Department (all three graduates of Johns Hopkins) worked together to recruit Carothers.

In the summer of 1927, when Carothers first learned of DuPont's new program from Bradshaw, his reaction was "very enthusiastic." Carothers perceptively commented that "it may do a good deal to vitalize academic research and to bring about a closer contact between the work of university and industrial laboratories to the mutual advantage of both."[65] In late September, after visiting DuPont's Experimental Station in Wilmington, Carothers was officially offered the position to head the organic chemistry group. Stine promised him a higher salary ($5,000 a year) than Harvard's ($3,200) and a staff of trained scientists to assist in working on any problems of his own choosing.[66] The position demanded only research, the results of which would be publishable. Although the job appeared fairly attractive, Carothers still felt it necessary to ascertain DuPont's policy on the nature of fundamental research. He therefore inquired as to whether, if he were to join DuPont, he could continue his present work at Harvard, exploring such subjects as the thermal decomposition of metal alkyls (to see if they might contain the alkyl anion) and the synthesis of succinic dialdehyde, both of which he believed to be of little practical use.[67] Stine replied quite positively:

We are interested in you because we believe you will not only do work of a high order, but also that you will select worth while problems, and therefore anticipate no difficulty with respect to decisions about the particular investigations you would want to undertake. We know of no reason why you should not continue work of this sort as long as you yourself consider it worth while — I mean worth while from a scientific point of view and not from the point of view of direct financial returns.[68]

It is noteworthy that neither party referred to polymerization, nor did Stine request that Carothers study it. The selection of problems was to be left entirely up to Carothers.

In early October, however, Carothers decided to decline the offer. He wrote to Stine, "In view of the complete balance in my mind between the advantages of my present and possible positions, it seems unwise to make the change." Yet he complimented Stine on his fundamental research program: "American chemists are to be congratulated that such an institution is to exist. No doubt it will prove to be an important factor in the future development of chemistry as a science."[69] A few days later he wrote to Tanberg explaining why it was difficult for him to leave Harvard's academic life. He said that money was of some consideration to him, but "the real freedom and independence and relative stability of a university position are sufficient" to compensate for the difference in salary. He also confided in Tanberg about his personal problem. He had had considerable difficulty adjusting to the change from Urbana to Cambridge. Now that the adjustment has been made, he did not like to make another change "unless there is a clear cut and indubitable advantage to be gained in so doing. . . . I suffer from neurotic spells of diminished capacity which might constitute a much more serious handicap there than here."

Carothers added, "I would have regarded the opportunity for participating in it [DuPont's program] as an extraordinary piece of good fortune," and showed his ambivalence by saying, "I am not altogether sure yet that my prospects are better at Harvard than they would have been at Wilmington."[70] By implication, he had left room for negotiations.

Reading hope between the lines, Bradshaw went to Cambridge to talk further and persuade Carothers to take the position.[71] This time, Bradshaw renewed the offer with an increased salary of $500 per month, namely $6,000 a year. Possibly James Conant cooperated in convincing Carothers as well.[72] As it turned out, Carothers accepted the offer in late October 1927.[73] Although the move to industry appeared to him "a fairly heavy gamble," the DuPont job looked sufficiently "alluring" to overcome his reluctance to leave the academic world.[74]

Most likely, Carothers made his decision to join DuPont only after finding a research problem with which he felt comfortable. That problem was polymerization. This was a field in which he had neither published nor performed any experiment, but which he came to consider worthwhile for DuPont's industrial research program. Only nine days after he notified them of his acceptance, Carothers wrote Bradshaw a long letter outlining in remarkably concrete terms what he would do for DuPont. This letter reveals that he had taken pains to select research topics possessing some practical significance. As for his interest in the thermal decomposition of metal alkyls, which he had previously described to Stine, Carothers now reconsidered, stating that it "is somewhat remote from any practical considerations, and I think it would be a mistake for me to devote all my attention indefinitely to such problems as these." Instead, he proposed polymerization as a far more promising research candidate. He had known about DuPont's concern with this problem since his first contact with Bradshaw in the summer of 1927. Carothers now wrote, "I have some appreciation of the commercial importance of this subject, because rubber, cellulose and its derivatives, resins and gums, and proteins may all be classified as large or polymerized materials." He wanted to attack the problem because of the need for theoretical exploration as well as its vast commercial implications. As he explained:

This is a class of substances about which relatively little is known in terms of structure. None of these substances is very amenable to the classical tools of the organic chemist, and no doubt some of the most important contributions in this field will be made by experts in colloidal chemistry. From the standpoint of organic chemistry one of the first problems is to find out what is the size of these molecules and whether the forces involved in holding together the different units are of the same kind as those which operate in holding the atoms in ethyl alcohol together, or whether some other kind of valence is involved — more or less peculiar to highly polymerized substances.

He said that he had been reading "with a great deal of interest lately" recent German articles on rubber by both Staudinger and his opponent, Rudolf Pummerer. From Carothers's perspective, Staudinger had "demonstrated rather convincingly" during the past few years its macromolecular structure. Nevertheless, Pummerer's latest paper "has succeeded in casting serious doubts on Staudinger's conclusions."[75] The unsettled German debate on polymers, particularly regarding the structure of rubber, greatly appealed to Carothers's theoretical interest. To resolve the issue, Carothers already had a method of attack in mind. Stine, Tanberg, and Bradshaw enthusiastically endorsed Carothers's new plans. In addition to his interest in the new research program, his scientific personality, a greater financial reward, recommendations from Conant and Adams, and DuPont's persuasiveness — besides all these factors, finding a research project suitable for industrial research was perhaps a strong consideration in Carothers's decision to move to DuPont. Without DuPont's forcible approach and constant contact with him during the fall of 1927, the young Harvard chemist might never have turned his attention to polymers, nor contemplated a new research program. It was also this very interaction between academic experience and industrial practice that would help to shape Carothers's successful method of polymerization, as will be discussed later.

American Response to Macromolecular Theory

What had been the state of polymer research in America up to the time Carothers embarked on his study? How did Americans respond to German theories of polymers in the 1920s? To place Carothers's position in its proper historical context, an examination of these problems is in order.

Between 1920 — the year Hermann Staudinger published his major work on polymerization, and 1927 — the year Carothers was initiated into the world of polymers, a number of Americans were exploring the problems of polymers, such as rubber, cellulose, proteins, and synthetic polymers. As we have seen in Chapter 1, America had a strong tradition in the polymer industries, as represented by the products Celluloid and Bakelite. The production of viscose rayon and cellulose derivatives also began during the 1920s. The automobile industry created huge demand for rubber tires, with American consumption of rubber quickly outdistancing the rest of the world (in 1920 for example 206,000 tons, more than eight times the amount used by Britain). As a result, rubber manufacturing companies, led by "the Big Four" (Goodrich, Goodyear, U.S. Rubber, and Firestone), grew and grew.[76] In an epoch when industrial research was booming, American polymer industries employed an unprecedented number of chemists to conduct research in their laboratories. In the U.S. rubber industry, for example, in 1920 there were nine research laboratories employing a total of 590 research personnel.[77]

During this period, the American Chemical Society founded three polymer-industry-related divisions: the Divisions of Rubber Chemistry (1919), Cellulose Chemistry (1922), and Paint and Varnish (1927). These divisions were created largely because of the demand of ACS members who worked in industry. Industrial chemists needed a mechanism for collecting and exchanging information. For instance, the Division of Cellulose Chemistry was an outgrowth of a cellulose symposium, the first being held at the 1920 ACS meeting in St. Louis. Minutes of the symposium recorded that it intended "to promote intercourse and cooperation between the chemists in the various cellulose industries," the field that "constitutes one of the largest and most important of American industries." The majority of the charter members consisted of industrial chemists, and the first three chairmen (1922–27) of the Division were industrial chemists.[78]

Despite increasing interest in polymers, Americans made few contributions to fundamental studies in this area. The National Research Council's *A Survey of American Chemistry*, published in 1927, reveals reviewers' frustration. Harry LeB. Gray of Eastman Kodak, who surveyed cellulose chemistry, wrote, "While great advances have been made in America in the technical application of cellulose and its derivatives, a review of the literature indicates that the major portion of the published results have appeared in foreign journals."[79] The surveyor of rubber chemistry, William C. Geer, who was a former vice president of Goodrich, expressed a similar sentiment:

In Europe such work [on rubber] as has been published has come largely from scientific and university circles, while in this country the laboratories connected with the industries have furnished a majority of the papers. One never can be quite sure, therefore, to what extent the published papers reflect the real progress of the science in the United States. And, as is natural under such conditions, our chemists lean in their publications toward the applied aspects of rubber chemistry.[80]

And regarding protein chemistry — a subject that should deal more with pure science — Phoebus A. Levene of the Rockefeller Institute for Medical Research commented, "It is noteworthy that American workers devoted little attention to the problem of the structure of the protein molecule, whereas European chemists centered their interest on this phase of protein chemistry."[81]

After all, American academia had engaged in little, if any, of the heated macromolecular controversy that went on in the German chemical community. Staudinger's theory of macromolecules attracted little attention from American investigators, especially in the first half of the 1920s. To understand the slow reactions of Americans to the macromolecular concept in the years before 1927, we must first consider the outlook of the chemical communities following World War I, and then the intellectual background which gave rise to American chemists' concern with polymeric materials.

During the early 1920s, the German scientific community was isolated from its counterparts in America, Britain, and France. An unsettled political situation in Germany, almost amounting to civil war, and a devastating inflation around 1920 greatly increased the difficulties of travel and exchange of information between Germany and the Allies. The postwar nationalistic mood on both sides extended to science. In early 1920, for example, Jacques Loeb of the Rockefeller Institute for Medical Research stated, "From all I hear the Germans are still on the whole in a very hostile attitude towards scientific work done in this country."[82] This sentiment is further illustrated by the actions of the International Union of Pure and Applied Chemistry, which was officially organized by the war victors in 1920. The Union excluded both Germans and Austrians from its conferences, because their scientists were considered to be tainted with war-guilt and not fit to attend international meetings. The antipathy towards Germans, of American chemists as well as of British and French chemists, continued well into the mid-1920s. The attitude of Theodore William Richards of Harvard, as reported by his colleague and son-in-law Conant, reveals the widening breach in scientific communications between the former wartime allies and Germany:

His [Richards's] condemnation of the Germans [around 1925] was as total as it had been in August 1914. He had made no move to reestablish communications with his former scientific friends in what he now considered to be a hopelessly barbaric land.[83]

Given the historical context, it is not surprising that the Allied chemical community was uninformed about advances in the German community during the postwar period. Consequently, the upheaval in German chemistry in the early 1920s, brought about by the emergence of the macromolecular theory, did not lead to immediate reactions in Britain, France, and the United States. Instead, the heated debate over Staudinger's theory remained, for the time being, a German phenomenon.

In the 1920s Staudinger's ideas were discussed on several important occasions in Austria, Switzerland, and Germany: in 1924 at the Innsbruck meeting of the Society of German Natural Scientists and Physicians; in 1925 at a meeting of the Zürich Chemical Society; and in 1926 at the Düsseldorf meeting of the Society of German Natural Scientists and Physicians. The macromolecular debate reached its peak on the occasion of the Düsseldorf meeting. But it was not until the early 1930s that Staudinger and other German scientists brought up the issue at scientific meetings outside German chemical circles.[84] Although Staudinger's views were available to Americans in such journals as *Berichte der Deutschen Chemischen Gesellschaft, Chemical Abstracts,* and *Journal of the Chemical Society, Abstracts of Papers,* they seldom were cited in American articles.[85] One therefore suspects that the infrequency of reaction to macromolecular theory in the United States between 1920 and 1927 was in large part due to Germany's postwar intellectual isolation.

Recent historical literature on the emergence of American science has focused attention on the influence of German scientific disciplines on higher education in the United States. A common pattern for this transmission was that American students who had been trained by influential professors at German universities imported a newly emerging specialty and subsequently founded research schools in American universities. In the chemical sciences one may well find this pattern in such areas as agricultural chemistry, biochemistry, physical chemistry, and, to some extent, colloid chemistry.[86] The question must therefore be raised whether the rise of macromolecular chemistry in America fits in with this stereotype.

At the Eidgenössische Technische Hochschule in Zürich, Staudinger introduced the subject of macromolecules to seventeen doctoral students, most of them Swiss or German (see Appendix Table 1). Although the American pilgrimage to German schools had passed its peak by that time, two American students studied this field under Staudinger in the 1920s. Herman Alexander Bruson, a native of Ohio, wrote his dissertation under Staudinger's direction in 1925. He published three articles on the subject in German and Swiss journals in 1926 and 1929.[87] After graduation, Bruson was employed by industry in the United States, including Goodyear Tire and Rubber Company (1925–28), Rohm and Haas (1928–48) and Industrial Rayon Corporation (1948–52). He worked on practical problems of synthetic polymers, such as plasticizers, polyurethane foam, and oil-soluble resins. In his long career as an industrial chemist, he published 28 papers and acquired about 400 patents in his name. Nevertheless, he made little substantial impact on the academic world in the late 1920s.[88] Another American, Avery Allen Ashdown, completed a Ph.D. dissertation at the Massachusetts Institute of Technology, followed by postdoctoral education as a Moore Fellow in Staudinger's laboratory from 1924 to 1925. He worked on the constitution of synthetic polymers such as polyindenes and poly-α-phenylbutadiene. The results of this study were published first in 1929 in a German journal (jointly authored by Bruson and Staudinger), and again in 1930 in a Swiss journal.[89] However, after returning to the United States in 1925, Ashdown did not pursue the subject further. He remained on the faculty of MIT, where his interest returned to the problem of organic reactions of ordinary low molecular substances.[90] Aside from their German articles, Staudinger's two American students did not exert any visible influence on polymer research in the United States during the 1920s. Hence the birth of American macromolecular chemistry cannot be ascribed to the role of Staudinger's American students.

In the late 1920s Staudinger's work was mentioned by certain English-speaking chemists. An English cellulose chemist, Walter N. Haworth, for example, first cited Staudinger's theory of macromolecules in 1928 in the *Annual Reports on the Progress of Chemistry*, published by the Chemical Society of London.[91] The British-born rubber chemist George Stafford Whitby, of McGill University in Montreal, occasionally made favorable remarks about Staudinger's views.[92] But, despite this information which was available to American readers, the macromolecular concept was still largely ignored in American scientific circles.

There were well-founded reasons why American chemists in the 1920s were, on the whole, not receptive to the idea of very large polymer molecules. Ironically, this "chemical *Zeitgeist*" had been to a considerable extent determined by prewar German influence. Emil Fischer's elaborate study of proteins through the syntheses of artificial polypeptides was widely recognized by American organic chemists. His famous 1913 lecture in Vienna was translated into English and appeared the following year in the *Journal of the American Chemical Society*.[93] In this lecture Fischer declared the molecular weight 4,021 of his synthetic compound to be the highest of all synthetic substances of known structure, and even of all natural proteins. As we have seen in the first chapter, this statement turned out to be an influential dictum by convincing chemists that compounds of molecular weight greater than 5,000 could not exist. It was this very dictum that Carothers would eventually challenge.

The influence of the Ostwaldian tradition on American colloid chemists no doubt played an even more important part in American reluctance to accept the concept of large molecules. Prior to 1910 only a small number of systematic studies of colloid chemistry had been carried out in America. In late 1913 to 1914, Wolfgang Ostwald, the German leader of colloid chemistry, came to America to give a series of lectures in this field. Highly successful, the lectures drew considerable attention from his American audience. "Originally invited by five universities," Ostwald noted, "I found the interest in the science [in this country] . . . so great that their number grew to sixteen while the actual number of lectures demanded of me during some seventy-four days was fifty-six."[94] He preached his scientific doctrine at many universities, including the University of Cincinnati, the University of Illinois, Columbia University, Johns Hopkins University, the University of Chicago, Ohio State University, the University of Pittsburgh, the University of Nebraska, the University of Kansas, and McGill University. He also lectured before the

National Academy of Sciences and the American Chemical Society (branches in Cincinnati, Indianapolis, and Washington). As Ralph Edward Oesper of the University of Cincinnati reported, "Those who heard his wonderfully interesting and well illustrated talks were infected with enthusiasm and went away with a fuller comprehension of what colloid chemistry was and could become."[95] These lectures resulted in his well-known book, *Die Welt der vernachlässigten Dimensionen* (1915), in which he noted, "I do not hesitate in consequence to designate this volume a propaganda sheet for colloid chemistry."[96] During the war the book was translated into English by his best and most trusted friend, Martin Henry Fischer, a German-born American and professor of physiology at the University of Cincinnati. The translation won wide readership well into the next decade.[97] Fischer applauded: "Wolfgang Ostwald's writings represent in colloid chemistry what those of Charles Gerhardt represent in organic, Justus Liebig in agricultural, and Wilhelm Ostwald in physical chemistry."[98] Through Ostwald's lectures and publications and Fischer's popularization of his ideas, a number of Americans were inspired to go into colloid chemistry, a branch of science with both theoretical and practical significance.[99] As American colloid chemistry arose, the Ostwaldian doctrine, which viewed colloids as a physical state of matter, penetrated the chemical community.

The mobilization of science during World War I greatly enhanced the status of chemistry in America, especially that of colloid chemistry, which formed an important area of wartime research. The Chemical Warfare Service, created in 1918, employed many chemists to work on colloidal problems of gas and masks. The National Research Council, established in 1916, organized a Committee on Colloid Chemistry at the urging of prominent colloid chemists such as Wilder D. Bancroft. During the postwar period, the Committee acted as the major organ to encourage and promote research and education in this field.[100]

The colloid chemistry boom reached a peak in America in the 1920s, as reflected by a flood of literature on colloids and the large number of practitioners. During this decade, according to a survey, over seventy English texts appeared on the subject; more than a hundred articles were published annually, in English alone; and in North America 20 to 25 percent of research chemists were focusing on colloids or related topics.[101] In 1925 the American Chemical Society's *Journal of Chemical Education* posted "Editor's Outlook," which urged:

> As teachers of chemistry, we are unmistakably confronted with the fact that colloid chemistry has not only taken its place as one of the major branches of our science, but that its rapid and continuing development is already making it a fundamental one.
> Even the specialist (in whatever branch of chemistry he may engage) is now

forced to recognize the fact that his field includes innumerable phenomena of a colloidal nature and that the pursuit of his specialty necessitates at least a general knowledge of colloidal principles. To the general chemist, the situation is even more obvious.

We trust that the teachers of chemistry will not be the last to interpret the handwriting on the wall.[102]

American colloid chemistry owed its institutional expansion largely to physical chemistry, a discipline American students of Wilhelm Ostwald had successfully introduced into U.S. universities around the turn of the century. In a country that before the war had no dyestuffs industry worthy of the name, organic chemists had failed to reach the academic heights their German counterparts enjoyed. Yet physical chemistry prospered on American institutional soil. As its universities expanded, with enrollments in chemistry courses growing rapidly between the 1900s and 1910s, physical chemists capitalized on the opportunity by occupying the bulk of the teaching positions in introductory or general chemistry.[103] As early as 1904, Arthur A. Noyes — one of Ostwald's former students and now an MIT physical chemist — in his presidential address before the ACS meeting held in Philadelphia had called special attention to colloids or an "important state of aggregation" as "a favorite hunting ground of physical chemists and physiologists."[104] Especially after the war, textbooks of physical chemistry often began including a chapter on colloids; and textbooks of colloid chemistry defined the study of colloids as an outgrowth of physical chemistry.[105] Also, a number of physical chemists entered physiological and biological fields via the study of biocolloids such as proteins. Symbolically, the Harvard Medical School in 1920 created a Department of Physical Chemistry primarily for the study of proteins.

Among the leading schools of colloid chemistry in the United States during the 1920s was one founded at Cornell University by the physical chemist Wilder Dwight Bancroft.[106] Bancroft, a student of Wilhelm Ostwald, was an advocate of his mentor's ideal of *allgemeine Chemie* that sought to unify the various branches of chemistry. Physical chemistry, in Bancroft's view, was not merely a branch of chemistry but encompassed the whole of chemical science. Furthermore, he claimed that other sciences, such as biology, medicine, and even physics, were "all subdivisions of chemistry." Bancroft felt that "It should be the aim of all chemists to have chemistry take its place as the fundamental science and that can only be done by and through the physical chemist."[107] In 1896 he founded at Cornell the *Journal of Physical Chemistry*, the first English-language periodical to focus exclusively on physical chemistry. During the 1910s and 1920s, Bancroft's interest increasingly centered on colloid chemistry, a field which he considered to be the most significant part of physical chemistry and as yet still in the early stages of development. By the mid-1920s, almost half the contents of his journal were concerned with topics in colloid chemistry.[108]

In 1921 Bancroft published a textbook, *Applied Colloid Chemistry*, of which two further editions appeared in 1926 and 1932.[109] In it, he asserted that colloid chemistry was the physical chemistry of everyday life. To justify this claim, he, like Wolfgang Ostwald, cited a long list of colloid-related industries and other human practices, ranging from cement, bricks, pottery, and glass to glue, starch, paints, lacquers, rubber, celluloid, leather, paper, textiles, dyestuffs, and drugs.[110] Also in this book, he, like Wolfgang Ostwald, strongly supported the view that colloids were not true chemical compounds but rather, a state into which any substance might be brought. As he explained:

> Graham believed that the distinction between a crystalloid and a colloid was fundamental and was due to some molecular condition. Though modern colloid chemistry begins with Graham, his distinction between crystalloids and colloids has been dropped. A colloidal substance is not necessarily amorphous. . . . We now speak of a colloidal state instead of a colloidal substance, and we call any phase colloidal when it is sufficiently finely divided or dispersed, without committing ourselves definitely as to what degree of subdivision is necessary in any particular case.[111]

Like Max Bergmann, Bancroft vigorously upheld the view that proteins were composed of aggregates of small molecules. Rejecting reportedly high molecular weights for proteins, Bancroft asserted that protein solutions formed colloidal suspensions consisting of two phases rather than one; for this reason they did not obey van't Hoff's laws of osmotic pressure, since the laws were to be applied to a true chemical solution. He stated:

> In general the apparent osmotic pressures of colloidal solutions are very small and this has led to absurd molecular weights. Such values as 30,000 for albumin, 700,000 for glycogen, over 48,000 for silica and "enormous" for Fe_2O_3 mean nothing whatsoever.[112]

For example, he did not accept the results of the Danish physical chemist Søren Peter Lauritz Sørensen (later known as the originator of the pH concept), who in 1917 calculated a molecular weight of 34,000 for egg albumin (an egg protein) on the basis of his measurement of its osmotic pressure.[113] Sørensen's study became known in the United States after the First World War, especially through the work of his American student Edwin Joseph Cohn. According to John Tileston Edsall, who worked in Cohn's Physical Chemistry Laboratory at Harvard Medical School during the 1920s:

At a scientific meeting shortly after 1920, Cohn had an exchange of remarks with Wilder D. Bancroft, the well-known colloid chemist, which went somewhat as follows:
Cohn: Sørensen has measured the osmotic pressure of egg albumin, and finds a molecular weight of 34,000.
Bancroft: Yes, yes. I understand. He is measuring a system of molecular aggregates. That is the molecular weight of the aggregate.
Cohn: But the tryptophan content and sulfur content of ovalbumin have also been determined, and they give a *minimum* molecular weight of 34,000.
Bancroft: Ah! Then in that case I would say that the aggregation factor is unity![114]

In Bancroft's view, as in that of many other colloid chemists, any effort to treat proteins as chemical individuals with definitive structure was a waste of time. Since the solutions of a protein or any other colloid represented merely a physical state of matter, they could not be substances of particular chemical structures.

Although colloid chemistry was booming in the mid-1920s, Americans believed that they still lagged behind European colleagues in fundamental contributions to this area. To activate chemical research and education, a number of American universities sought to invite foremost foreign colloid chemists to join their faculties. Thus the University of Chicago planned to offer a chair of colloid chemistry to either The Svedberg of the University of Uppsala or James W. McBain of the University of Bristol.[115] MIT's Department of Chemical Engineering showed strong interest in having Ernst Alfred Hauser (an Austrian-born chemist of the Colloid Chemical Laboratory of the Metallgesellschaft, A.G.) "to develop at the Institute the outstanding colloid research center of the country."[116]

As early as 1923 the University of Wisconsin had succeeded in bringing Svedberg to Madison as a visiting professor. Joseph Howard Mathews, a colloid chemist and chairman of its Chemistry Department, arranged the invitation in response to the Dean of the Graduate School's admonishment that "we ought to do something to pep up research at Wisconsin."[117] At Madison, Svedberg directed eight graduate students, including two future DuPont researchers, Elmer O. Kraemer and James B. Nichols. It was there that Svedberg began his famous experimental work on centrifugal methods for determining the size of colloidal particles. He and Nichols built a small centrifuge that yielded a force of only 150 times that of gravity, but this was sufficient to establish the prospect for further study.

While in the United States, Svedberg was frequently invited by the American Chemical Society to give lectures around the country. However, in marked contrast to Wolfgang Ostwald, Svedberg proved to be a failure as a speaker. His presentations were always poorly organized and perplexing. As one observer commented: "His large audiences generally went away disappointed with his uninspired presentations."[118]

Nevertheless, his stay in America culminated in the First National Colloid Symposium, held in June 1923 on the Madison campus. The event was organized by Mathews, aided by the National Research Council's Committee on Colloid Chemistry. A member of the committee, Bancroft used his influence in supporting the symposium. Svedberg was chosen to be the special guest of the symposium. It drew 175 participants, and twenty-five papers were presented — a clear indication of the phenomenal recent growth of the colloid community in the United States.[119] The creation of a Division of Colloid Chemistry in the American Chemical Society, in the year 1926, was an outgrowth of this symposium.[120]

After his return to Uppsala, Svedberg constructed an improved ultracentrifuge that could generate a force 100,000 times that of gravity. This machine allowed him to measure very high molecular weights of proteins. He and Robin Fåhraeus, professor of pathology at the University of Uppsala, published the first paper on proteins, entitled "A New Method for the Determination of the Molecular Weight of the Proteins," in the *Journal of the American Chemical Society* in 1926, the same year that Svedberg was awarded the Nobel Prize.[121]

However, so staunch was Bancroft's belief in proteins as molecular aggregates that he remained unconvinced by Svedberg's results demonstrating the molecular weight of proteins to be extremely high. In Bancroft's eyes, Svedberg was only confusing particle size with molecular weight.[122] What had appeared to others, such as Staudinger, Meyer, and Mark, to be a decisive step toward the establishment of the macromolecularity of proteins did not convince him at all.

Jacques Loeb was one of a few outspoken critics of the colloid chemistry movement, both in America and in Germany. His physico-chemical experiments on gelatin, Loeb claimed, demonstrated that proteins were primarily crystalloid molecules like the amino acids from which they were built, and that the laws of classical chemistry would explain the chemical and physical behavior of proteins. As he stated, "what has been called colloid chemistry is in reality only a part of classical chemistry and finds its complete explanation from classical chemistry."[123] By "classical chemistry" he referred to physical chemistry, especially the ionic theory of Svante Arrhenius. An evangelist for physical chemistry as the key to biology, the German-born American physiologist grew resentful towards colloid specialists such as Wolfgang Ostwald, Richard Zsigmondy, and Wolfgang Pauli, as they paraded the claim in Germany that since colloids were subject to special laws, colloid chemistry should be declared an independent discipline. In 1923 Loeb wrote bitterly to Arrhenius:

I have read the Leipzig transactions of the new "Colloid-Abteilung." It is poor stuff and I do not quite understand what has become of the critical scientific spirit in Germany that such empty verbalisms and speculations based on purely qualitative experiments or no experiments at all are accepted into the society of scientific production. "Colloid chemistry," or rather the adsorption theory, was an offshoot from the temporary error into which Wilhelm Ostwald had fallen that molecules have no real existence and that stoichiometrical behavior is an artifact. . . . With the limited means available in Germany for scientific research they have no right to waste money on futilities like "colloid chemistry."[124]

Wolfgang Ostwald had spent two years as a postdoctoral student at Loeb's laboratory in Berkeley some fifteen years earlier. But now Loeb condemned even this former student most vehemently:

. . . colloid chemistry in Germany has fallen into the hands of men like Wolfgang Ostwald, who, to say the least, are not scientists of the first rank; in fact, I consider Wolfgang Ostwald, whom I know personally, as a rather incompetent man. His leadership will prove an expensive luxury to the prestige of German science as well as to protein industries in Germany, because his methods and his theories are antiquated.[125]

Yet when it came to the American situation, Loeb was too optimistic. He believed that the U.S. protein industries had accepted his own work. Loeb added:

Of course, the medical quacks probably will continue to stick to the old fashioned "colloid chemistry" which they find quite remunerative. One occasionally finds also a physical chemist like Wilder Bancroft sticking to that confusion, but I do not think Bancroft has ever done an experiment in his whole life, and since he is neither a mathematician nor capable of rationalistic thinking, I think he can be ignored.[126]

Despite his strong sentiments, it was Loeb who remained ignored. He died in 1924 at the height of the colloid chemistry movement in America, which he had underestimated. He never lived to see the acceptance of his views. Loeb argued for the molecularity of proteins on the basis of the premises and techniques of physical chemistry rather than the principles of molecular structure. As Edsall noted, Loeb's work made almost no contribution to the determination of the molecular size and structure of proteins.[127] Typical of a physiologist and physical chemist of the time, he did not care about the recent German upheaval in organic chemistry; nor was he familiar with Staudinger's theory of macromolecules.

Despite Bancroft's attack, Cohn continued Sørensen's line of research on the molecular weights of proteins. One of the arguments in favor of the aggregate structure of proteins, made by Reginald O. Herzog and others in the mid-1920s, was that proteins, when dissolved in phenol, fell apart into units with molecular weights of only between 200 and 600. Proteins were therefore considered to be aggregates of these small units held together by secondary valences.[128] In 1926 Cohn and his colleague James B. Conant, who was then teaching in the Department of Chemistry at Harvard, demonstrated by measuring freezing points of proteins in phenol that the low values previously obtained were due to the presence of water as an impurity in the phenol. When the water content of phenol was eliminated, the freezing-point depressions produced by the addition of proteins were extremely small, which indicated that proteins had very high molecular weights.[129] In this way, they supported Sørensen's position and rejected the prevailing argument which favored the aggregate theory.

While working with Cohn on proteins, the organic chemist Conant had maintained a cautious and rather skeptical attitude to Staudinger's belief that polymeric compounds with a colloidal nature were made up of macromolecules. As early as 1925—at a time when antipathy to German chemists was still strong among his colleagues—Conant had taken a private trip to Germany to visit postwar German laboratories and to become personally acquainted with a number of outstanding chemists.[130] In Switzerland he met Staudinger, then professor of chemistry at the ETH in Zürich. Staudinger, together with his American student Ashdown, explained Staudinger's "new and revolutionary point of view" to Conant.[131] The meeting occurred during the time when Staudinger's theory was under strong attack from many of his contemporaries in the Swiss-German chemical community. According to Staudinger's memoirs, when Conant visited his laboratory in Zürich, "My co-workers and I told him our arguments in favor of the macromolecular structure of these polymeric compounds. On his visit to Germany which immediately followed he was told not to believe a word of Staudinger!"[132] In Germany he, in fact, visited Max Bergmann, director of the Kaiser Wilhelm Institute for Leather Research in Dresden, and learned from him a version exactly opposite to that of Staudinger. Conant first related this story three decades later after Staudinger had received the Nobel Prize.[133] The climate of opinion of the time, one might well suspect, led Conant in his reluctance to accept the macromolecular view of polymers.

The context in which Conant and Cohn launched their cooperative study of proteins, mentioned above, was well explained by George Scatchard, who worked at the Harvard Medical School in the mid-1920s:

In 1925 Jim Conant came back from Europe sold on the idea that the real molecular weights of proteins are 200–600 and that what was studied in aqueous solutions were colloidal aggregates. This sat badly with Edwin Cohn, who was then determining minimum molecular weights from the amounts of some rarer elements or amino acids. So they agreed to settle it experimentally. The most precise measurements in the world were being made in T. W. Richards' laboratory where Cohn had worked with Richards' brother-in-law during the war, and Conant was Richards's son-in-law. So they went to Richards' laboratories. They started with the freezing point depression of zein in phenol. For a while it was one of those researches which gave one answer on Monday, Wednesday and Friday, and another on Tuesday, Thursday and Saturday. I saw them both frequently. On the days when I saw Conant the molecular weight seemed very small, when I saw Cohn it seemed large, when I saw them both together the answer was indefinite. Then they decided that the phenol must be splitting water out of the protein. . . . They added calcium chloride and its lowest hydrate to give a constant water activity, and they found no measurable depression. They published a paper concluding that the molecular weight of zein in phenol is certainly greater than 10,000 and probably greater than 100,000, and showing no indication that there had ever been any disagreement between them.[134]

Although he acknowledged the apparent high molecular weights for proteins, Conant did not immediately conclude that proteins were composed of long-chain molecules held together by normal valences. In his well-received textbook, *Organic Chemistry*, which was published in 1928, he wrote:

The question arises, are the polypeptide units in the protein molecule held together by amide-like linkages or some other way? There is at present no conclusive answer to this question. Many believe that cyclic systems are involved and even the forces which unite the polypeptide residues are different from the usual valences which hold atoms together in the simple organic compounds. This is one of the important questions now facing the chemist.[135]

Concerning the structure of rubber, he expressed more explicitly a view favoring the aggregate theory:

If one obtains as pure a sample as possible of crude rubber and examines it in the laboratory, one finds that its analysis corresponds to the formula C_5H_8. The material dissolves in only a few organic solvents, forming a colloidal solution. It is not possible to determine the molecular weight because the substance cannot be vaporized without decomposition, and the freezing point and boiling point methods are not applicable to colloidal solution. The formula $(C_5H_8)_n$ is often written for it. It is probable that n has a value of 10 to 15.[136]

The polymerization of isoprene molecules would yield rubber-like materials $(C_5H_8)_n$. "But we are," he continued, "still uncertain how the isoprene molecules are united in the polymer; it is possible that the ten or more molecules combine in one or several large rings."[137] Thus Conant, like many chemists of the time, was inclined toward the aggregate theory developed by Carl Harries and Rudolf Pummerer, although he was well acquainted with Staudinger's ideas.

It was at Harvard about this time that Conant's peer, Carothers, developed his interest in the field of polymers. As the former observed, Carothers's "first thinking about polymerization and the structure of sub-stances of high molecular weight began while he was at Harvard."[138] Before moving to DuPont, Carothers had "quite a long talk" with Conant about polymers and polymerization. Conant did endorse some of his views.[139] But clearly, Carothers departed from Conant's basic interpreta-tion of the polymer structure. "He was never content," recalled Conant, "to follow the beaten track or to accept the usual interpretation of or-ganic reactions."[140] From the outset, Carothers accepted Staudinger's concept of macromolecules, and developed it into his own research proj-ect on polymerization. Despite the skepticism of Conant and other aca-demic colleagues, he arrived at DuPont with a research program based on this idea.

Origins and Development of Carothers's Macromolecular Synthesis

In early February 1928, Carothers started work at a new laboratory in the DuPont Experimental Station in Wilmington, Delaware. In recogni-tion of its devotion to pure science, DuPont employees quickly dubbed the new building that housed his laboratory "Purity Hall" and the re-searchers there "Virgins."[141] Eight days after he began working at the company, Carothers wrote cheerfully to a friend, John R. Johnson, about his new life as a DuPont research chemist:

Regarding the funds, the sky is the limit. I can spend as much as I please. Nobody asks any questions as to how I am spending my time or what my plans are for the future. Apparently it is all up to me. So even though it was somewhat of a wrench to leave Harvard when the time finally came, the new job looks just as good from this side as it did from the other. According to any orthodox standards, making the move was certainly the correct thing.[142]

At Purity Hall, Carothers soon launched into research on polymers. His study was, he later wrote, "first initiated at a time when a great deal of scepticism prevailed concerning the possibility of applying the usually accepted ideas of structural organic chemistry to such naturally occur-ring materials as cellulose."[143] He found that the tendency to evade or ignore the classical molecular concept in dealing with polymers was the source of such skepticism and confusion among chemists.[144] Unlike many of his contemporaries, he adopted Staudinger's macromolecular concept from the start. It is to be emphasized that Carothers's direct motives to adopt the concept were neither interest in the recent studies of X-ray crystallography nor Svedberg's ultracentrifuge work on proteins. As an organic structuralist, he shared with Staudinger the belief in the molecularity of polymers on the basis of the classical structural theory. Like Staudinger, Carothers maintained that the chemical molecule was the entity from which stemmed all physical and chemical properties of the substance. Polymers were not an exception. Hence the unique properties of polymeric substances, such as colloidality, could be explained only by the large molecules themselves, and not by physical forces. In this respect, he followed the organic-structural approach, on which Staudinger based his arguments, as opposed to the prevailing physicalist view of polymers.[145]

However, to demonstrate the existence of macromolecules and also examine their properties, Carothers took an initial step in his research that differed significantly from Staudinger's methods. His letter of November 9, 1927 to Bradshaw reveals that Carothers had by then already devised a synthetic approach:

> For some time I have been hoping that it might be possible to tackle this problem from the synthetic side. The idea would be to build up some very large molecules by simple and definite reactions in such a way that there could be no doubt as to their structures. This idea is no doubt a little fantastic and one might run up against insuperable difficulties, but after all [Emil] Fischer synthesized an octadecapeptide [$C_{220}H_{142}O_{58}N_4I_2$ with a molecular weight of 4021] and it behaved like a simplified protein.[146]

Carothers intended to expand the synthetic approach developed by Fischer, who had attempted to elucidate the protein structure by attaining the synthesis of artificial polypeptides of known structure. His objectives were first to synthesize macromolecules of definitive structure by use of established classical organic reactions, and then to examine the relation between the properties of polymers and their chemical structures. "The largest molecule of known structure either natural or synthetic is," Carothers said at DuPont, "one synthesized by Fischer and having a molecular weight of 4021."[147] Whereas Fischer's structural scheme limited the size of the *Riesenmolekül* to less than 5000 in molecular weight, Carothers felt certain about the possibility of making far larger molecules using his method. As he wrote to Johnson, "It would seem quite possible to beat Fischer's record. . . . It would be a satisfaction to do this."[148]

In order to prepare artificial macromolecules, he decided to use what he called the "bifunctional reaction." By the fall of 1927 Carothers had developed a general interest in this type of reaction, and, in all likelihood, hit upon its application to the macromolecular synthesis in mid-December 1927, one and a half months before he moved to DuPont.[149] Two important lines of thought were to merge into his concept of bifunctional reaction; the first concerned Grignard reagents, the second,

Glyptal resins. While still at Harvard, he worked on Grignard reagents, or alkylmagnesium halides (as represented by RMgX), named after the French discoverer, Victor Grignard. By using electrolytic cells, Carothers was able to examine the formation and structure of a series of these reagents.[150] Grignard reagents reacted readily with such chemicals as aldehydes, ketones, and esters, reactions which were known as "Grignard reactions." When Carothers came across one such reagent, acetylene (di-)magnesium bromide ($C_2(MgBr)_2$), which was bifunctional (having two reactive atomic groups one on each end of the molecule), it occurred to him that it might react readily with bifunctional peers, such as dialdehydes, diketones, and diesters to produce long chain molecules. These Grignard reactions involved condensations, that is, reactions proceeding with the elimination of simple substances like water. According to Carothers's recollection:

> I first became interested in what I have since called bifunctional condensations during the autumn of 1927 in connection with some work that one of my students was doing on acetylene (di-)magnesium bromide. It seemed to me inevitable that such condensations would, in many cases, lead to long chains. In a rather vague way I planned some experimental work.[151]

He now gave much thought to a link between condensations and the making of long chain molecules.

"Early work at Cambridge," said Carothers, "was unproductive, and a successful method of attack was first suggested when I came into contact with some of the resin work at the [Experimental] Station."[152] In mid-December 1927 he paid a second visit to Wilmington.[153] At that time Arthur Tanberg, director of the Station, told him about Glyptal (glyceryl phthalate) resins.[154] Originally discovered by the British chemist Watson Smith at the turn of the century, glyceryl phthalate was first produced commercially in 1914 by General Electric, under the trade name "Glyptal."[155] Glyptal resins were used for coating transformer coils and other electrical appliances. In the mid-1920s chemists in DuPont's Chemical Department became interested in Glyptal resins, since they were considered promising candidates to improve "Duco" automotive finishes. Having entered into a patent-licensing agreement with General Electric in early 1927, DuPont was ready to exploit the modified Glyptal resins commercially under the trade name "Dulux."[156]

Glyptal resins were formed by the action of phthalic acid and glycerol, but their precise structure and formation mechanism remained unclear.[157] In late 1927, while visiting the Experimental Station, Carothers immediately recognized that the reactants had respectively two carboxyl groups (bifunctional) and three hydroxy groups (trifunctional) and that the only reaction involved in this process should be esterification. He recalled, "This, it seemed at once, must be a polyfunctional intermolecular esterification. For purposes of theoretical exploration, a purely bifunctional type seemed more interesting and appropriate, and esterifications more suitable than Grignard reactions."[158] Hence, through his encounter with the commercial product, bifunctional esterifications flashed across Carothers's mind as a more simplified, ideal method of attack on the problems of polymerization.

In this context, Carothers must have recalled an experiment he had conducted at Adams's Illinois school in 1926. "Before leaving the University of Illinois," Johnson remarked "he [Carothers] had started an investigation of the reaction of ethylene glycol [HO–C_2H_4–OH] with succinic acid [HOOC–C_2H_4–COOH]."[159] Although he had no intention to synthesize polymers at that time, this reaction was, in retrospect, certainly a particular case of bifunctional esterification.

Carothers now decided to extend the reaction mechanism to the building up of giant molecules. As was long known, an alcohol (R–OH, where R is an organic radical) reacts with a carboxylic acid (R'–COOH, where R' is another organic radical) to form an ester (R'–COO–R) by the elimination of water (H_2O). Likewise, both a bifunctional alcohol (HO–R–OH) and a bifunctional carboxylic acid (HOOC–R'–COOH) would yield an ester (HOOC–R'–COO–R–OH) through the condensation process. He realized that a continuous condensation reaction of these small molecules would produce long-chain molecules of known structure (HOOC–R' $\cdots$ COO–R–OOC–R'–COO–R–OOC–R'–COO $\cdots$ R–OH), so long as the reaction did not end with the formation of large rings.

By the time he joined DuPont, Carothers had developed his ideas into a coherent program for the synthesis of macromolecules. In the above-mentioned letter to Johnson of February 14, 1928, Carothers summarized in general terms the bifunctional reaction. He outlined his plan,

to study the action of substances xAx on yBy where A and B are divalent radicals and x and y are functional groups capable of reacting with each other. Where A and B are quite short, such reactions lead to simple rings of which many have been synthesized by this method. Where they are long formation of small rings is not possible. Hence reaction must result either in large rings or endless chains. It may be possible to find out which reaction occurs. In any event the reactions will lead to the formation of substances of high molecular weight and containing known linkages.[160]

The reaction of compounds xAx with yBy, or the bifunctional reaction, remained essential to his study of polymerization. It was an application of a simple classical organic reaction, namely, condensation, as represented by esterification.

Staudinger had overlooked condensation reactions in his vision of polymerization. He even suspected that condensations would not lead to macromolecules, for the end group reactivity of a molecule would decrease rapidly with the increase of chain length.[161] But from his prior experience with organic reactions Carothers was confident that this would not be the case. The reactivity of functional groups, he believed, would not be affected by an increase of chain length insofar as increased viscosity did not impinge molecular mobility.[162]

In early March Carothers proposed to Stine his research project at DuPont. In the proposal, he stated that esterification was especially susceptible to experimental study of polymerization, because (1) the product, whatever its structure, must be an ester; (2) esterification would not be complicated by side reactions; (3) it is quantitatively reversible, which would allow kinetic studies of the formation and saponification; (4) the structural unit was known; and (5) materials were available for the study of a large number of reactions of this type. The polymers that resulted, he argued, would justify the study at DuPont for three reasons. First, he stated, "Commercially important resins probably belong to this class." Second, known substances of this class presented some of the peculiar properties of cellulose and its derivatives. Hence, there might be an analogy between condensed polymers and cellulose. Finally, the study would furnish general insight into how the nature of the structural unit affected physical properties of polymers as a whole.[163]

Early in 1928 Roger Adams and Carl (Speed) Shipp Marvel, an assistant professor of the University of Illinois, were engaged as consultants to DuPont. They began alternate monthly visits to Wilmington.[164] Seeking advice, Carothers informed them of his research project. Adams replied, "The problem . . . of a general study of large molecules is an enticing one, though naturally it will be a difficult one." At Illinois, Adams had incidentally encountered bothersome, highly condensed products which were "atrociously insoluble in every type of solvent" so that it was "almost impossible to get a means of attack." Yet Adams encouraged Carothers to tackle this new challenge, and Marvel was able to assist the latter by obtaining such materials as various acids and alcohols.[165]

Meanwhile, the DuPont colloid group was busy constructing and testing an ultracentrifuge, a costly apparatus that even Staudinger could not obtain in Germany. DuPont was the first company in American industry to obtain one. This and the apparatus of the University of Wisconsin were the only two ultracentrifuges in the United States during the mid-1930s.[166] For the colloidal chemical study, DuPont had recruited The Svedberg's two students, Elmer O. Kraemer and James B. Nichols.

Kraemer had received his bachelor's degree from the University of Wisconsin in 1918. After studying at Uppsala and Berlin between 1921 and 1923, he returned to Wisconsin to write a dissertation on colloids under Mathews and visiting professor Svedberg, which was completed in 1924. While teaching at Wisconsin as an assistant professor, he served as a DuPont consultant until they hired him fulltime to head the colloid group in September 1927.[167]

As an undergraduate at Cornell, Nichols had written a senior thesis under Wilder Bancroft, who assigned him the subject of molecular weight determination of nitrocellulose using the boiling and the freezing point methods.[168] Moving to Wisconsin in 1923, Nichols came under the influence of Svedberg, who introduced him to the ultracentrifuge. From 1925 to 1927, he continued his graduate work with Svedberg at Uppsala, where he had occasion to see his teacher win the Nobel Prize. Joining DuPont as Kraemer's assistant in November 1927, Nichols helped install the Svedberg ultracentrifuge. With this instrument, they began to investigate the particle size of pigments and the molecular weights of cellulose derivatives.[169] This machine, together with other tools of the colloid group, would help determine the molecular weights of Carothers's synthetic polymers.[170]

Now the stage was set for Carothers to commence work on polymers. Despite Adams's apprehension, Carothers's synthetic work started favorably. In March 1928 he began with the preparation of poly(ethylene sebacate). Later that year and into the next, three assistants, J. A. Arvin, Frank Van Natta, and George L. Dorough, joined him in synthesizing a series of polyesters. In December 1928 he wrote to Adams:

Our work is progressing rather slowly but still favorably — that is to say, in the expected direction. . . . We have made quite a large number of esters. . . All those in the aliphatic series are crystalline and quite soluble, so we can purify them and get molecular weights in various solvents. The molecular weights don't check very well but they are in the region from 3 to 4000. . . . So far everything goes along to check up my original guess.[171]

By that time "the conviction was firmly established in my own mind," Carothers later recalled, "that bifunctional condensations involving unit length beyond 7 were, under appropriate conditions, almost perfectly intermolecular, and they were therefore functionally capable of almost indefinite structural propagation." It became evident to him that "most of the few statements in the scientific literature concerning such reactions were wrong." Moreover, he soon became impressed with the idea that the linear condensation polymers he was developing provided a close analogy in structure to cellulose or silk.[172]

In 1929 Carothers presented a general theory based on his results in the landmark paper, "An Introduction to the General Theory of Condensation Polymers," published in the *Journal of the American Chemical Society*.[173] In this article he classified polymers into two groups: addition polymers (or A-polymers) and condensation polymers (or C-polymers), according to the type of polymerization. Addition polymers were those produced by the self-addition reaction of monomers. The molecular formula of the monomer was therefore identical with that of the recurring structural unit of the polymer. Polystyrene and polyoxymethylene, which Staudinger had studied, belonged to this group. Condensation polymers were those formed by a polyintermolecular condensation reaction of the polyfunctional monomers through the elimination of such simple molecules as water.[174] The formula for the monomer of this type differed from that of the structural unit of the polymer. Polyesters, polyanhydrides, and polyamides represented the linear condensation polymer. "Polymerization then is," he wrote, "the chemical union of many similar molecules either (A) without or (C) with the elimination of simpler molecules."[175]

Carothers's classification of polymers was thus created solely from the synthetic standpoint. Moreover, he applied this classification not only to synthetic polymers but also to naturally occurring polymeric substances. For example, using the structural analogy, he identified rubber with linear addition polymers and silk and cellulose with linear condensation polymers.[176] In Carothers's mind, there was no breach between natural and artificial processes. As he later put it:

The idea that natural high polymers involve some principles of molecular structures peculiar to themselves and not capable of being simulated by synthetic materials is too strongly suggestive of the vital hypothesis, which preceded the dawn of organic chemistry, to be seriously considered.[177]

Carothers continued to cultivate the realm of bifunctional condensation polymers. Together with co-workers, he extended his synthetic study from polyesters to polyanhydrides and polyamides, beginning with various possible starting materials. By analytically demonstrating the presence of end groups (e.g., hydroxyl and carboxyl groups) in the products, he showed that they consisted not of molecular rings but of open-chain molecules. High molecular weights calculated from end-group determinations were in agreement with those determined by physical methods, such as the boiling-point method. In this way, he was able to conclude that condensation polymers were indeed made up of linear macromolecules. These results were published in a series of articles entitled "Studies on Polymerization and Ring Formation," which would total twenty-nine by 1936.[178] Carothers later evaluated the series as "an unusually significant contribution to the science of organic chemistry."[179]

Figure 8. Carothers in his DuPont laboratory, ca. 1930. Courtesy Hagley Museum and Library.

In a paper published in 1930, Roy Herman Kienle, a chemist at General Electric, had accounted for the formation of alkyd resins such as Glyptal in terms of "the random bonding of poly-reactive molecules that is responsible for heterogeneity, polymeric complexity, and colloidal properties of high polymers."[180] He acknowledged Carothers's 1929 paper, "An Introduction to the General Theory of Condensation Polymers," as well as subsequent papers, though he did not adopt the term "functionality" to explain essentially the same phenomena. The similarity between Carothers's and Kienle's views would lead some chemists, particularly coating chemists, to regard the latter as the originator of the functionality concept. For example, forty years later the editor of the *Journal of the Oil and Colour Association* pointed out that even Kienle's earlier paper, presented at a meeting of the American Chemical Society in September 1928 and published in 1929, "clearly foreshadowed" the 1930 paper, and therefore suspected: "Carothers *wrote* it [the concept of functionality] before Kienle, but . . . he [Kienle] may very well have been the first to *think* it."[181] Whatever the editor's sympathy for Kienle, Kienle's 1929 paper lacked a clear picture of the formation of macromolecules. Moreover, as we have seen, Carothers had first thought of polyfunctional reactions as early as late 1927. While Kienle's interest was limited to alkyd resins, Carothers concerned himself with the formation of polymers in general. At any rate, in his paper "Polymerization," published in 1931 in *Chemical Review*, Carothers, now aware of Kienle's work, made it clear that the formation of synthetic resins, such as Glyptal resins and Bakelite, should be classified as polyfunctional condensation reactions to form non-linear, three-dimensional macromolecules.[182]

In many ways, Carothers's paper, "Polymerization," laid the foundation of the new polymer chemistry. In this work, he provided an extensive survey of the existing literature, including works by Staudinger, Meyer, and Mark; gave clear definitions of terms and a new classification of a wide variety of polymers; and also presented systematic interpretations of polymerization based entirely on the macromolecular concept. Thus, he now endorsed the cellulose formula developed by Meyer and Mark, "at least in its essential outlines."[183] Following Meyer and Mark, he also suggested, "The vulcanization of rubber probably also involves the cross-linking of the long chains through the agency of the unsaturated linkages present."[184]

The impact of Carothers's review paper may well be illustrated by Marvel's comment: "After that article, the mystery of polymer chemistry was pretty well cleared up, and it was possible for less talented people to make good contributions in the field."[185] Indeed, Marvel himself, originally working on classical organic compounds, soon turned to research on polymers and was to become a prominent contributor to the field. Now James B. Conant, too, was led to accept Carothers's views. In a revised edition of his textbook on organic chemistry (1933), Conant quietly altered his earlier description of the aggregate theory of synthetic rubber:

. . . it seems probable that they [isoprene molecules] are united in a very long chain thus:

$$—CH_2-C=CH-CH_2—CH_2-C=CH-CH_2—CH_2-C=CH-CH_2—$$
$$\underset{CH_3}{|}\qquad\qquad\underset{CH_3}{|}\qquad\qquad\underset{CH_3}{|}$$

isoprene unit

This chain must be imagined as extended until the molecular weight is at least a hundred thousand.[186]

What Carothers's investigations continued to impress on contemporaries was that polymerization was a normal organic reaction: "polymerization is chemical combination involving the operation of primary valence force, and . . . the term polymer should not be used (as it frequently is by physical and inorganic chemists) to name loose or vaguely defined molecular aggregation."[187] On the same grounds, he did not accept the micelle concept of Meyer and Mark, who assumed that cellulose was composed of micelles, that is, aggregates of "primary valence chains" held together by special micellar forces. Well aware of the recent controversy between Staudinger and Meyer-Mark, Carothers favored Staudinger's view, stating that "micelles" were nothing but macromolecules. As Carothers explained, "so far as the minor differences in the views of these two groups of investigation are concerned, our own experiments on polyesters incline us to favor those of Staudinger. That is, we can find no real objection to referring to primary valence chains as molecules."[188]

Carothers's synthetic approach contrasted with Staudinger's analytical approach. Staudinger had attempted to show the macromolecularity of polymers primarily by analyzing and examining final products such as natural rubber and cellulose. He converted such products into their derivatives through hydrogenation, methylation, and nitration, and showed the similarities in properties between the original products and their derivatives, as exemplified by his 1922 experiment on the hydrogenation of rubber. As was discussed in Chapter 2, this method formed the basis of Staudinger's important evidence establishing the long-chain structure of polymers. Such an analytical approach was by no means exclusively Staudinger's. In order to examine the structure, his opponents — all advocates of the aggregate theory — also focused their attentions on the analysis and chemical treatment of naturally occurring polymers, such as rubber and proteins. In this regard, everyone debated the issues on the same ground.

Staudinger did, however, use a few synthetic models as a means of demonstration, a method which to some degree inspired Carothers's synthetic study. But, according to Carothers,

> the products studied by him [Staudinger] (polyoxymethylene, polystyrene, polyacrylic acid, etc.) although unquestionably simpler than naturally occurring polymers, were produced by reactions of unknown mechanism, and their behavior, except in the case of polyoxymethylene, was not sufficiently simple to furnish an unequivocal demonstration of their structure. On the other hand, the development of the principles of condensation polymerization [proposed by Carothers] . . . has led to strictly rational methods for the synthesis of linear polymers.[189]

Time and again, Carothers distinguished his own approach from Staudinger's use of synthetic model substances, which clearly seemed to complicate the demonstration, as indicated in the following passage:

> The obvious importance of simple synthetic models as an aid in studying macromolecular materials has been emphasized repeatedly by Staudinger. . . . Our own researches on condensation polymers were started with the idea that the fact of a proposed model's being synthetic is of little value unless the method of synthesis is rational, *i.e.*, unless it is sufficiently clear-cut to leave no doubt concerning the structure of the product. Polystyrene may, for example, serve as a simplified model of rubber, but it has the disadvantage that the method used in its synthesis . . . furnishes no certain clue to its structure. The independent demonstration of its structure presents the same difficulties as does rubber; in fact today the formula of rubber can be written with more assurance than that of polystyrene.[190]

In sum, Staudinger's synthetic models — addition polymers, according to Carothers's classification — were produced in particular by spontaneous polymerizations of unknown mechanisms. Therefore their structures had to be inferred from analysis of the final products and from examination of their properties. They could hardly be inferred from the reaction mechanism. Then Staudinger boldly used these synthetic polymers as models in his demonstration of the macromolecular structure of such natural polymers as rubber and cellulose. However, Carothers remained unconvinced by this approach because the structures of the synthetic models themselves appeared never to have been rigorously proven.

Whereas Staudinger was concerned with *final products*, Carothers attached greater importance to their formation *processes*. Carothers's method was strikingly simple and straightforward in contrast to Staudinger's roundabout method: Combine normal small molecules one by one by the use of well-established organic reactions, and you will eventually get long-chain molecules of definitely known constitution. Thus the structure of the polymerization product was beforehand predictable theoretically from the reaction mechanism, and afterward confirmable experimentally through, for instance, the demonstration of the presence and nature of end groups. This method appeared to leave no doubt when it came to demonstrating the structure of the product. In this regard, Carothers considered his own synthetic method to be "strictly rational," as distinguished from Staudinger's approach.[191]

A logically minded perfectionist, Carothers was critical even of Staudinger's writing style and his opinionated manner. For instance, reviewing Staudinger's book, *Die hochmolekulare organischen Verbindungen*, Carothers commented:

Unfortunately, this exposition is not as concise or as well-proportioned and organized as it might be. It is in parts unnecessarily repetitious. Abandoned views of erstwhile opponents are again set forth and again re-demolished; but aside from this the contributions of other investigators are, by comparison, rather inadequately recognized. The author is inclined to be dogmatic concerning certain speculative points of interpretation where, in fact, only slender and dubious evidence is available.[192]

For his part, Staudinger surely was familiar with Carothers's publications, as indicated by references to them in *Die hochmolekulare organischen Verbindungen*.[193] Yet he continued to maintain that condensations were less than essential to the study of polymerization.

Yukichi Go, a Japanese polymer chemist who had trained in Germany between 1930 and 1933, was particularly impressed by Carothers's balanced scientific style, which "at first sight, looks like a Germanized orderly system. But free from awkwardness and stiffness common to German academic style, it exhibits a remarkable character of flexibility and liveliness."[194] Paul J. Flory, an eminent DuPont co-worker, saw Carothers's approach to polymers as "the epitome of clarity of concept and of simplicity. He set out to synthesize polymers of high molecular weight by procedures free of ambiguity."[195] This approach was, Flory continued:

rigorous science, and he [Carothers] imbued that into his people [at DuPont] — at least if they were at all receptive. If they weren't, they didn't stay in his group very long. His conviction [was] that these seemingly complicated substances — polymers — could be approached scientifically.[196]

It is important to note that Marvel, Conant, Flory, and many other American chemists were led to accept the macromolecular theory not through Staudinger's work but by Carothers's work.

Industrial Fruits: Nylon and Neoprene

The month of April 1930 marked a turning point for Carothers's group as well as the company. Two years had elapsed since he had initiated his fundamental research on condensation polymers, which was winning a growing reputation both inside and outside DuPont. At the beginning of 1930, Stine proudly reviewed their progress: "Publication of results has occasioned favorable comment from numerous sources, and several of our men are earning increased recognition in the scientific world."[197] Carothers now usually had over ten men conducting experiments along the lines of the problems he had set forth. In April, within a short span of two weeks, his team made two separate discoveries that would lead to DuPont's unprecedented commercial products, nylon and neoprene.

Although much of the literature on the discovery of nylon has endorsed the view that Carothers went to DuPont in order to make synthetic fibers, that claim cannot be substantiated.[198] He did not intend to produce an artificial fiber when he started his study of polymerization in 1928, for, as he later remarked, "I do not think we . . . had definitely in mind at all the idea of making a fiber, but we did want to make a molecule as large as we could get."[199] His early structural analogy between synthetic polymers and natural fibers (such as silk and cellulose) signaled the first step, yet even this effort did not advance beyond a loose reasoning.

The key idea of the synthesis of artificial fibers occurred in the spring of 1930, and was incidental to Carothers's study of making giant molecules. Between 1928 and early 1930, Carothers's group had failed to synthesize high polymers with a molecular weight of more than 5000, although some of them slightly exceeded Fischer's record of 4,021. Polymers prepared by the Carothers team were generally brittle, opaque solids that melted at a low temperature (100°C or less); they dissolved readily in certain solvents, such as chloroform, and could be crystallized in the form of white powder.

Despite his conviction that they were built up from very large molecules, these polymers showed no signs of inherently colloidal behavior — a property peculiar to naturally occurring macromolecular compounds.[200] He therefore thought it necessary to make larger molecules in order to examine further the relationship between molecular size, the molecule's chemical structure, and its properties. Was it that molecules of a molecular weight more than 5000 did not exist, as the Fischerian dictum suggested? Carothers never thought that way, but believed that he could make far larger molecules by solving certain technical problems in his experiment. He eventually became determined "to develop more drastic conditions for the displacement of equilibrium in reversible bifunctional condensations."[201] Since experimental techniques used thus far by his group appeared inadequate for the purpose, he decided to employ another means, namely, the molecular still, which had recently been developed by several investigators, including Edward Wight Washburn, chief chemist at the National Bureau of Standards, for the separation of chemical mixtures. Carothers conceived the idea, inspired by

Washburn's report, "Molecular Stills," presented at the 1928 meeting of the American Chemical Society held in Swampscott, Massachusetts.[202]

The molecular still was a distillation device in which the distance between the distilling and condensing surfaces was shorter than the mean free path — the mean distance traveled by a molecule between successive collisions with other molecules — at very low pressures. Under this condition the molecule escaping from the distilling surface usually reached the condenser without colliding with another molecule; hence the molecules of the evaporated substance could be promptly removed from the system of reaction. The reaction was therefore forced to completion in accordance with Le Chatelier's law of mobile equilibrium. Thus this apparatus offered a method for displacing the equilibrium of the condensation of polymers towards a more complete reaction and a higher molecular weight by distilling off simple reaction products, such as water, as soon as they were formed.[203] In April 1930 Carothers's co-worker Julian Werner Hill, a 1928 MIT graduate, constructed the still, and soon was able to synthesize what Carothers called "superpolymers" by this means. The superpolymer, an exceedingly long polymeric chain having a molecular weight of 10,000 or more, exhibited physical properties different from those of a polymer of lower molecular weight, although the superpolymer closely resembled the initial polymer in its analytical composition and chemical properties. The product was a tough, horny, and elastic mass, and it displayed colloidal behavior in solution.[204]

Synthesis of superpolymers was soon followed by discovery of a phenomenon peculiar to these materials. Twelve days after the first superpolymer synthesis, Hill observed that the superpolymer (a polyester derived from hexadecamethylene dicarboxlylic acid and trimethylene glycol) could be mechanically drawn out from a molt or dry-spun from a solution into fibers or threads.[205] However, this mechanical operation profoundly changed the physical properties of the original superpolymers. The drawn filaments exhibited properties such as tensile strength, pliability, elastic recovery, transparency, and luster similar to those of natural fibers such as cellulose and silk. X-ray diffraction patterns indicated that superpolymers in the undrawn state were crystalline, but that the crystals had a random orientation. Drawn filaments, on the other hand, gave a fiber pattern in which the long-chain molecules seemed to be in an ordered array with the fiber axis. The character of this pattern was similar to that obtained from natural silk fibers or rayon filaments under tension. The condensation polymers also resembled cellulose and silk in their basic chemical structures. Together, these analogies stemming from the so-called "cold-drawing" phenomenon led Carothers to realize the possibility of making artificial fibers from linear-condensation superpolymers.

Earlier, Carothers had believed that the strength and elasticity of natural fibers depended on their macromolecular structure. From a chemical standpoint, it had been reasonable to assume that sooner or later a way would be discovered to prepare artificial fibers from synthetic macromolecular compounds. Carothers's study now disclosed that although a particular macromolecular structure was a necessary condition, it was not in itself sufficient to account for the physical properties of fibers. Since properties depended not only on the chemical constitution but also on the physical treatment (or what he called "physical history") of high polymers, the action of the mechanical stress, namely the cold-drawing, was the essential step in constructing fibers. The phenomenon had never before been observed with a synthetic material of any kind. Hill recalled, "The only effect known at the time that at all resembled it was the 'cold-drawing' of the silk glands of silk worms."[206] After questioning why this mechanical operation yielded a permanent high strength of the substance, Carothers concluded that it was due to the great size of the linear condensation polymer. He observed that "the property of cold-drawing does not appear until its molecular weight reaches about 9000 . . . a useful degree of strength and pliability in a fiber requires a molecular weight of at least 12,000 and a molecular length not less than 1000 Å." When the superpolymer received external mechanical tension, the long molecular chains were arranged in a highly ordered array parallel with one another. In this state, the mutual cohesive force of the very long chain would act fully; hence the drawn fibers exhibited the maximum possible strength.[207]

Staudinger had argued that a structural analogy existed between polyoxymethylene and cellulose. In 1929 he reported photomicrographs of fibrous crystals of polyoxymethylene formed directly from formaldehyde vapors, and claimed that these crystals were the first fibers to be prepared by synthesis.[208] However, they were only a few millimeters in length and too fragile for Carothers to consider fibers. In his opinion, a useful fiber must satisfy certain mechanical requirements.[209] While Staudinger predicted that sooner or later a way would be found to prepare useful artificial fibers from the addition polymers he had created, Carothers claimed that addition polymers, especially those produced from vinyl compounds, were less suitable for the fiber. To form oriented fibers, a compound must be capable of crystallizing, for crystallization would make possible the parallel arrangement of macromolecules. Unlike condensation polymers, addition polymers were often amorphous and rarely crystalline. According to Carothers, the presence of side chains (such as methyl or phenyl groups in the structural units) in the addition polymers diminished the tendency towards crystallinity. However large, the macromolecules of such addition polymers were less subject to the formation of oriented fibers. For the same reason, "three-dimensional polymers are obviously unsuited for fiber orientation, and synthetic materials of this class are besides invariably amorphous."[210] By contrast, condensation superpolymers (such as polyesters and polyamides) were usually crystalline because of the high degree of linear symmetry of their molecular shape. Thus they could easily be drawn out into strong, oriented fibers.[211]

After establishing a theoretical basis for making synthetic fibers, Carothers first announced his findings publicly at the Buffalo meeting of the American Chemical Society, held on September 1, 1931:

> While the method of preparation and the raw materials of the compounds so far studied are too costly to make them of any immediate practical application, the results clearly demonstrate for the first time the possibility of obtaining useful fibers from strictly synthetic materials.[212]

The next day the *New York Times* covered Carothers's presentation with the headline, "Chemists Produce Synthetic 'Silk.' "[213] His study evoked the dream of the seventeenth-century microscopist Robert Hooke, who had conceived a man-made silk in the *Micrographia* (1665). In the manuscript of his paper on artificial fibers, published in 1932, Carothers quoted a passage from Hooke, although it was deleted for publication:

> And I have often thought, that probably there might be a way found out, to make an artificial glutinous composition, much resembling, if not full as good, nay better, then that Excrement, or whatever other substance it be out of which, the Silk-worm wire-draws his clew. If such a composition were found, it were certainly an easie matter to find very quick ways of drawing it out into small wires for use.[214]

With Carothers's superpolymers and Hill's cold-drawing method, realization of the long-cherished Hooke's dream seemed near at hand.

Apparently Carothers was excited by the new findings. Samples of superpolymers were taken to DuPont's Rayon Department in Buffalo, New York for testing. "The members of the staff there," recalled Carothers, "were favorably impressed and they encouraged us for further work."[215] Hill overheard a conversation that took place at the time: "Showing one of the drawn fibers, Carothers cheerfully said to the staff of the Rayon Department, 'Well gentlemen, this is very interesting. I hope you will tell me every two or three years how you are coming along.' "[216] Carothers forecast the coming of the age of synthetic fibers and subsequent fate of rayon, although admittedly he believed, "Commercialization still seemed very remote."[217]

From then on, Carothers's basic research group shifted its aim to the more practical goal, that is, the search for polymers which could be drawn into fibers for commercial use. Elmer K. Bolton, the new Chemical Director who had taken over Stine's position in June 1930, strongly encouraged this shift. In 1931 Carothers filed two patents on synthetic fibers from condensation superpolymers, which were to become the first basic patents of fiber-forming polymers.[218] He hoped that the broad claims made in this patent application were "likely to dominate any practical developments that are made in future."[219] In Bolton's eyes, Carothers's fibers were still only of theoretical interest. Besides the high cost of raw materials, they were deficient in certain properties. For example, the melting points of the polyester fibers were too low for textile purposes and their solubilities were too great. They therefore had little utility as commercial fibers.[220]

During the period from 1930 to 1933, Carothers and his group systematically investigated various types of linear condensation superpolymers, including polyesters, polyanhydrides, polyacetals, polyamides, and polyester polyamide mixtures, which were synthesized by his co-workers from hundreds of possible combinations of starting materials. As Thomas Edison had done while searching for a filament for the light bulb from thousands of materials, Carothers's group — though proceeding more rationally and taking theoretical considerations into account — examined the properties of drawn polymers by the trial-and-error method. But each was found to be deficient in one or more textile properties. "At one stage of the work," Bolton recalled, "the outlook was so dark that investigations along this line were actually halted for a time."[221]

Under company pressure, however, Carothers continued his search, thinking that "it was worth one more effort."[222] Early in 1934, after a survey of his progress so far, he decided to resume work on the superpolyamides. In addition to their structural similarity to silk, the polyamides — prepared from amino acids and also from dibasic acids and diamines — had high melting points and high tensile strength. The amides he had previously obtained showed melting points too high to deal with. He reasoned that "if we were to go to longer unit lengths [of acids] we could probably reduce the melting point to any desired degree."[223] This time, he decided that by using a carefully purified ester of an amino acid (instead of a simple amino acid) he could make a superpolyester without the molecular still. Following this scheme, his co-worker Donald D. Coffman succeeded in May 1934 in synthesizing a superpolyamide from aminononanoic ester, an exceedingly expensive raw material. It was spun at 210–215°C into the most silk-like fiber ever prepared in terms of physical and chemical properties.[224] Meanwhile, another co-worker, Wesley R. Peterson, successfully prepared a spinnable polyamide from sebacic acid at ordinary pressure without using the molecular still. From the summer of 1934 to the following spring, his collaborators prepared eighty-one polyamides along this line, including 2-10, 5-6, 5-10, 5-6, 5-12, 5-18, 6-6, 10-1, 10-2, 10-6, 10-10, 11-10, and 12-6 polymers (numbers indicating those of carbon atoms respectively in amine and acid molecule).[225] Among these, 5-10 and 6-6 polymers proved most promising for commercial fibers. Carothers recommended for commercial fiber 5-10 polymer, which could be polymerized and spun easily. But, after due consideration, Bolton in the summer of 1935 decided to select 6-6 polymer for manufacture "because it had the best balance of properties and manufacturing cost of the polyamide then known."[226] This polyamide had been first synthesized by a co-worker, Gerard Jean Berchet, in late February 1935 from hexamethylenediamine ($NH_2(CH_2)_6NH_2$) and adipic acid ($HOOC(CH_2)_4COOH$).[227] It melted at 263°C, and its cold-drawn fibers exhibited a high strength and an elasticity greater than any natural fibers. Above all, the two raw materials could be made from benzene, readily available from coal. The fiber, later named "nylon," was to become the most successful commercial product in DuPont's research and development history.

The discovery of a new addition polymer, also to become an important commercial product, namely neoprene rubber, had taken place in April 1930 — at the same time as the discovery of a fiber-forming polymer. Again, the discovery was an unexpected event. Whereas nylon turned out to be a direct application of Carothers's work on condensation polymerization, the initial discovery of neoprene stemmed from a different line in his investigations.[228]

In the late 1920s, a group at DuPont's Jackson Laboratory in Pennsville, New Jersey had been working on a conversion of acetylene to its dimer (monovinyl acetylene) and trimer (divinyl acetylene) by use of a cuprous chloride catalyst, on the basis of the recent study by Father Julius Arthur Nieuland at the University of Notre Dame. The research was carried out as part of Bolton's synthetic rubber project, inaugurated in the mid-1920s. Bolton had hoped that the study on acetylene chemistry would provide a basis for another on synthetic rubber, since the chemical structure of monovinyl acetylene was similar to that of butadiene, a possible alternative for isoprene. Despite the efforts of DuPont chemists over a span of several years, the study yielded no rubbery material.[229]

At Bolton's suggestion, Carothers began in early 1930 to review the chemistry of the low polymers of acetylene.[230] In this case, his initial target was not synthetic rubber, but rather a fundamental study of the reactions of these compounds. To Carothers, "it was an attractive field even from the standpoint of pure chemistry."[231] Since his emphasis in any polymer experiment was placed on the purity of chemicals, Carothers first assigned his co-worker Arnold Miller Collins, a graduate of Columbia, to purify crude divinyl acetylene, which was prepared from acetylene in the presence of cuprous chloride. Collins carried out a very careful fractional distillation of the substance for this purpose. In this process he obtained a constant boiling main fraction of pure divinyl acetylene, along with a lower-boiling fraction that seemed to be an impurity. Several days later, he observed that

strange things had happened. The low boiling liquid fraction which had been collected in an attached test tube, had solidified, not to a resin but to something that had never been seen before in this work—a ball with a lively bounce and other characteristic physical properties of natural rubber.[232]

Thus a rubber-like material was obtained from the impurity Collins had removed from pure divinyl acetylene.

Carothers immediately interpreted this phenomenon in terms of his macromolecular concept.[233] He had known that rubber belonged to the class of addition polymers, and that this type of polymer could be obtained by spontaneous polymerization, unlike condensation polymers. Chemical analysis of the rubbery material showed that it contained chlorine atoms, obviously derived from the cuprous chloride catalyst. From this, it was reasonable to assume that, with the exception of the presence of chlorine, the rubber-like product had a structure similar to that of natural rubber, and that the original fractional liquid was analogous to isoprene. Subsequent experiments supported the analogy: addition of hydrogen chloride to monovinyl acetylene directly yielded the same liquid, which in turn transformed spontaneously into the elastic solid. Thus, Carothers identified the new liquid as 2-chloro-1, 3-butadiene, or what he named "chloroprene." Its high addition polymer resulted in a rubbery material, the long-chain structure polychloroprene, later called "neoprene."[234] The company recognized that as a synthetic rubber neoprene held great potential for commercial use since it exhibited some physical properties far superior to those of vulcanized natural rubber (e.g., resistance to oxidation, to heat, and to many chemicals).

The results were reported at a Rubber Division Meeting of the American Chemical Society, held in Akron in November 1931, and published the same year in the *Journal of the American Chemical Society*.[235] After establishing a practical method for the polymerization of chloroprene, DuPont started the commercial production of neoprene, under the trademark "Duprene" in June 1931. In the face of the low price of natural rubber, DuPont shrewdly presented Duprene as a wholly new material with special properties, rather than just as a substitute for natural rubber. This marketing strategy would help neoprene rubber sell for a higher price than natural rubber.[236] The rubber chemist G. Stafford Whitby remarked that the discovery of neoprene rubber "represents one of the most striking achievements of industrial chemistry in recent years."[237]

The discovery of neoprene led Carothers to cultivate the subject of acetylene polymers, which formed his second major field. "From the

Figure 9. Carothers stretching neoprene rubber, early 1930s. Behind him is a molecular still that enabled him to synthesize superpolymers. Courtesy Hagley Museum and Library.

theoretical standpoint," he noted in 1932, "it is very interesting since the mechanism by which either natural or synthetic rubbers are formed is still very obscure, and the rapidity with which chloroprene polymerizes greatly facilitates the study of this problem."[238] Between 1931 and 1934, along with co-workers, he published nineteen scientific papers on the synthesis and formation mechanism of acetylene polymers in a series entitled, "Acetylene Polymers and Their Derivatives."[239] Although chemists such as Whitby spoke very highly of the series, Carothers himself rated this work lower than his study of condensation polymers when he stated, "the scientific results were abundant in quantity but perhaps a little disappointing in quality."[240]

Leaving behind the company preoccupation with fiber and rubber, Carothers traveled to England in the fall of 1935 to give a paper at a meeting of the Faraday Society, where he was a principal speaker together with Staudinger, Mark, and Meyer. His outstanding presentation at the meeting rounded off a tragically short but productive career as one of the founders of macromolecular chemistry, as will be discussed in the next chapter.

Chapter 4
Triumph and Struggles of Two Giants

With all his fine physique he [Carothers] had a super sensitive nervous organism—too finely strung—from which he always suffered. . . . [H]e hated his shyness and the dark moods that assailed and engulfed him. I suffered too, realizing how alone he was and seeing the struggle that was going on in his mind much of the time.

He made a valiant effort to be a normal person to adjust himself to the ways of the world and the hard headed business men who were his chiefs.

— Mary E. M. Carothers to Roger Adams, 1937

Staudinger was, I admit, rather brusque toward Meyer and me back in the twenties, but that is just his manner. . . . Besides, he was involved in some personal troubles at the time, and they may have made him a bit sharp. One must take such things into consideration. He stayed on in Germany under the Nazis, so in the end he had more difficulties than I ever did.

— Herman F. Mark, quoted in Morton H. Hunt, "Polymers Everywhere," 1958

There always emerge such wrong evaluations, when—as was the case in the Nazi period—the scholar's attitude to the party has much more decisive influence on the appreciation of his work than his accomplishment itself does.

— Hermann Staudinger, "Einfluss des Nationalsozialismus auf die Entwicklung der makromolekularen Chemie," ca. 1945

The postwar intellectual isolation of Germany had apparently disappeared by the opening of the 1930s. For the first time, German and Austrian chemists were guests at the 1928 meeting of the International Union of Pure and Applied Chemistry, held at the Hague. At the meeting, Fritz Haber told Charles Lee Reese, a member of DuPont's board of directors and a vice president of the Union, "the difficulty was that many German scientists had never been out of Germany since the war and they still had the bitter feeling which was created by their defeat, but those who have traveled see the light of day and have no prejudices of this kind."[1] German scientists appeared on the international scene more frequently in order to exchange scientific results with foreign colleagues during the first half of the 1930s.

Consequently, the significance of the issue of macromolecules received growing recognition in scientific communities outside Germany during this period. As a proselytizer for the macromolecular theory, Hermann Staudinger wasted no time spreading his theory throughout Europe. In 1931, for example, he lectured on the macromolecular structure of polymers at the International Solvay-Congress in Brussels and also before the Chemical Society of France (Société Chimique de France) in Paris.[2] The following year, England's Faraday Society organized a symposium, "The Colloid Aspects of Textile Materials and Related Topics" in Manchester, where Staudinger, Herman Mark, and other German chemists were invited to present papers.[3] The subject of high polymers was again discussed by Staudinger at the ninth Congress of the International Union of Pure and Applied Chemistry, held in Madrid in the spring of 1934.[4]

Concurrent with Germany's return to the international scientific community was the rise to power of an extremist political group. The ensuing political upheaval, in turn, began to make a profound impact on German academic circles. The Weimar Republic ended in 1933 when Adolf Hitler came to power with his anti-Semitic campaign. In April of that year the Nazis were able to pass the Civil Service Law (Gesetz zur Wiederherstellung des Berufsbeamtentums) under which no employment was to be given to persons of non-Aryan descent. Leading authorities on polymers who were Jewish, including disputants of Staudinger's chemical theories, were forced to resign their positions at the state-controlled Kaiser Wilhelm Institutes and at universities.[5] In 1933 Max Bergmann, director of the Institute for Leather Research in Dresden, and a Jew, fled to America. It was some consolation to find a good post there; he served as a member of the Rockefeller Institute for Medical Research in New York until his death at the age of fifty-eight in 1944.[6] Bergmann, as a scientist, considered the United States to be "the best country on this globe to live in."[7] By contrast, the case of Reginald O. Herzog was tragic. He was accused by the National Socialists of cheating his German assistant, Helmut Hoff-

mann, a member of the Nazi Party, of the scientific reward due his position. Herzog's Institute for Fiber Chemistry was cited as an example of
"parasites on the body of the German people" and of "vampires at the
highest scientific cultural center."[8] In 1934 Herzog fled to Turkey and his
institute in Dahlem was subsequently closed. Despite the efforts of Bergmann and others to secure a suitable industrial post for him in America,
he suffered a nervous breakdown. The despondent Herzog committed
suicide while in exile in Zürich in early 1935.[9]

In the summer of 1933 Fritz Haber resigned, under pressure, from the
Institute for Physical Chemistry and Electrochemistry. He moved to En-
gland, and early in 1934 he died of a heart attack in Switzerland. Herbert
Freundlich, chief of the Division of Colloid Chemistry and Applied Physi-
cal Chemistry at Haber's institute, relocated in 1933 to London to join
University College as an honorary research associate. After moving to
America in 1938, he become a professor of colloid chemistry at the
University of Minnesota, but died in Minneapolis three years later. Michael Polanyi, another eminent physical chemist at Haber's institute, left
Germany in 1933 and taught at the University of Manchester, where he
would eventually give up chemistry and become a philosopher and social
scientist. Hans Pringsheim, associate professor at the University of Berlin,
left for Paris in 1933 and died in Geneva in 1940.

The impending success of Nazism also forced Kurt H. Meyer and Mark
at I. G. Farben to leave Germany in 1932. Meyer succeeded Amé Pictet as
head of the Laboratory of Organic and Inorganic Chemistry at the University of Geneva, Switzerland, where he continued research on natural
polymers.[10] Mark, who was part Jewish, moved to Austria, accepting the
directorship of the First Chemical Institute at his alma mater, the University of Vienna. There, together with a number of his young colleagues, he
zealously injected polymer chemistry into the traditional chemical curriculum and was able to advance his investigations on polymerization.[11]
While the non-Jewish Staudinger remained in Nazi Germany, the intellectual migration of the country's chemists led to a renewed stimulus in
the field of polymers within European scientific circles. By the mid-1930s,
the major arena of heated discussions on macromolecules had moved
outside the walls of the Third Reich.

The Faraday Society, 1935

The Faraday Society arranged a symposium on "The Phenomena of Polymerisation and Condensation" at the University of Cambridge in September 1935. Founded in 1903 "to promote the study of Electrochemistry, Electrometallurgy, Chemical Physics, Metallography, and Kindred
Subjects," the Faraday Society had brought together pioneers in these

fields, from all parts of the world, to hold intensive discussions on important new research topics.[12] Physical chemistry and colloid chemistry were dominant areas of interest for Faraday Society discussions during the 1920s and 1930s. These discussions, including both formal papers and ensuing verbal debates, were published in the *Transactions of the Faraday Society*, which provided "a source book of new knowledge having a profound influence on subsequent developments."[13]

The 1935 Cambridge discussion was the first international conference devoted exclusively to general studies on polymers. It was at this meeting that the two champions of macromolecules, Staudinger and Carothers, met for the first time. Other overseas guests included Meyer (Geneva), Mark (Vienna), Pringsheim (Paris), Johann R. Katz (Amsterdam), and Karl Freudenberg (Heidelberg). Among the British speakers were Eric Keightley Rideal and Harry Work Melville—both at Cambridge—who had embarked on their research in this field. The large size of the conference was indicated by the total of thirty-three papers to be presented during the three-day session.[14]

Appearing before a European audience for the first time, Carothers gave a lecture on "Polymers and Polyfunctionality," which dealt mainly with his major field, condensation polymerization.[15] Staudinger presented a paper on "The Formation of High Polymers of Unsaturated Substances" that was devoted to addition polymerization.[16]

Let us remember that, by striking a blow against the physicalist conception of polymerization, Staudinger had revived Berzelius's classical notions on polymers. In his comments at this meeting, he again stressed that "a polymerisation is a process in which a substance of low molecular weight is transformed into a substance of equal composition but of higher multiple molecular weight."[17] In this process, a monomer and a polymer ought to have the same composition, namely, the polymer is simply made up of monomer units. However, this definition could be applied only to the addition polymerisation of unsaturated monomer molecules, which tended to combine without changing their composition. It followed that condensation polymers, such as polyesters and polyamides, were by definition not "polymers," since they were formed by elimination of secondary compounds (such as water) and not by pure self-addition. Consequently, Staudinger's group directed chemists' attention exclusively to the "true" polymerization products, addition polymers. But Carothers, at the meeting, suggested that the definition of polymerization needed to be altered. He stressed:

Professor Staudinger's point of view has considerable historical justification, but it presents certain logical and practical difficulties. Apparently it involves the necessity of making a distinction between polymers and "real polymers" and of admitting that polymers can be formed by reactions that are not polymerisation. . . . It is true that large molecules are in some cases built up from small ones by reactions that appear, at least, to consist in pure addition, while in other cases they are formed by reactions that are demonstrably condensations. Staudinger proposes to call the latter type of reaction polycondensation . . . but I contend that we may as well give in to the logic of the situation and admit that such reactions constitute one type of polymerisation: the products are polymers.[18]

Although the issue was one of wording, clarification of terminology was particularly important during a period of transition in chemists' concepts of polymers and polymerization. Now that the old definition did not match practical usage, creation of a new definition involving both addition and condensation polymers seemed indispensable.

Carothers's claim demanded sufficient consideration among his contemporaries, and an agreement was to be postponed until the Council of the International Union of Pure and Applied Chemistry issued its "Report on Nomenclature in the Field of Macromolecules" in 1951. The report provided the following general definitions:

A *high polymer* is a macromolecular substance which . . . consists of molecules which are, at least approximately, multiples of a low molecular unit. In agreement of present-day usage, a high polymer need not consist of molecules which are all of the same size, nor is it necessary that they have *exactly* the same composition or chemical structure as each other or as the corresponding monomer. . . .

Polymerization is the process of formation of polymer molecules from small molecules, with or without the production of other small molecules not entering into the composition of the polymer.[19]

The issues under discussion at the 1935 Faraday Society meeting centered on details of polymers and polymerization, including the mechanism of polymerization reactions, the distribution and determination of molecular weights, the viscosity law, the nature of polymers, the shape of macromolecules, and potential areas of application. As Melville recalled two decades later, the 1935 meeting "embraced every branch of polymer chemistry as it is now defined. The gathering was truly international and the vigor of the discussion clearly showed that new ideas were afoot and that they were not readily accepted by those who had just began to take an interest in the subject."[20]

Despite conflicts of opinions between lecturers, the outcome of the 1935 Faraday Society symposium was, on the whole, impressive enough to convince the audience to recall it as one of the "milestones of modern chemical history."[21] Absent from the gathering was the familiar debate revolving around the macromolecular theory and the aggregate theory, which had dominated so many previous scientific gatherings. Pringsheim, a former exponent of the aggregate view and now present at

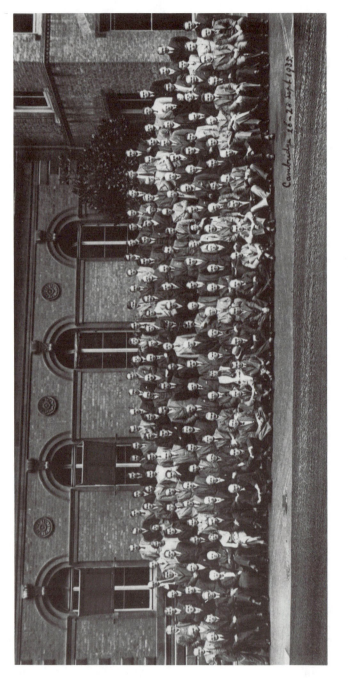

Figure 10. Participants in the 1935 Faraday Society meeting, Cambridge, England. Bottom row (seated on chairs, right to left), Roelof Houwink (no. 4), Eric K. Rideal (no. 6), Herman F. Mark (no. 7), Hermann Staudinger (no. 14), Hans Pringsheim (no. 16), Kurt H. Meyer (no. 18), Herbert Freundlich (no. 24), Johan Katz (no. 28). Wallace Carothers appears not to be in this picture. Courtesy Chemical Heritage Foundation.

the Cambridge meeting, no longer invoked his initial theory against the macromolecularity of polymers. Referring to recent developments brought about by the work of Staudinger, Carothers, Meyer, himself, and others, Mark was able to declare in his lecture that "the chain-structure of the polymerization products can now be considered as a fairly well-established fact."[22] The large attendance and high tone of the symposium, Mark noted,

proved the enormous progress which the young branch of polymer science had made during the last decade. There was no question any more about the *existence* of macromolecules. . . .
    At the end of the symposium, everybody was convinced that polymer chemistry had grown into a full-scale science with unexpected new vistas for intensification of understanding and expansion of application.[23]

The international Cambridge meeting on polymerization thus vividly illustrated the end of the decade-long controversy over a fundamental principle concerning the macromolecular structure of high polymeric substances. With the shift in scientists' minds from the aggregate theory toward the macromolecular theory, polymer chemistry now became identical with the chemistry of macromolecules.

Following the Cambridge symposium, Staudinger and Carothers conferred privately at the DuPont office in London, a meeting arranged by William F. Zimmerli, the company's European technical representative. Carothers took this opportunity to explain to Staudinger more thoroughly his study of condensation polymers, especially his work on polyesters, and probably some of the recent developments in synthetic fibers at DuPont. At last, Staudinger was convinced of the significance of condensation polymers, which he had underestimated since initiating his research in this area. He later humbly recalled, "In retrospect, I admire Mr. Carothers's extraordinarily critical remarks, which, as our discussion [in London] proved, above all raised the interest of the clarification of his polycondensation products in the light of organic chemistry."[24] Carothers, for his part, acknowledged that he had been much inspired by Staudinger's 1927 paper on polyoxymethylene as a model for cellulose.[25] Staudinger must have found in Carothers a genial comrade in the fading battle against the aggregate and the new micelle theory: "on the ground of these experiments by Mr. Carothers, we both came to the conviction that macromolecules truly exist and that the view [by Meyer and Mark] of a micelle structure, which was advocated at this meeting of the Faraday Society, must be rejected."[26]

Despite Staudinger's and Carothers's shared feeling of triumph in 1935, the former would do further battle with Meyer and Mark, on the one hand, and with the last academic opponents in Nazi Germany, on the other. Carothers would suffer a fatal depression, following a struggle with the dilemma created by working in industry. In many ways, the scientific careers of the two giants were in jeopardy.

Carothers's Industrial Dilemma and Fate

By the mid-1930s, years before he became known as the inventor of nylon, Carothers had earned a successful reputation as an organic chemist in the American scientific community. "Carothers is, in my opinion," Roger Adams remarked in 1933, "the best organic chemist in the country. . . . Practically all of my closest personal friends in organic chemistry rate him as I do—at the top."[27] Elmer K. Bolton, Carothers's chief at DuPont, considered him "the smartest organic chemist that the DuPont Company ever had."[28] On April 29, 1936 the National Academy of Sciences elected Carothers into its membership, with an unusual "overwhelming vote."[29] Leo H. Baekeland, the inventor of Bakelite, won membership at the same time.[30] The two men became the first organic chemists associated with industry to be so honored. They were the sixteenth and seventeenth industrial scientists of a total of 291 members.[31] Celebrated by colleagues, Carothers responded, "I feel very highly honored to join so distinguished a group." He thanked his mentor Adams, then a member of the Academy's Council, in particular "for all that you have done to provide the occasion for these congratulations."[32] Rumor had it that Carothers would be a future candidate for the Nobel Prize.[33]

Despite his growing fame as a successful industrial scientist, Carothers had for years been caught in a conflict between academic and industrial values. Apparently his primary interest lay in basic science. Like many academics, he was perennially concerned with publication. Within his industrial organization, he was, by and large, able to fulfill this desire, and published an exceptionally large number of scientific papers, many of which were candid enough that the readers could realize what research DuPont was currently engaged in. Yet the papers were submitted for publication only after meticulous censorship by the company and, when necessary, application for patenting. This tendency grew more pronounced after the company's management recognized the feasibility of making commercial fibers on the basis of superpolymer research by Carothers.

In retrospect, synthetic fiber was an unforeseen consequence of Carothers's basic research. In an attempt to make molecules as big as possible, he serendipitously gained an insight into the possibility of preparing artificial fibers. Realizing its practical implication, a DuPont official delayed publication of Carothers's paper on superpolymers and recommended that "our position should be thoroughly protected by a well planned patent program."[34] Carothers began to sense that the restriction

would abridge the "academic freedom" which Charles Stine had be-
stowed upon the fundamental research program at its inception.

The Great Depression affected DuPont's policy regarding its fundamental research program. In the summer of 1930 Bolton replaced Stine
as Chemical Director, the latter being promoted to the company's Executive Committee. An outspoken critic of Stine's new program four years
earlier, Bolton now revitalized the traditional mode of research management. He knew that industrial research needed science, but insisted that
it should aim at profitable targets. When he induced the fundamental
research staff to set up more practical goals, the "academic era" at Purity
Hall virtually came to an end.[35] Accordingly, Carothers was encouraged
to transform his group's findings into commercially feasible products.

Carothers provided thoughtful criticism of this reorientation. In a review of his past efforts at DuPont, Carothers wrote that fundamental
research should be pure research with objectives of increasing the body
of knowledge as Stine had envisioned: "Its success would be gauged by
the significance of the scientific contributions produced, and any financial profit that might accrue would be so much gravy." Gauged as such,
his own fundamental research work in organic chemistry would, to date,
be overall "very successful." Now that fundamental research was expected "to pay its own way," however, the new policy brought forth entirely different criteria to estimate whether his work was successful or not.
Although his fiber syntheses then seemed to have "considerable possibilities for cash return in future," he still had little to show "in the way of
direct cash benefits." Therefore under the new pragmatic criteria his
scientific work would be judged unsuccessful. He complained that, had
he initially been invited to do research of this nature for DuPont, he
would never have accepted the job:

I never had any confidence in my ability to initiate and carry on research of this
kind, and I still haven't any. There are certainly people that do have this ability,
but I think that they are rather rare, and I doubt that there are any on the present
fundamental research staff. . . . [D]uring the present transition period practically the only guide we have for formulating and criticizing our own research
problems is the rather desperate feeling that they should show a profit at the end.
As a result, I think that some of our problems are being undertaken in a spirit of
uncertainty and skepticism without any faith in a successful outcome, or even
without any clear idea of what would constitute a successful outcome.

He stressed that anything of practical utility that might come out of
fundamental research must be regarded as "an accidental by-product."
DuPont should, he argued, continue fundamental research as a pure
scientific pursuit even if economic conditions forced the company to
reduce the expenditures devoted to it.[36]

The economic depression made some of Carothers's co-workers anxious lest they should be laid off.[37] As a group leader, he needed to protect his men from that anxiety. However, DuPont's management neither eliminated the fundamental research program nor reduced the budget for Carothers's group. Instead, Bolton's new policy prevailed. In 1934, for example, a quarter of the time of his group was charged to nonfundamental research.[38] Aside from fiber searching, he was busy making patent applications for synthetic polymers (for which a total of sixty-nine U.S. patents would be granted), most equaling scientific papers in terms of their content. He wrote to a friend that

Research lately has been rather foul on account of the depression. . . . We are still spending money like nothing at all; but an atmosphere of anxiety has arisen which causes us to scrutinize topics from a misty standpoint of what may ultimately be practical. As a result most of my research seems to be of the unclassified kind—neither theoretical nor practical. I don't complain about that; since everybody is more or less in the same boat nowadays; but if I have a chance to get back into a really good university appointment it will not require very long to decide to leave this place.[39]

His first chance to consider an academic position came in May 1933, when James B. Conant became the new president of Harvard University.[40] This news stimulated Carothers's desire to take over the vacant professorship in chemistry that would be created by Conant's promotion. In a letter to Adams, he confided:

I have lately had a yearning to get back into university work. . . . As a matter of fact I haven't any confidence about practical problems; but enough nice theoretical ones have turned up during the past two years to suffice for a long time. . . . Also the mere idea of going to *a* university does not particularly appeal to me. But if there were a chance to go to Harvard that would be a different matter.[41]

Shortly afterward, however, Carothers changed his mind and asked Adams to disregard that "rather silly letter":

It was the result of a brain storm which blew over rather promptly. It would be difficult to suddenly pull up stakes here anyway. There are too many loose ends, especially in the matter of getting things ready for publication. Perhaps the situation will be different generally in a year or two; and for the present it might be better to say nothing to the people at Harvard.[42]

In addition, Carothers had decided to buy a house in Arden, near Wilmington, where he planned to bring his parents from Iowa, as they had suffered financially due to the depression. But Adams had already "dropped a line to some of the people in Cambridge, merely mentioning in a general way what the situation was."[43]

In early 1934 Harvard began to seek Conant's successor to the Sheldon Emery Professorship of Organic Chemistry: someone who was expected to be interested in "developing the biological and natural product side of organic chemistry from either the synthetic or the constitutional point of view" in cooperation with Harvard's Biology Department.[44] Full professors of the Chemistry Department, including James B. Conant and Elmer P. Kohler, did not choose Carothers, but instead voted unanimously to invite Roger Adams with "the top salary at Harvard [$12,000]."[45] Mrs. Adams wanted her husband to accept the offer. To the disappointment of Harvard's staff, however, Adams turned it down.[46] As it turned out, an English chemist, R. P. Linstead, filled the position.[47]

During the summer of 1934, Carothers felt "a deflated interest in chemistry," a symptom of the mental depression from which he had suffered periodically since his Illinois days. He consulted a Baltimore psychiatrist, who diagnosed his condition as "nervous collapse," and was then sent to a nursing home for a temporary stay. After taking a "forced holiday" in August with Julian Hill's family at Martha's Vineyard, Massachusetts, Carothers apparently recovered: "Chemistry and many other things begin to seem important again," he wrote to Adams.[48]

Carothers encountered another chance for a "good" academic position in the fall of that year when the University of Chicago conducted a search for an able chemist to develop its chemistry department. Learning from Conant that Carothers was a prospective candidate, Robert Maynard Hutchins, president of Chicago, offered him an invitation to consider the post.[49] This time Carothers's initial response was negative because recent developments in his laboratory had "very lately reached an exciting stage. It would be impossible to carry them elsewhere and difficult to abandon them."[50] This was about the time when spinnable 5-10 polymer was being prepared in his laboratory. Nevertheless, the offer gave him an opportunity to "assess the ultimate future possibilities" as a DuPont researcher. He replied to Hutchins:

> The pros and cons are numerous, but one outstanding factor is that we are limited in the choice of problems and methods of approach: we now have to regard scientific contributions as an occasional and accidental by-product, not as a primary objective; and there is no prospect of change in this situation. Also for considerable financial rewards we are dependent upon the ultimate industrial outcome of our own researches.

Against these disadvantages of industrial research, he weighed the advantages of the academic post, that is, "complete freedom in the selection of problems and the aiming of work directly toward scientific contributions, and [an] immediately adequate salary." These features, he said, "would make an academic position definitely very attractive."[51] Confidentially,

Hutchins requested that Adams and Conant sent recommendations to Chicago. Conant, as well as Adams, emphatically endorsed him: "I regard Carothers as by far the best American organic chemist who is likely to be available for the position at the University of Chicago. . . . If you succeed in getting Carothers you are very much to be congratulated."[52] On the other hand, Adams cautioned that Carothers's outstanding ability lay as a researcher rather than an administrator: "He is not the man to hire if he is to be expected to spend from 60–75 per cent of his time in departmental detail. He is too good a man to waste on that sort of job."[53] Conant also alluded to Carothers's "certain personal peculiarities which do not make him quite as good a bet from a university point of view as from a research institute point of view. . . . [H]e might find a certain human problems a little difficult to handle."[54]

Meanwhile, Carothers, with Chicago's offer in mind, took up the question with Bolton of DuPont's general policy on fundamental research. Consequently, Carothers reported to Hutchins that the differences of opinion between them had "in part been adjusted. In a conversation with Dr. Bolton . . . I indicated my intention of staying in my present position. This decision seems rather conclusive."[55] Whatever the adjustment, Hutchins's further efforts to persuade him were futile.[56]

Carothers no longer considered leaving the industrial research environment to which he had become so accustomed. A few months after his return from the Faraday Society, he said that Wilmington was "my adopted city where everything seems to be in approximately the order. . . . [T]he chance for a real chemical argument was a real boon . . . I miss here where we have no real chemical fanatics or if we do I am not a sufficiently genuine one myself to inspire debate on purely academic subjects."[57]

The desperate efforts of Carothers's group to find fibers of practical use were rewarded in the summer of 1935, when Bolton decided to produce 6-6 polymer commercially. Once this decision was made by the company, development of processes for manufacturing intermediates and for fiber spinning and processing was launched in various divisions of the company. In order to reduce to a minimum the "time between the test tube and the counter," some 230 chemists and engineers were at one time or another engaged in the "D" of the R & D, in other words, developing technologies of ingredients manufacturing, melt spinning, high-speed cold drawing, heat-setting, and sizing.[58]

Between 1930 and 1935, Carothers always had ten to thirteen co-workers, but after his return from Cambridge, England, in the fall of 1935 the number fell rapidly (see Appendix Table 5). While most of the applied work on the fiber was transferred to other groups, "minor aspects of a rather theoretical or exploratory nature" were left with Carothers and his three assistants.[59] Left out of the mainstream of the R & D fiber project, as Adams observed, "His mental sensitiveness came to play a more and more important role in his life and led to not infrequent physical upsets which made his mental difficulties even worse." He did not feel free to talk over personal problems even with his most intimate friends.[60] In February 1936 Carothers, now near forty, abruptly and quietly married a twenty-five-year-old worker in the Patent Division of the Chemical Department, Helen Everett Sweetman, a marriage that some of his colleagues saw as a mismatch. Carothers commented to a close friend, "No doubt this properly calls for condolences to the lady."[61]

Carothers's attacks of depression grew most pronounced in the summer of that year. And as they grew, he came to depend increasingly on alcohol.[62] Worried DuPont executives sent him to the Institute of the Pennsylvania Hospital in Philadelphia, what Carothers called "an especially elegant, large, and elaborate semi-bug house."[63] Kenneth E. Appel, an acknowledged authority on psychiatry at the Institute, found in Carothers symptoms typical of depressed individuals: he "was obsessed with the idea that he was a failure."[64] Feeling himself worthless, Carothers, as Adams put it, "could not hope to continue the accomplishments already completed which he looked upon in his modest way as not being due to his ability but to good fortune and the opportunity which had been provided for him."[65] Appel's treatment consisted in conversation and rambling that were, Carothers complained, "inconsequential, pointless, and sometimes so repetitious and puerile as to be the source of laughter, amazement, or anger."[66]

After being hospitalized for five weeks, Carothers visited Europe to attempt to relax. In Munich he joined Roger Adams, who was in Europe to attend an international conference.[67] Adams observed that Carothers "looked rather dejected—had been drinking heavily."[68] But, as they hiked the Austrian Alps, he became "a changed man—quite normal" and enjoyed climbing.[69] After Adams left for the conference, Carothers remained and explored more of the surrounding valleys. After a solitary tour of Vienna and Baden, Carothers made "an exasperatingly slow return" by ship to the United States.[70] But apparently, when he was back by himself, his obsession that he was a failure resurfaced. His friends, and Adams in particular, increasingly felt that they "could do nothing to help him out of the mental state."[71]

Back in Wilmington, Carothers began searching for new research problems, "especially in the field lying between organic and inorganic chemistry," but he confessed, "I have become rather a fish out of water."[72] The unexpected death in January 1937 of his beloved sister Isobel further shocked him so staggeringly that he was unable to completely reconcile himself.[73]

Early in April 1937 Carothers appeared to recover his spirits. After filing a patent on "Synthetic Fiber," which would form the important basic patent on nylon,[74] he wrote an unusually long and forward-looking letter to his mother, in which he stated that one of the old problems of the Arden days — the period in which he lived briefly with his parents — was now "bursting into real flower" but his role in the fiber development was finished and he was therefore looking for something entirely new.[75]

However, typical of the depressed person who finds life unbearable, tragedy suddenly occurred at a time when Carothers appeared to be recovering rather than in his worst condition. Very early in the morning of April 29, Carothers registered a room at the Hotel Philadelphian, several blocks from the institution where he had once been hospitalized. After remaining in his room all day, Carothers took his life with cyanide. According to an account in the *New York Times*, "Groans early in the evening led guests in near-by rooms to notify . . . the [hotel] manager, who forced his way in and found the chemist lying dead on the floor." "A few grains of crystalline poison and a squeezed lemon near his body led police to believe that he had drunk poison in lemon juice."[76] Carothers's suicide took place two days after his forty-first birthday and precisely the same day as his election to the National Academy of Sciences a year earlier. His ashes were buried in his hometown, Des Moines. Seven months later, his daughter Jane was born.

Most newspaper obituaries referred to Carothers simply as the DuPont co-inventor of synthetic rubber. Our popular image of him as the inventor of nylon was then nonexistent, since DuPont's nylon project had yet to be made public. It was not until 1938 that the name Carothers became associated with nylon. The year marked the beginning of a new epoch in DuPont history, as the scaling-up effort of nylon neared completion. By August the company had christened its polyamide fiber with the generic name "nylon" after examining some 400 names submitted by DuPont employees.[77] In September Carothers's basic patent was disclosed. While DuPont officials continued to maintain silence, the news was leaked — though without details — into chemical circles, and rumors about the new fiber spread rapidly. Newspapers began mentioning Carothers's patent and DuPont's "synthetic silk" as a potential replacement for natural silk, America's largest import from Japan. As the *Chicago Tribune* put it:

You have a substance which historians of the future may find caused Japan to lose in Delaware all the dreams of empire she pursued in China. . . . It was patented recently in the name of the late W. H. Carothers, chemist for the du Pont Company. It is possible that it may have as revolutionary an effect on world trade as did the artificial dyes which ruined indigo planters [in India] or the fixation of atmospheric nitrogen which broke Chile's virtual monopoly of nitrates.[78]

Representing the company, Charles M. A. Stine finally unveiled the new product, nylon, to the public in late October. Addressing an audience at the Herald Tribune Forum in New York, he said:

> I am making the first announcement of a brand new chemical textile fiber. This textile fiber is the first man-made organic textile fiber prepared wholly from raw materials from the mineral kingdom. . . .
> Though wholly fabricated from such common raw materials as coal, water, and air, nylon can be fashioned into filaments as strong as steel, as fine as the spider's web, yet more elastic than any of the common natural fibers and possessing a beautiful luster.[79]

This historic speech, purposefully entitled "What Laboratories of Industry Are Doing for the World of Tomorrow," was simultaneously Stine's declaration of the triumph of industrial research. Far exceeding his own expectations, Stine's fundamental research program proved to bear exceptional industrial results during a ten-year span. DuPont had invested a total of $27 million for nylon from basic research through the startup of the first plant, a cost that would eventually bring a profitable enough return to double the size of the company.

Production of nylon textile yarn started in a new $8,000,000 plant in Seaford, Delaware in December 1939. The new product promised a wide variety of uses, ranging from women's hosiery and toothbrush bristles to military parachutes and airplane tires. When nylon stockings first went on sale in New York City in May 1940, four million pairs were purchased within a few hours. Depression would inevitably hit Japan's silk market. What has been called "the fourth textile revolution" then followed, in the wake of earlier inventions of mercerized cotton, synthetic dyes, and rayon. In July 1940 a leading U.S. magazine announced:

> ON THE U.S. FRONTIER   The Giant Molecule is a greater fact of history than Adolf Hitler, although it may take vision to believe it. Nylon, a product of the Giant Molecule and the fourth basic revolution in textile chemistry in four thousand years, is less a substance than a group of substances, all unlike anything found in nature. They didn't just happen — they were made to happen.[80]

Carothers lived neither to hear the name nylon nor to see its grandiose commercial production and subsequent impact on culture. As David Hounshell and John K. Smith have argued, nylon could not have been discovered without Carothers's basic research; at the same time, had he been left entirely on his own, nylon might not have been invented either. Nylon emerged from the creative tension between "the pure-science idealist Carothers and the pragmatic Bolton."[81] The story of nylon vividly illustrates a simple paradox: successful industrial research needed pure science but could not go entirely with science for its own sake. After Carothers's death and the subsequent success of nylon, it became a common practice for DuPont's researchers to shift back and forth between fundamental and applied research.[82]

The suicide of the inventor was in such striking contrast to the sensational nylon drama that it posed an enigma and aroused much public interest. Naturally, DuPont was sensitive to the matter. Six months after Carothers's death, in the course of gathering biographical information in preparation for the National Academy of Sciences's memoir of his former student, Roger Adams corresponded with Carothers's colleagues, friends, and family. Carothers's mother Mary wrote to Adams that her son "had a super sensitive nervous organism — too finely strung — from which he always suffered. . . . He made a valiant effort to be a normal person to adjust himself to the ways of the world and the hard headed business men who were his chiefs."[83] Adams quoted these sentences verbatim in his draft of the memoir, which was submitted to Bolton for review before publication. Bolton requested that Adams revise the last sentence because of "unanimous objection" to it by the company.[84] Adams deleted it in the published version.[85]

Whatever the expression, Mary Carothers and Adams were right to point to Carothers's personal difficulties and his struggle in adjusting to the industrial world. However, his dilemma and unfortunate demise may obscure the positive aspects of the industrial scientist. Carothers was by no means a naive, single-minded pure science idealist. Nor did he consider himself a misfit or a hapless victim of industry. He surely knew where he belonged and what his role was. While desiring to remain a "Virgin" at "Purity Hall," Carothers never denied the importance of applied research. He was critical to Bolton's policy not because he objected to the practical application of his science, but because he wanted to restore the DuPont program to its original form and thereby retain his own role as a pure scientist in industry. He unmistakably realized that he could show himself at his best only in this capacity. Restrictions on "academic freedom" provoked concern for a period, but this factor alone was not sufficient to cause Carothers to return to academia.

Perhaps, as he himself admitted, Carothers lacked real talent in applied research and business-oriented judgments.[86] But this does not mean that he lacked sympathy with the purpose of applied research. "Carothers was a pure scientist per se, at the same time he was a practical man," explained Julian Hill, his chief collaborator for over eight years.[87] In retrospect, as an admirer of Duncan in his formative years and a top student in the Adams school, Carothers maintained a more or less dualistic approach to chemistry, a blend of pure and practical science ideals.

Similarly, the DuPont program had emerged from a Remsenian dualistic outlook, combining academic and industrial interests, at the dawn of American industrial research. Carothers welcomed its creation. As we have seen in Chapter 3, the shaping of Carothers as a polymer chemist began with DuPont's initiative. Had it not been for this contact, it is unlikely that the young Harvard instructor would have set out on a journey into the frontier of polymer research. Or he might have remained a conventional organic chemist, but probably in the role of a less than successful teacher. The new chemistry of macromolecules which appeared to Carothers to embrace academic as well as practical possibilities for exploration; DuPont's — or more precisely Stine's — new mode of basic research organization as the site for its pursuit; and a new role of professional scientists in American industry — all these elements promised to satisfy the scientific personality and ethos of the young Midwesterner. While he was at DuPont, Carothers's study of macromolecules was bound to give rise to both academically and industrially significant knowledge.

Industrial practice certainly stimulated Carothers's basic study of polymers. Just as DuPont's technological innovations benefited from Carothers's science, so in turn did his science benefit from technology. In fact, his scientific pursuits went back and forth between theory and practice: from his encounter with Glyptal resins to his concept of functionality; from his synthetic study of macromolecules to the idea of making synthetic fibers, and to his theory of fiber formation; from the chemistry of acetylene to synthetic rubber; and from neoprene back to the theoretical work of addition polymers. Above all, the nylon venture represented a large-scale industrial test that proved the validity of his theory of condensation polymers. His fifty-two scientific papers on polymers and polymerization, published in the short space of nine years at DuPont, owed a great deal to this "reciprocal approach" in which scientific principles were advanced by constant contact with industrial practice. It was this approach that made Wallace Carothers's study of polymers so cogent both in industry and in the academy.[88]

Staudinger's Political Struggles Under the Nazis

During the 1930s, Hermann Staudinger's troubles were political; he struggled against the oppression of Nazism.[89] His wife Magda was barred from accompanying him to the 1935 Faraday Society meeting in Cambridge, England, because Nazi officials feared that if she were with him they might defect.[90] And once he arrived in Cambridge, the Baden Ministry of Culture and Education (Baden Ministerium des Kultus und

Figure 11. Martin Heidegger (with black mustache, seated near center) attending a university rectors' meeting in support of the Nazi regime, Leipzig, November 1933. From *Illustrierte Zeitung,* 23 November 1933.

Unterrichts) compelled Staudinger to explain her absence in a way that would not hurt "the prestige of the German Reich."[91] That meeting signaled his last international participation until the end of World War II.

Staudinger's distress began in 1933 when his "political past" became a Nazi target. The first threat came from the rector of his university. In April 1933 Wilhelm von Möllendorff, a professor of anatomy, was elected rector of the University of Freiburg. Two weeks later, however, the Baden Minister of Culture and Education forced him to resign on the grounds that he was a member of the German Social Democratic Party and that he had prohibited the posting of placards for an anti-Jewish campaign on campus. Martin Heidegger, the eminent existentialist philosopher and professor of philosophy at Freiburg, replaced Möllendorff as rector. In a well-publicized interview given some thirty years later, Heidegger claimed that he had unwillingly joined the National Socialist Party due to circumstances of the time, and further that he had opposed a Nazi scheme to purge Jewish professors at Freiburg, including Georg von Hevesy, a renowned inorganic chemist and future Nobel laureate.[92] However, in view of his lectures, presidential address, and articles for the student newspaper around 1933, there can be little doubt that at the time he admired Nazism.[93] That Heidegger's apology was contradictory is especially evident when it comes to his action toward Staudinger, an event disclosed by the historian Hugo Ott, who in 1983 discovered confidential documents dating from the Nazi period.[94]

In September 1933, Heidegger secretly reported details about Staudinger's political past to a Nazi official. Although not Jewish, Staudinger was alleged to fall under an article (Paragraph 4) of the Civil Service Law in which those who had not constantly demonstrated their loyalty to the Reich were to be removed from public service. According to Heidegger, during World War I Staudinger had publicized his anti-war ideas and opposed German militarism. Going on this information, the Gestapo (Geheime Staatspolizei) in Karlsruhe began an investigation, secretly named "Aktion Sternheim."[95]

Within a few months, a large amount of documentation was gathered as evidence from Karlsruhe, Berlin, Bern, and Zürich. This information confirmed that Staudinger attempted to obtain Swiss nationality and that he had exchanges with such pacifists as Frederic Nicolai, who denounced German militarism. In early February 1934 the Baden Minister of Culture and Education revealed the documents, asking Heidegger to proceed quickly with "Aktion Sternheim" since the statutory applicability of Paragraph 4 was to expire at the end of March. Quoting from a report by the German Consulate General in Zürich, Heidegger dashed off a reply to the Ministry:

St. [Staudinger] "made no secret of his keen opposition to the national drift in Germany and over and over again declared that he would never support his fatherland with the weapon or other services." . . .

These facts certainly demand the application of paragraph 4 of the Civil Service Law. These have been widely known in German circles since the discussions on Staudinger's call in 1925–26, and the prestige of the University of Freiburg too calls for legal steps, all the more as Staudinger today brags that he is a 110 [sic] % friend of the national uplift. Dismissal, rather than retirement, may be suitable. Heil Hitler![96]

In mid-February Staudinger was asked to report for questioning by the Ministry of Culture and Education at Karlsruhe. Unaware of the rector's tip-off and hence unprepared, Staudinger could barely clear himself of the charge. Waiting outside the interrogation room, Magda Staudinger worried that she might not see her husband again.[97] Although he was released, Staudinger could only explain that he had not been a pacifist in the same sense as a Quaker or conscientious objector, and that he had expressed his anxieties only in view of the significance of technology in modern warfare. He had to admit to his wartime actions, but emphasized, however, that he had long since retreated from his earlier political views.[98] Shortly after, probably to make his current position publicly clear, he wrote an article on the significance of chemistry for the German people that appeared in a Düsseldorf newspaper, and personally sent a copy to the Minister at Karlsruhe. The article tactfully conveyed the impression that he, as a chemical authority, could help contribute to the self-sufficiency, or autarky, of a new Germany.[99]

In March Heidegger retracted his initial proposal, recommending to the Ministry that Staudinger resign. That month Staudinger was invited as a guest speaker at the Congress of the International Union of Pure and Applied Chemistry in Madrid. He had been honored a year before with the Cannizzaro Prize, awarded by the Royal Academy of Lincei (Reale Accademia Nazionale dei Lincei) in Rome. Heidegger could not ignore Staudinger's international fame. Stressing that it indicated no change in his judgment, Heidegger wrote to the Baden Ministry:

After due consideration of Staudinger's case, it appears to me advisable to seek an appropriate way in consideration of his scientific reputation abroad. . . . It is simply a question of avoiding a new burden of foreign policy as far as possible.[100]

The Baden Ministry of Culture and Education conceived a face-saving solution that would again be humiliating to Staudinger. They ordered Staudinger to request resignation on his own initiative but offered a probationary period with a stipulation: if new suspicions did not arise about him in the next six months, the Ministry would decline the request.

Since by October no suspicions had surfaced, the request for discharge was turned down.[101]

To the end of his life, Staudinger remained ignorant of the fact that the incident sprang from Heidegger's accusation. According to Magda Staudinger, her husband had no occasion to converse with Heidegger, although he had seen the rector at a distance on campus.[102] Among the Staudinger Papers, housed at the Deutsches Museum, there is a five-page typed memo entitled "Einfluss des Nationalsozialismus auf die Entwicklung der makromolekularen Chemie" (The Influence of Nazism on the Development of Macromolecular Chemistry), which he wrote immediately after the end of World War II. "The year 1933 brought a complete change of my scientific position in Germany because it was well known that I had disagreements with the party that wanted to remove me from the official position," Staudinger wrote.[103] This memo, however, in which he named Nazi scholars who he believed had offended him, made no mention of Heidegger. Clearly, Heidegger kept secret his actions against Staudinger for the rest of his life.

Heidegger resigned the rectorate within a year. Yet the Nazis continued to harass Staudinger in various ways. The Gestapo had the Staudingers' housemaid inform them as to what the couple said and did.[104] Friedrich Metz confessed later that during his rectorate at the University of Freiburg between 1936 and 1938 the Reich Ministry of Economy (Reichswirtschaftsministerium) sought to fire Staudinger several times "because of his antagonism to National Socialism."[105] Continually under pressure, Staudinger considered seeking refuge in the United States, as had his younger brother Hans.

A Social Democratic representative in the Reich Parliament in Berlin and Privy Council of the Reich Ministry of Economy (1919–27) as well as State Secretary of the Prussian Ministry of Trade and Industry (preußisches Ministerium für Handel und Gewerbe) (1927–33), Hans Staudinger was dismissed from office on the basis of Paragraph 4 of the Civil Service Law. His publications on economics were burned. Moreover, he was charged with treason and arrested in June 1933 by the Reich police and threatened with hanging. After a six-week imprisonment, he was liberated only because of the desperate efforts of his Jewish wife, Else, who wisely used the influence of her friends to gain his release. In September 1934 Hans and Else fled to the United States by way of Belgium, France, and England. Hans was appointed a professor of economics and later dean at the New School for Social Research (or "the University in Exile," as it was aptly called) in New York. Together with the New School's president, Alvin Johnson, Else founded the American Council for Emigrés in the Professions, an organization that would help thousands of refugee professionals and other highly trained men and women in meeting America's educational and economic needs.[106]

After due consideration, however, Hermann Staudinger chose to stay in Germany in order to continue his polymer research, which was just then reaching an exciting stage. He had already invested a large amount of his own resources (approximately 80,000 Reichsmarks) in the Freiburg laboratory and had just found several able young collaborators, so dogged was his determination to establish macromolecular chemistry on his own terms.[107]

Whereas Jewish disputants of Staudinger's theories were compelled to leave Germany, the few antagonists who remained in Germany gained influential positions in the Nazi regime. Among them was Kurt Hess, a cellulose chemist at the Kaiser Wilhelm Institute for Chemistry. Staudinger's memo, mentioned above, reveals Hess's determination to harass him:

Also important [for my decision to stay in Freiburg] was the fact that, as I heard, Kurt Hess at Berlin schemed to become my successor [at Freiburg]. Owing to his leading position in the SA [Nazi Storm Troopers], this man at the time, and in the subsequent period, had good connections with various governmental offices in Berlin. Later too, I head again and again that he was framing a plot against me in the hope of getting my position here.[108]

By the early 1930s, Staudinger's macromolecular theory had gained significant support from abroad, notably from the work of The Svedberg in Sweden and Carothers in America. While a few German supporters of the aggregate theory began to change their minds, Hess continued to oppose the macromolecular theory. According to the memoirs of Ichiro Sakurada, Hess's student from Japan, Hess confessed to him, as he was about to leave Berlin for Kyoto in 1931, that

If I now give up this research [on cellulose], the public would think that the macromolecular theory of cellulose has been established by the German scientists. But the truth is not so simple as they have claimed. There are still many tasks to be done and many problems to be solved. Unlike Karrer and Bergmann, I will not give up this problem and will continue thoroughly my study from the point of view which I have maintained for years.[109]

After many of his academic colleagues fled Nazi Germany, Hess remained, and continued as one of Staudinger's most formidable opponents in German scientific circles. The rumor that Hess, as an important member of the Nazi SA, was trying to discharge Staudinger and even to have himself appointed to Staudinger's position at the university must have caused the latter much fear and agony. Perhaps the rumor was not groundless; Hess had long cherished a desire to be a university professor,

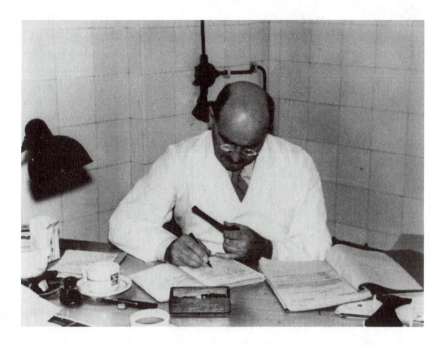

Figure 12. Kurt Hess in his laboratory at the Kaiser Wilhelm Institute for Chemistry, Berlin-Dahlem, 1932. Courtesy Archiv zur Geschichte der Max-Planck-Gesellschaft.

especially at Freiburg, where he had taken his *Habilitation* and served as a non-tenured associate professor during World War I. In a 1934 letter to a Dresden chemist, Staudinger wrote:

I think that his [Hess's] interpretation on cellulose is wrong and his series of experimental studies is wrong as well. Because he continuously insists on his peculiar views, which are not demonstrated experimentally, the development of cellulose study in Germany has been severely hampered. My personal attitude toward him was changed recently by his method of debating. . . . In my view, this debate would damage the German scientific reputation abroad.[110]

In any event, Hess remained associate professor at the University of Berlin, and his move to Freiburg did not take place. While holding his university position, Hess remained as head of the Guest Division (Gastabteilung) of the Kaiser Wilhelm Institute for Chemistry. According to Otto Hahn (then director of the Institute and a non-Party member), Hess not only suggested to the Nazi Reich Ministry that they dismiss Hahn's Jewish assistant, Lise Meitner, but he also conspired to take Hahn's post.[111]

Wolfgang Ostwald at the University of Leipzig was another academic opponent, as well as Nazi Party member, who annoyed Staudinger. Well before the popularity of National Socialism, Ostwald had displayed an antipathy to Jewish scholars, especially "the Berlin Jews" who occupied many of the important positions at the Kaiser Wilhelm Institutes in Dahlem. He welcomed the 1933 political upheaval, because, as he wrote to Martin Fischer,

No one could avoid our political recrystallization. The students in my lab are almost all active in politics and in uniform. At first with hesitation, and then complete conviction, I was thrust into this process. . . . I am convinced that the revolution was necessary for us and completely a blessing, and will be shown so in coming years. In such a revolution, foreign bodies, like Jews, etc., will be gotten rid of—it belongs to the definition of the concept.[112]

Despite his efforts to preach the importance of his science, and despite the success of colloid chemistry, Ostwald himself had been unable to obtain a secure post for nearly three decades, which no doubt made him an even more outspoken propagandist of his science. Twelve years earlier, when he failed to be appointed director of the Kaiser Wilhelm Institute for Leather Research, Ostwald discontentedly referred to its new director, Max Bergmann, as "a completely unknown young organic chemist, naturally a Jew."[113] While attached to the Physical Chemistry Institute headed by his father at the University of Leipzig, Ostwald had to wait long to be given a full professorship of colloid chemistry. Then, after joining the Party in May 1933, he was promoted to full professor and was considered a promising candidate as Herzog's successor at the suspended Kaiser Wilhelm Institute for Fiber Chemistry.[114] Ostwald also became an arbitrator (*Schiedsrichter*) of Nazi academic affairs.

As Staudinger saw it, Ostwald, who had maintained a distance from him throughout the macromolecular debate in the 1920s, embarked on a new course of open criticism of Staudinger's work only after the Nazis came to power. One aspect of his behavior that Staudinger found inscrutable was Ostwald's persistent protest against Staudinger's use of the word *"Eukolloide."* As early as 1925 Staudinger had used *"Eukolloide"* (meaning true colloids) to refer to the group of colloids in which primary colloidal particles represented the macromolecules themselves. Admittedly, he had overlooked the fact that two years earlier Ostwald had used the same term to designate colloidal substances as distinguished from colloidal states. Although silent about the issue up until 1933, Ostwald now asked Staudinger to "return" the word to him. In his journal, *Kolloid-Zeitschrift,* Ostwald twice criticized Staudinger's books, such as *Organische Kolloidchemie,* in an "exorbitantly derogatory" manner, while excessively praising works by Paul Karrer and Kurt Hess.[115] Ostwald's

Figure 13. Wolfgang Ostwald at Leipzig, 1930s. Courtesy Wilhelm Ostwald Gesellschaft and Chemical Heritage Foundation.

animosity lasted until the 1940s, well after Staudinger's theory appeared to have won wide acceptance. "Since he could have supported my theory on the existence of macromolecules in previous years [especially because of his concept of *Eukolloide*], his later actions were so much the less understandable," Staudinger speculated in his memoirs.[116] The timing, unreasonableness, and far from scientific nature of Ostwald's criticism aroused Staudinger's distrust of this leading colloid chemist.[117]

In 1940 Staudinger took over the editorship of the *Journal für praktische*

Chemie, which was published in Leipzig. Taking advantage of this position, he transformed the oldest existing German chemical periodical into a new journal with the sub-title, *Unter Berücksichtigung der makromolekularen Chemie.* The revised issue started in February with Staudinger's introductory essay, "Über niedermolekulare und makromolekulare Chemie."[118] Although the journal continued to cover traditional subjects in the field of organic chemistry, its primary emphasis was the new field of large molecules. Clearly, Staudinger intended to use it as a vehicle for the establishment and dissemination of his science, an attitude that made academic opponents like Ostwald uneasy. As wartime conditions limited the supply of paper, the Reich Union of German Journals (Reichsverband der deutschen Zeitschriften) advised journal editors that journals dealing with similar subject matter merge. Ostwald seized this opportunity to request that Staudinger annex the *Journal für praktische Chemie* to his *Kolloid-Zeitschrift,* a request that did not materialize due to Staudinger's strenuous objections.[119] Staudinger's negative response angered Ostwald, who complained, "It was a major concession on my part to propose the merger to such a narrow-minded man."[120] In November 1943, the confrontation ended, however, with Ostwald's sudden death at the age of sixty.[121]

Staudinger experienced many other difficulties. He was no longer invited to lecture by the major scientific societies in Germany, such as the Society of German Natural Scientists and Physicians, the German Chemical Society, and the Society of German Chemists. Nor did the Berlin Ministry for Science and Education allow him to attend international meetings after 1937, something other scholars were still permitted to do.[122] In vain, Staudinger entreated the Cultural Ministry for overseas travel privileges, expressing concern that in his absence from international congresses individuals like Kurt H. Meyer and Herman Mark, now residing outside Germany, were freely appropriating "my scientific results in the field of rubber, cellulose, plastics, etc. for their own purposes without appropriate citation."[123] Meanwhile, Staudinger's research funds were cut, and his publications (including the fourth edition of *Organische Kolloidchemie*) blocked. His assistants, such as Werner Kern and Günther V. Schulz, found it extremely difficult to obtain academic positions. These problems, he realized, resulted largely from the actions of Nazi academic opponents. Staudinger therefore believed that German polymer chemistry was victimized by Nazism.[124]

Staudinger made no overt political commitments during the Nazi regime. Whatever the rhetoric used by opponents, it is unlikely that the difficulties Staudinger encountered in his scientific activities stemmed solely from his "political past" during World War I. Although politically voiceless, he was no quitter in scholarly debates. In papers and lectures, he continued to criticize the research of his scientific opponents, including Hess and Ostwald.[125] Without a doubt, this provocative stance further stirred their antipathy. Losing the scientific domination they had once held, some of Staudinger's opponents who remained in Germany found a point of counterattack by appealing to their last resort, the Nazi officialdom. These opponents were in a position to use that political power to attack Staudinger's chemical ideas and professional activities. Hence the later phase of the macromolecular debate in Germany not only reflects a picture of a theoretical conflict or priority dispute, but also political tensions of the time.

Unlike "Aryan physics," there was no "Aryan polymer chemistry"; we find few traces in the chemistry of Hess and Ostwald of efforts to combine the content of science with their political ideology.[126] Apparently, their anti-macromolecular stance had little to do with National Socialist ideology. Instead, they opportunistically used their political affiliations with the Party, in part to secure and even improve their own academic positions and status. Not a few Nazi scientists participated in such a political game, in which some became winners while others gained less than their expectations. Unlike Hess, Ostwald seems to have belonged to the latter group. Though a loyal Nazi sympathizer, Ostwald also expected that his political ties with the party and the departure of his "Jewish enemies" "will put me somewhat up the ladder."[127] He certainly won a belated promotion at Leipzig. But he was crestfallen on learning that Peter Adolf Thiessen, Zsigmondy's student and a staunch National Socialist who sat more securely with the Berlin Ministry, had succeeded the purged Freundlich at the Kaiser Wilhelm Institute for Physical Chemistry and Electrochemistry. Ostwald apparently coveted this independent post for colloid chemistry.[128] Moreover, the Nazi government named Thiessen as director of the institute. There he oversaw work related to the goals of the Four Year Plan, an economic outline officially announced by Hitler in 1936 to make Germany self-sufficient in the development of raw materials and thereby to save on foreign exchange. In fact, by 1940 Thiessen's institute was admiringly called a "National Socialist model institute."[129] As we have seen, Ostwald was also a leading candidate to succeed Reginald O. Herzog as director of the Kaiser Wilhelm Institute for Fiber Chemistry. But again he failed, as financial problems and the outbreak of war nullified the society's plans for reopening that institute.[130] It is not surprising that Ostwald turned out to be a severe critic of the Party in the years before his death.[131]

Finally, the question needs to be asked why Staudinger was able to retain his position and allowed to work throughout the Nazi regime despite mounting opposition by his colleagues. Ironically, the Nazi period was concurrent with the height of Staudinger's scientific productivity.

Between 1933 and 1945, he published over 140 papers on macromolecules (52 percent of his entire output on the subject) as well as four textbooks, and trained thirty-eight doctoral students (nearly half of the Ph.D.s in the field).[132]

A non-Party member, Staudinger had to find ways to survive the swastika. One strategy for adapting to the changed political environment was to appeal to the value of his chemistry for German self-sufficiency. As mentioned above, he used this approach in his newspaper article. In March 1937 he ventured further by proposing the creation of a "Kaiser Wilhelm Institute for Wood and Cellulose Research" (Kaiser Wilhelm Institut für Holz- und Cellulose-Forschung)" in Freiburg.[133] However, Peter Debye, who at the time directed the Institute for Physics, and Otto Hahn, who oversaw the Institute for Chemistry, advised the president of the Kaiser Wilhelm Society, Max Planck, that there was little need for the society to create a new institute, since Staudinger's intention was "only an expansion of the chemical institute of the university."[134] That fall Staudinger proposed an alternative, this time using the name "Kaiser Wilhelm Institute for Macromolecular Chemistry (Kaiser Wilhelm Institut für makromolekulare Chemie)." Stressing how appropriate the study of macromolecular materials, such as wood, cellulose, rubber, and plastics, was for Hitler's Four Year Plan, he made a case for the establishment of such an institute at Freiburg:

Today, the chemical industry is assigned many great tasks to produce synthetic fibers as substitutes for cotton and wool, artificial rubber, artificial petrol, lubricating oil, and many other synthetic materials. According to the Four Year Plan, the German people should not be dependent on importation of these products which constitute 1 to 1½ billion [Reichsmarks], namely about 30% of the whole imports.

In order to produce synthetic substitutes, it was expected here, as in the field of dyestuffs, to elucidate the constitution of these products and also to investigate scientifically the mechanism of their formation. Staudinger continued:

It was a ten-year work—resulting in over 200 scientific publications—at the Freiburg Laboratory that has succeeded in elucidating structures of the most important of these materials. . . . Therefore, this new branch of organic chemistry should be completed by creating a Kaiser Wilhelm Institute for Macromolecular Chemistry in Freiburg i.Br., a special fostering place.[135]

Hess did not fail to challenge Staudinger by also proposing a polymer institute. Shortly after Staudinger's lobbying, he proposed a strikingly similar "Institute for Wood and Cellulose Research" (Institut für Holz- und Zellstoff-Forschung) in Berlin. Hess, for his part, requested that the Society transform his Guest Division at the Kaiser Wilhelm Institute for

Figure 14. Staudinger with his assistants and students at Freiburg, ca. 1935. Front row (left to right), Emil Dreher, Hermann Staudinger, Magda Staudinger, Erwin R. Sauter, Werner Kern; second row, K. Feuerstein, Elfriede Husemann, Kuno Wagner, Günther V. Schulz, Rolf Mohr, A. E. Werner; third row, K. Rössler, H. Frey, I. Jurisch, Günter Daumiller, Friedrich Reinecke; top row, Hans-P. Mojen, H. Schwalenstöcker, H. von Becker, L. Hildenstab. Courtesy Magda Staudinger.

Chemistry into an independent Kaiser Wilhelm institute.[136] But the Office for German Raw Materials (Amt für deutsche Roh- und Werkstoffe), a division created under the Four Year Plan's apparatus—in agreement with the Kaiser Wilhelm Society—decided to support the expansion of Karl Freudenberg's Chemical Institute at Heidelberg and Walter Brecht and Georg Jayme's Institutes at Darmstadt as special cellulose institutes that were expected to contribute to Germany's Plan. The proposals of Staudinger and Hess were therefore both turned down.[137] Yet Hess's desire to have his own independent institute was eventually fulfilled in 1943 when he became head of the Institute for Rubber Research (Institut für Kautschukforschung) in Löwenberg.[138]

There was a more important reason why Staudinger managed to survive Nazism. Magda Staudinger believes, "It was industry that saved him."[139] As we have seen in Chapter 2, since the mid-1920s certain German industrial leaders had recognized the industrial significance of Staudinger's work and tried to incorporate it into their research on and development of synthetic rubber and plastics. In all likelihood, these individuals helped him escape dismissal or arrest by the Gestapo.

Staudinger continued to maintain a good relationship with industry throughout the Nazi regime. At Freiburg, between 1928 and 1944, he trained forty-eight Ph.D.s in the field of macromolecules; all but one went into industry immediately after graduation (see Appendix Table 1). Besides doctoral students, there were several associates who became industrial researchers as well, including Erich Konrad, I. G. Farben's chief rubber chemist at Leverkusen.[140] Staudinger endeavored in vain to secure academic positions for some of his most able students. In addition to his own strained relationship with the Party, the peculiar Nazi employment system for faculty positions, which required ideological training and loyalty to the Party, discouraged most of his students from pursuing academic careers.[141] As a result, this situation opened up opportunities to propagate his theory in industry even earlier than in academia.

Staudinger gained support from Emil Wilhelm Tscheulin, who, as he later acknowledged, brought "an improvement of my situation" after 1936. The president of the Freiburg Chamber of Industry and Commerce (Industrie- und Handelskammer Freiburg) and a Reich Commissioner, Tscheulin encouraged Staudinger's study of macromolecules and sought to raise funds from industry when Staudinger proposed the creation of an independent institute at Freiburg. While the Reich Ministry and the Kaiser Wilhelm Society turned down Staudinger's proposal for a new Kaiser Wilhelm institute, Tscheulin supported Staudinger in early 1940 in his efforts to establish an institute under the name "Research Division for Macromolecular Chemistry" (Forschungsabteilung für makromolekulare Chemie) within the Chemical Laboratory at the University of Freiburg.[142]

Staudinger's connection with I. G. Farben merits special attention. This tie was noticed with fear and envy by Hess, whose Guest Division at the Kaiser Wilhelm Institute for Chemistry had been sponsored by I. G. Farben since 1931. In 1937 Hess cautioned Max Planck that Staudinger "has tried to influence authoritative persons of the I. G. Farben Industrie" to block its support for his division.[143]

Georg Kränzlein was among the influential Farben officials who maintained contact with Staudinger. The head of polymer research at I. G. Farben Hoechst, he was in a position to help preserve Staudinger's academic career. As mentioned in Chapter 2, impressed by Staudinger's lecture at the 1926 Düsseldorf symposium, Kränzlein became a firm believer in the macromolecular theory and maintained links with the Freiburg school in the hope of applying Staudinger's work to industrial undertakings. Aware of Staudinger's scarce state funding, Kränzlein continuously provided the Freiburg laboratory with financial support. He hired

Staudinger's 1928 doctoral student, W. Starck, to work in his Hoechst group on the polyvinylacetate R & D project. It was also Kränzlein who found employment for Werner Kern—Staudinger's only doctoral student who had stayed in academia upon graduation during this period—when Kern, under Party pressure, was discharged in 1939 from the newly planned German Research Institute for Plastics (Deutsches Forschungsinstitut für Kunststoffe) in Frankfurt.[144] Aware that Staudinger was under close observation by the Party, Kränzlein was enough of an enthusiastic Nazi supporter to secure himself and to protect Staudinger and his students against political slanders. He expected that Staudinger's research activities would soon "be acknowledged by the National Socialist State," because of Staudinger's "co-operation in the new development of the autarkic chemistry in Greater Germany."[145]

In 1936, the editor of the *Berichte der deutschen chemischen Gesellschaft* declared an official end to the debate between Staudinger and Kurt H. Meyer, a Jewish chemist and then professor at Geneva, by inserting a notice in the former's article that had responded bitterly to Meyer's recent criticism.[146] Kränzlein cautioned Staudinger—who he believed shared his racist assumptions—concerning the debate:

In my opinion, you make a mistake to contend continuously with Jews. . . . You need not get involved in polemics with Jews, because by doing so you give them far too much honor. Avoid and ignore this gang. . . . We systematically keep our distance from Jews, as the Nürnberg law shows. Thereby we bring them back to the place from which they came. Why don't you keep your distance in science? Here too, they must return to the spiritual ghetto from which they came. . . . Today, it is your duty no more to mention Jews at all, let alone to get involved in polemic with them.[147]

To Kränzlein, science had to conform to Nazi politics, a point he made far more explicitly than Hess and Ostwald had. Although the views of this loyal Nazi somewhat embarrassed him, Staudinger quietly followed his advice. Although political conditions did affect the nature of the controversy on macromolecules, political tensions in the debate cannot be seen simply as a dialectical conflict between the macromolecular school and adherents of Nazism.[148]

Whatever the reasons, Staudinger made no mention of the political aspects of his scientific activities in his published memoirs. As we have seen, however, his pacifist activities during World War I, his controversy with Haber over chemical warfare, his move to Freiburg, the Düsseldorf symposium on polymers, his struggles under the Nazi regime, and the prolonged domestic debate on macromolecules during the 1930s and 1940s—all these events, in large measure, reflect a series of vital interactions between Staudinger and political forces of the time.

Chapter 5
Restoration of the Physicalist Approach

Macromolecular chemistry is the youngest branch of organic chemistry.
— Hermann Staudinger, "Nobel Lecture," 1953.

Polymer chemistry was becoming a new branch of organic and physical chemistry [around the mid-1940s]. It needed and deserved the attention that would be cultivated through an organization devoted to teaching and research.
— Herman F. Mark, *From Small Organic Molecules to Large,* 1993.

In physical chemistry, as in other sciences, progress usually occurs by a series of rather discontinuous steps separated by periods of consolidation. Thus certain topics and branches of a subject become popular fields of activity once the pioneering work has defined the field of endeavour. Thereafter papers flow in ever-increasing numbers.
— Harry W. Melville, "High Polymers," 1953.

In this study, I have called particular attention to the role of conceptual and epistemological reasoning in the birth of macromolecular chemistry, which was of some significance in the historiography of chemical science as well. The dominance of physics over other natural sciences has been one of the general features of modern science. Concepts and methods of physics indeed exercised a wide influence on the ways chemists, biologists, and scientists of other fields pursued their distinctive research areas. In chemical science, physical chemistry had emerged as a boundary science by the late nineteenth century, and was to provide a deeper methodological insight into the study of inorganic, organic, and biological chemistry. Polymer chemistry proved no exception. Contemporary students of this specialty know how much our understanding of macromolecules owes to physical chemistry and physics. Modern textbooks of polymer chemistry are full of discussions of rigorous mathematical and physico-chemical treatments of polymers. The term "polymer science" today designates the entire field of study, encompassing both polymer chemistry and polymer physics. Understandably, this climate has led to a conviction that any chemical science could be deduced from the principles of physics. Not surprisingly, this conviction is sometimes applied a priori to the history of chemistry. Modern chemistry textbooks and even some historical studies seem to have fallen under this influence, viewing the birth of modern chemical sciences simply as linked with the victory of the physicalist view of nature or the "mechanization of the world picture."

However, history and logic are not necessarily one and the same. The story of the origins of macromolecular chemistry offers a counterexample to this historical pitfall. Macromolecular chemistry arose as a new field of organic chemistry from a conceptual conflict between the physicalist and the organic-structuralist traditions. The physicalist approach to polymer-colloid substances, as represented by the colloid theories and the aggregate theory, flourished in the early part of this century and reflected the rise of physical chemistry. Yet the success achieved by Staudinger and Carothers in the 1920s and early 1930s, following a decade-long controversy over macromolecules, proved to be a manifestation of the conceptual power of orthodox organic chemistry in the face of the current physico-chemical trend. Both conceptually and epistemologically, the notion of macromolecules represented an anti-physical deductivism, in that it dismissed concepts which reduced chemical phenomena to the smallest units of matter and interacting physical forces. As we have seen, the new theory was firmly grounded in the traditional Kekuléan molecular approach to organic compounds, which viewed the molecule as the unit from which stemmed properties of matter. Even Paul John Flory, a pioneer in the physical chemistry of polymers, was sensitive about the historical foundation of macromolecular theory. His cautionary view merits quoting:

> . . . to hold that all chemistry follows deductively from physics and dismiss the matter therewith is to overlook the central role of science in erecting constructs for representation of physical reality in terms rational to the human mind. . . . [T]o present chemistry as a deductive science is to conceal the historical foundations and conceptual framework of the science of molecules and molecular behavior. This viewpoint could conceivably lead ultimately to denial of the rightful existence of chemistry as a separate discipline.[1]

The macromolecular view carried a holistic or emergent conception that the whole was not a mere sum total of its constituent parts, but something more than that. In short, a macromolecule was not a mere collection of atoms. This acknowledgment was an epistemological ground essential to the macromolecule concept, which Staudinger stressed time and again in opposing the physico-mechanical reductionism of proponents of the aggregate theory. Defining colloids as "physical compounds" and not as chemical ones, Victor Cofman, a physicalist, noted in 1927:

> Simplicity and symmetry should be among the chief aims of a scientific theory. It is probable that the same laws which regulate the movement of electrons within the atom also determine the paths of planets in their orbits; a complete understanding of the simplest phenomenon may enable us to explain the Universe.[2]

Cofman worked as a member of the colloid group in DuPont's fundamental research program, but left the company shortly after Carothers joined DuPont.[3] In contrast to Cofman's view, the following statement by Flory, who was Carothers's loyal assistant at DuPont in the mid-1930s, illustrates the viewpoint of the macromolecularist:

> The reductionist attitude has encouraged chemists to focus their investigative efforts on the simplest molecules to the virtual exclusion of all else, as if full knowledge gained at the simpler level would suffice to explain the more complex by straightforward deduction. This is demonstrably false. . . .
> To be sure, a great deal can be learned by investigation of the simplest systems, but comprehension of those of higher complexity cannot be achieved through processes of deduction alone. . . . [F]ull knowledge concerning simple molecules did not pave the way for comprehension of macromolecules. Further creative effort, formulation of new concepts with appropriate abstraction, and so forth, were required.[4]

As a whole, the large molecule exhibited its own properties, which could not be deduced from those of smaller molecular or atomic units, and which could not be predicted even by a thorough study of the low molecular weight substances. Hence emphasis was placed by macromolecularists on the superstructure of molecules: the molecular size and shape that determined unique properties of polymers, such as colloidal phenomena, elasticity, and fibrousness. These new conceptions associated with the organic-structural approach to macromolecules expanded the theoretical outlook of classical chemistry, that is, the science of molecules. Thus it may well be argued that organic chemistry at its inception became a prime mover of the field on the basis of its own principles and by eradicating the shadow of the old physicalist tradition. Armed with this conviction, Staudinger maintained throughout the debate that macromolecular chemistry was a new branch within organic chemistry.

To be sure, the history of the study of macromolecules does not end here. The now familiar outcome followed: the coming of age of the physical chemistry of macromolecules. Physical chemistry and physics did greatly help macromolecular chemistry mature as a science. How did this transition take place? Who were the cultivators and how did they enter the field? What were the reactions of Staudinger and Carothers to the trend? We shall briefly examine here the historic dynamism of this new development.

The Physical Chemistry of Macromolecules

By the late 1930s, it had become increasingly clear that organic chemistry alone could not solve the whole problem of macromolecules and that Staudinger's organic-structuralist approach had certain limits. Missing from his intellectual scope was a physico-chemical insight into polymers, as had been pointed out by Herman F. Mark and others from an early stage. Once the macromolecular concept was firmly established by the mid-1930s, physical chemists and physicists alike found plenty of room left to apply their methods and theories to polymers. By the end of the war, physical chemistry — for which Staudinger showed a strong distaste — had become a vital, even pivotal part of polymer research. The new physico-chemical studies expanded the scope and understanding of the dynamic behavior of macromolecules and even disproved some of Staudinger's initial concepts. Physical chemistry thus was proven to be not an inveterate foe of organic chemistry but its essential complementary partner in perfecting the science of macromolecules.

It was an irony of fate that Staudinger's viscosity law turned out to pave the way for the rapid growth of the physical chemistry of macromolecules. Viscometry — the only and "primitive" physical method that the organic chemist Staudinger had introduced into his polymer research — did not remain merely a technique of measuring molecular weight, but posed the essential question of what macromolecules were really like. What did his viscosity law, namely, that viscosity was in direct proportion to molecular size, imply? Staudinger reasoned that linear macromole-

Figure 15. Staudinger's wooden stick model of macromolecules stored in the Deutsches Museum. Photography by author. Courtesy Deutsches Museum.

cules must have a thin, rigid rod shape like a glass fiber in solution. Long, rigid rods of macromolecules would move across a flowing liquid, rotating as they moved in a disk-like plane. The intrinsic viscosity was considered to be proportional to the volume of disk-like cylinders swept out by macromolecules. In 1932 he explained that his viscosity law

can only be understood on the assumption that the molecules possess a rigid fiber shape. . . . In order to demonstrate my opinion about these molecules by a comparison, I would compare such a molecule with a thin flexible glass fiber, and not with a wool fiber, which is capable of assuming any shape. Of course, these molecules cannot be compared with absolutely rigid rods. But a rigid shape of the molecules seems to me to be necessary from very general experiences in organic chemistry. The large number of organic compounds is only to be understood if their molecules are rigid.[5]

For many years, Staudinger in his lectures used wooden sticks as models of macromolecules (see Figure 15). At the 1934 Madrid meeting of the International Union of Pure and Applied Chemistry, for example, he explained the process of lowering of viscosity by snapping off these thin sticks before the audience.[6]

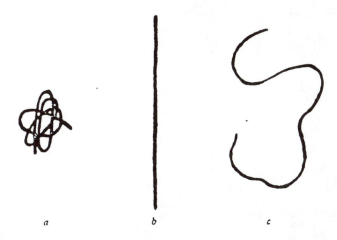

a   *b*   *c*

Figure 16. Mark's sketch of macromolecules. From *Trans. Faraday Soc.* 32 (1936), Pt. 1, p. 312.

Staudinger's concept of rigid macromolecules elicited considerable opposition from physical chemists. The 1935 Faraday Society meeting witnessed the controversy between Staudinger and Mark over this problem. Mark claimed that Staudinger's concept contradicted the basic requirements of physical chemistry:

The chain-like macromolecules, which must be investigated in extremely diluted solution, seem not to be compact, more or less sphere-like clusters (compare for instance, Fig. a [Figure 16a]). They are, moreover, not quite extended and stiff with elastic vibrations (as shown in Fig. b [16b]), but they are in a state, as it is shown in Fig. c [16c], that is to say, they are bent but not rolled entirely together. If one assumes the form shown in Fig. c one gets in fairly good agreement with all experimental evidence . . . and remains at the same time in concordance with fundamental statistical considerations and with the principle of the free rotation round the single carbon bond. I think that the shape shown in Fig. c may be regarded as a close approximation to the real form of the long molecules in solution.[7]

Mark's position was supported by the latest statistical study of the Swiss chemist Werner Kuhn, who assumed that the macromolecule was partially rigid but as a whole flexible; that is, the separate rigid links of the molecular chain could rotate freely around single chemical bonds in relation to each other.[8] Kuhn showed his picture of macromolecule as a flexible, coiled chain-molecule analogous to a pearl necklace (see Figure 17).

Staudinger had responded sarcastically, "If you [physical chemists like

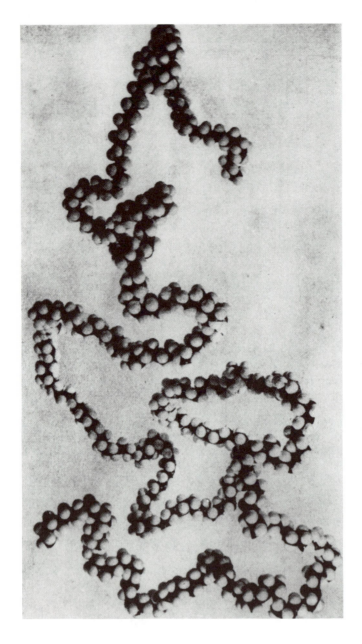

Figure 17. Werner Kuhn's "pearl necklace" model of macromolecules. Courtesy Hans Kuhn and Chemical Heritage Foundation.

Mark and Kuhn] insist, why don't you prove the viscosity law which we had obtained from experiments, by you guys' skillful calculations on the basis of your assumption of rotations around the chemical bonds?"[9] Indeed, Mark and Kuhn immediately did so. The concept of flexible-chain molecules gave rise to a new interpretation of the mechanism of viscosity. Because of the highly extended configuration, the effective hydrodynamic volume swept by the chain molecule in solution was considered to be much larger than the molecular volume. By 1940, refined general formulas had been proposed by Kuhn, Mark, Roelof Houwink in Eindhoven, the Netherlands, and Ichiro Sakurada in Kyoto, Japan.[10] They indicated that Staudinger's empirical viscosity-molecular weight relationship was valid only for a few specific systems, not for many important high polymers such as rubber, polystyrene, and polyamides. Staudinger, however, would continue to refuse to endorse any such formulas by simply claiming them to be the product of "theoretical arguments," rather than experimentation.[11]

Once physical chemists accepted the concept of macromolecules in which atoms were held together by covalent bonds, they were far from bothered, but rather fascinated by the complexities peculiar to polymers. Polymers were, as Paul Flory put it, "non-homogeneous substances consisting of mixtures of chemically similar molecules which are distributed in size about an average, some of them being much smaller and some much larger than this average."[12] Here physical chemists found a golden opportunity fully to exploit probability and statistics—mathematical tools which would be less applicable to the realm of ordinary small molecular substances. As Flory summarized:

The complexities of high polymers are far too great for a direct mechanistic deduction of properties from the detailed structure of the constituent molecules; even the constitution of polymers often is too complex for an exact description. The very complexities which make the task of rational interpretation of polymers and their properties appear formidable actually provide an ideal situation for the application of statistical procedures.[13]

The study of the distribution of molecular weights, sizes, and shapes, and of the behavior of macromolecules appeared amenable to statistics, mathematics, kinetics, thermodynamics, and hydrodynamics. This was indeed virgin territory left for the physical chemist. The period between the late 1930s and 1940s represented an exciting formative stage "when polymer science looked easy," as American polymer physical chemists Walter H. Stockmayer and Bruno H. Zimm reminisced years later.[14]

The concept of flexible chains, along with the statistical and kinetic approaches, opened up new explanations of certain physical properties of polymers, such as the exceedingly high viscosity of dilute polymer solutions, double refractions of flow, and elasticity. For example, by the early 1940s the cause of rubber elasticity was thermodynamically and statistically clarified by Kurt H. Meyer, Eugene Guth, Mark, Kuhn, and others.[15] The kinetic units of macromolecules were best explained as segments rather than whole molecules (which could only move sluggishly because of their great sizes). Due to the frequent, free movements of many thousands of individual segments ("microbrownian motions"), the macromolecule could take a wide variety of irregularly contorted configurations. The probabilities of the random motions and shapes of the macromolecule could be statistically calculated. The probability of the stretched state of the macromolecule should be nearly zero; higher probabilities would lie in a thread-ball-like state. Rubber, when relaxed, tends to return to its original state because the stretched rubber macromolecule tends to restore to its innate thread-ball shape. This observation is also in harmony with the second law of thermodynamics: in nature, entropy increases from the molecularly ordered system (stretched state) to the molecularly disordered system (crumpled, rolled state).

Rubber elasticity increases as the temperature increases, because the higher the temperature, the greater the microbrownian motions. Johan Katz's finding that rubber, when stretched and kept cool, exhibits a fibrous X-ray crystallographic pattern could now be adequately expounded as a phenomenon in which the stretched macromolecules are arranged in an orderly manner in one direction and held by the inter-macromolecular force to take a crystalline form like a fiber. Yet when heated the rubber immediately loses its crystalline form and shrinks because heat increases the microbrownian motions enough to break the intermolecular force, and the macromolecules shrink to take the original thread-ball-like shape. Vulcanized rubber has a strong elasticity and stability that is less temperature dependent than natural rubber. The former forms three-dimensional networks of macromolecules without losing much of the microbrownian motion of its macromolecules, so that the molecules hardly slip off from each other and exhibit a high elasticity and strength. These contorted configurations and the dynamic behavior of the macromolecule were an aspect which the structural organic chemist Staudinger had failed to take into account. He had explained properties of macromolecules only in terms of their chemical constitutions and fixed shapes, without considering their microbrownian movements. The new study of physical chemistry thus exposed the limits of Staudinger's static structural approach and promised vast possibilities which remained in the renewed physicalist approach to macromolecules.

Toward an Interdisciplinary Science

In the process, polymer organic chemists came to perceive the signifi-cance of mathematico-physical treatments in the field, although their initial studies had made little use of such methods. However uncomfort-able this made him, Staudinger, who had trained a large number of polymer chemists along the lines of organic chemistry, also could not deny their importance. The assistance of physical chemists was needed. Taking the macromolecular concept for granted, physical chemists, for their part, learned a great deal from the accomplishments of polymer organic chemists, by exploiting the data they had compiled and using the polymers they had synthesized. A new generation of polymer physical chemists, which emerged in the 1930s, typically began their studies by collaborating with organic chemists, before establishing themselves as independent polymer researchers. For example, Kuhn worked with Karl Freudenberg, Günther Viktor Schulz with Staudinger, and Flory with Carothers. Their collaborations proved to be remarkably fruitful. Even Mark, a physical chemist who was also adept at organic chemistry, invited an able physicist, Eugene Guth at the University of Vienna, in 1932 to help in his work. As Mark recalled,

Since I was not mathematically skilled enough to handle such a problem [statisti-cal treatment of polymer conformations], I asked the famous Viennese theoret-ical physicist H.[Hans] Thirring [at the Department of Physics] for help. He recommended to me Dr. E. Guth, one of his best collaborators. In fact, in a very short time he developed an important equation which represents the entropy gain during the coiling of linear flexible molecules.[16]

Rivaling Kuhn, Guth later claimed himself and Mark to be "the founders of the physics and physical chemistry of polymers," because of their 1934 paper that pioneered in the clarification of rubber elasticity by means of statistical mechanics of flexible chain molecules.[17] With this capable col-laborator, Mark was swift to seize the opportunity to design an ambitious and farsighted teaching program in polymer chemistry at Vienna. In contrast to Staudinger's Freiburg school, which clung to organic chemis-try, Mark's core curriculum consisted of organic chemistry, physical chemistry, and physics. His organic chemistry course dealt with the syn-thesis of monomers and polymers, the physical chemistry course with characterization of polymers, and the physics course with their proper-ties. As he later recalled, "It was really an interdisciplinary activity, as it is now. Now you start with organic chemistry, you go into physical chemis-try, and you end up with physics." By 1938 Mark had managed to assem-ble a group of enthusiastic co-workers and doctoral students, including

Anton Wacek in organic chemistry; Frederick R. Eirich, Philip Gross, Otto Kratky, Franz Patat, and E. Suess in physical chemistry; and H. Dostal, and Robert Simha, as well as Guth, in physics.[18]

A versatile theoretician, Kuhn entered the polymer field in 1930. Born in Switzerland, he had studied chemistry in the technological chemistry section of the chemistry department at Zürich's Eidgenössische Technische Hochschule in 1917, while Staudinger was teaching as head of its general chemistry section. After receiving a degree in 1921, Kuhn moved to the neighboring University of Zürich, where he obtained his doctorate in physical chemistry two years later, with a dissertation on the photochemical decomposition of ammonia. He then entered Niels Bohr's Institute of Theoretical Physics in Copenhagen to study quantum mechanics. With this background, Kuhn accepted an invitation in 1927 from Freudenberg at the University of Heidelberg to collaborate in research on optical activity, a partnership that served to evoke Kuhn's keen interest in polymers. Freudenberg, a cellulose organic chemist, was one of the few early supporters of the long-chain molecular concept, which Kuhn adopted from the outset.[19] His study benefited from Freudenberg's data as well as his broad knowledge of cellulose. For instance, Kuhn applied statistics to the molecular weight distribution of degraded cellulose by assuming that cellulose molecules were broken up randomly.[20] In 1939, he left Nazi Germany for the University of Basel, Switzerland, to become director of its Institute of Physical Chemistry, where he continued work in polymer physical chemistry.[21]

Schulz was not Staudinger's doctoral student, but had studied colloid chemistry under Herbert Freundlich at the Kaiser Wilhelm Institute for Physical Chemistry and Electrochemistry until his Jewish teacher fled Germany. Schulz then fell under the influence of Staudinger, when the macromolecularist accepted the young physical chemist's application to join the Freiburg school as *Dozent* in the fall of 1933. There Schulz embarked on a physical and mathematical study of macromolecules, although at first he was often puzzled by Staudinger's bias against the physical approach. Schulz found no difficulty in accepting "an ingenious idea" in Kuhn's recent work that explained rubber elasticity in terms of an entropy effect of thread-like molecules, an idea which "Staudinger did not like at all."[22] By re-examining "somewhat confusing" experimental data compiled by one of Staudinger's former assistants, Schulz was also able to arrive at his idea of molecular weight distribution as early as the beginning of 1935. He conducted an experiment using polymerides of isobutylene and managed to confirm the validity of his idea. Soon he prepared a paper on his findings for publication. However, "Staudinger was very suspicious of my calculation," recalled Schulz, "and left long my manuscript on his desk." But Staudinger's attitude changed abruptly

Figure 18. Staudinger's physical chemist, Günther V. Schulz, ca. 1935. Courtesy Deutsches Museum.

when he heard from Carothers at the Faraday Society meeting at Cambridge that "in America very similar considerations were being made." This priority concern caused Staudinger to publish Schulz's paper immediately in a German journal.[23] Schulz eventually gained Staudinger's respect and supported his mentor's work by conducting extensive research on the molecular weight and the kinetics of polymerization reactions.[24] He served as the sole physical chemist at the Freiburg school for ten years, until he moved to the University of Rostock in 1943.

Carothers had a more flexible attitude to the physico-chemical approach than Staudinger. DuPont's research management encouraged interdisciplinarity, an environment in which organic chemists, colloid chemists, and physical chemists frequently exchanged opinions and worked closely together.[25] The industrial undertaking necessitated such a teamwork atmosphere, whereas disciplinary boundaries were kept more rigid at the university in which the holder of a specialty chair dominated the department. Carothers's draft manuscripts were often internally reviewed before publication by such physical chemists as Elmer O. Kraemer of the colloid chemistry group. While this system of internal review partly served the company's purpose for protecting technological know-how, the cross-disciplinary check, which would not have been customary at a university, probably benefited Carothers's organic-chemical study.[26] Carothers himself had ample knowledge of physical chemistry, as illustrated by his early ambitious study of the electronic theory of the double bond. Roger Adams confirmed that Carothers "has a better knowledge of physical chemistry and mathematics than any of the other organic chemists" in the country.[27] Nevertheless, his mathematical capacities were reportedly limited, and in the summer of 1934 he hired Flory to work as his physical chemist in the organic chemistry group at DuPont.[28] Flory had just acquired a Ph.D. in physical chemistry at Ohio State University. Like Kuhn's, Flory's doctoral research concerned photochemistry (photochemical dissociation of nitric oxide), which provided him valuable experience in chemical kinetics.[29] Still, he had known very little about polymers. He later recalled that Carothers was "the influence, the person, that interested me in polymers." Flory went on to say:

I was not a synthetic organic chemist and he [Carothers] didn't try to make me one. He felt that he wanted to have a physical chemist in his [organic chemistry] group: I was his physical chemist. It was an extraordinary opportunity, I realize now in retrospect. He came to me one day during my first year there. . . . He said, "You know, the polymer field is an area where it is my belief [that] mathematics could be applied." Now he had very limited capabilities in mathematics — mine were limited, too, but his even more. But he had the appreciation of this, you see, and he conveyed this appreciation to me, that youngster in his lab.[30]

Carothers encouraged Flory to carry out mathematical investigations of polymerization. Materials to be explored — condensation polymers such as polyesters and polyamides — were abundantly at hand. As he remarked, "The polycondensation incisively investigated by Carothers and his coworkers during the early 1930s provided ideal examples for examination of the effects of molecular size and of viscosity on the rate of chemical processes such as occur in polyesterification and polyamidation."[31]

Figure 19. Carothers's physical chemist, Paul J. Flory, ca. 1935. Courtesy
Hagley Museum and Library.

In many ways, Flory extended and elaborated Kuhn's physico-chemical
study of polymers to the point of perfection. His first study at DuPont
concerned molecular size distribution in linear condensation polymers.
Before the work was published in 1936, Carothers had presented a preliminary account of this portion in a paper given at the Cambridge meeting
of the Faraday Society, concluding: "The study of molecular weight dis-

tributions and their relation to properties thus represents a field which should ultimately be of great importance to the subject of polymerization."[32] It was this announcement that compelled Staudinger to publish a similar work of Schulz's on addition polymers. Flory shared with Carothers the assumption that functional groups of molecules were equally reactive regardless of size. Later, he was able to confirm experimentally this principle of "equal reactivity," thus undermining Staudinger's and others' long-held belief that reactivity would be reduced as the molecule grew larger.[33] Flory's study extended to the kinetics of addition reactions or chain reactions, such as vinyl polymerization. In this study he introduced the important concept of "chain transfer," whereby a growing chain macromolecule would stop its growth by being saturated with an atom from another molecule (that might be a monomer, a polymer, or a solvent molecule), transferring the active center to that molecule.[34]

After leaving DuPont in 1938, he successively developed pathbreaking physico-chemical theories throughout the 1940s. These included a theory of three-dimensional macromolecular network formation which explained the formation of gels (1941); a theory of polymer solutions which explained the thermodynamic activity of macromolecules in solution (1942) (known as the "Flory-Huggins theory" since it was proposed independently by him and Maurice L. Huggins at Eastman Kodak); and a theory which extended the concept of "excluded volume" (a concept Kuhn had proposed in 1934 to account for the volume occupied by the real polymer molecule in solution and to see how that volume would depend on the chain length) to polymer solutions and polymer configurations (1949).[35] Flory was also among the first to use infrared spectrophotometry as a physical technique to examine the structure of polymers. He worked out virtually every aspect of polymer physical chemistry. To quote the words of a contemporary, "There are few areas within the discipline that have not been advanced and enriched by his efforts."[36] In 1974 Flory was awarded the Nobel Prize in chemistry "for his fundamental achievements, both theoretical and experimental, in the physical chemistry of macromolecules."

A novel physical technique for molecular weight measurement was introduced during the war when Peter Joseph Willem Debye, a Dutch Nobel laureate, studied the phenomenon of light scattering. A former director of the Kaiser Wilhelm Institute for Physics, Debye fled Germany to the United States in 1940, where he was appointed chairman of Cornell University's chemistry department. Debye, who had worked on the frontier between physics and chemistry (he could be called a chemical physicist), became involved in polymer research rather incidentally when he joined the U.S. wartime synthetic rubber program.[37] The government program provided chemists and physicists unique opportunities to work closely together on polymers.[38] Debye's study of light scattering was one of the fruits.

At the time the rubber program was launched, highly reliable methods for determining the molecular weights of polymers were meager. Researchers considered the available techniques, such as osmometry and the freezing point method, to be inaccurate. The ultracentrifuge was not only costly but also too complicated to handle, and data reduction was extremely time-consuming. Viscometry, then the predominant method, generally gave too low weights, and its fundamental theory was still controversial. Thus in the 1940s the issue of how to obtain the most accurate molecular weight of polymers, especially synthetic rubbers, was a subject of fundamental interest among American investigators.[39]

Debye's study had forerunners. The scattering of light in a gas (Rayleigh scattering) as well as that by colloidal particles suspended in a liquid medium (the Tyndal effect) had long been known. More recently, chemists at Goodyear attempted in 1937 to determine the size of the rubber molecule by measuring the light scattered by a rubber solution. Assuming that colloidal particles of rubber took the form of "rods" (as Staudinger thought), they arrived at too low a molecular weight for rubber.[40]

In 1943, when Debye took up the subject in order to solve the Goodyear chemists' problems, he was a neophyte in the polymer field. But the physicist was quick to recall Albert Einstein's 1910 paper, which suggested that fluctuations of the solute concentration in a solution caused light scattering. The relationship between the concentration and the refractive index of the solution could be known empirically; the molecular weight of the solute could then be obtained by measuring the solution's turbidity at different concentrations. In collaboration with industrial chemists, including those at Bell Telephone Laboratories and U.S. Rubber, he was able to develop mathematical equations on light scattering to calculate molecular weights of various polymers.[41] This method permitted a quick and accurate molecular weight determination of polymers, a method later further improved both theoretically and technically by Bruno H. Zimm at Brooklyn Polytechnic Institute and Walter H. Stockmayer at the Massachusetts Institute of Technology.[42] Debye accepted the concept of random flexible chain molecules and worked on polymer viscosity as well. This work, together with Flory's alteration, provided a neat solution to the problem of a general viscosity formula.[43]

Thus, whether by conscious arrangement, or contingent interaction, or through unique institutional channels, cooperation took place between organic chemists, physical chemists, and physicists in the study of macromolecules. The result was a rapid expansion of the theoretical framework of macromolecular chemistry. By the end of the 1940s, the new physicalist wave had greatly changed the outlook of the field previously dominated by organic chemistry. At the same time, the polymer community was able to enlarge its size and scope of professional activity by winning over to its side physical chemists and physicists who could share a disciplinary identity as polymer scientists.

Chapter 6
The Legacy of Staudinger and Carothers

By the immediate postwar period, the chemistry of macromolecules had taken recognizable shape as a new scientific discipline in an institutional setting. This noteworthy achievement was indicated by the large number of academic and industrial researchers, an even larger quantity of papers and books, the creation of several new journals devoted exclusively to macromolecules, and the growth of university teaching in this area. In the years between the 1950s and 1970s, four Nobel Prizes in chemistry were awarded to this field of study. Because of its theoretical significance in the science of molecules, its biological implications, and its phenomenal industrial applications (such as synthetic fibers, synthetic rubber, and plastic materials), macromolecular chemistry rapidly matured as a science during the postwar period. Practitioners' voices commenting about the state of affairs were indicative of this development. For example, in 1932, while working on polymerization, Wallace Hume Carothers wrote, "The chemistry of macromolecular materials is still in its infancy."[1] In 1953 Hermann Staudinger confirmed that "Macromolecular chemistry is the youngest branch of organic chemistry."[2] Ten years later the Italian pioneer of polymer chemistry, Giulio Natta, announced to the public, "Macromolecular chemistry is a relatively young science."[3] By 1980 chemists did not hesitate to declare: "The field of polymer chemistry has already reached a high degree of excellence and a certain maturity."[4] To conclude this book, we shall examine the institutional and industrial growth of the field up to the postwar period in light of the legacy of Staudinger and Carothers.

American Polymer Chemistry and Carothers's Legacy

Unlike the academic Staudinger, the industrial researcher Carothers did not train students in polymer chemistry. Yet Carothers introduced a number of able research chemists at DuPont into the world of polymers. From June 1928 to his death in April 1937, he had a total of twenty-five co-workers, most of them only slightly younger than the group leader, including twenty-one chemists with Ph.D. degrees. Five of those Ph.D.s graduated from the University of Illinois; four from Ohio State; two each from Harvard, Johns Hopkins, and MIT; and one each from Colorado, Columbia, Iowa, Iowa State, McGill, and Michigan (see Appendix Table 4).

A number of historical studies have focused attention on the problem of "research schools." These studies have expounded various factors (in general terms) that might contribute to the success or failure of scientific research schools. The existing literature tends to consider the research school as a peculiar phenomenon associated only with the university. A common picture of successful or influential research schools brings to mind the image of a charismatic university professor with a distinguished research reputation and institutional power directing a host of faithful students along the lines of his innovative research program, controlling publication outlets, and exerting strong influence on the scientific community. Indeed, such a picture does correspond with Staudinger's university-based school of macromolecular chemistry. Carothers's group in industry does not fit with this image of research schools. But if a "research school" designates a group of mature scientists pursuing a coherent, innovative program of research under the direction of one or a few creative leaders in the same institutional context, then the Carothers group, however peripheral in an academic sense, can rightly be considered to be a research school.[5] Like others, it was a laboratory-based research school with a coherent new program. It did indeed carry out basic research, which was to make DuPont a pioneer in American polymer chemistry. After Carothers's death, DuPont chemists continued to cultivate the polymer field, dominating American industrial research in the field.

The Wilmington school bequeathed a research tradition of polymers not only within the company but also to American academia. Under Carothers's influence, a new generation of polymer chemists emerged in American universities. Paul J. Flory, Elmer O. Kraemer, and Carl Shipp (Speed) Marvel were among the influential scholars who inherited Carothers's legacy. We have seen the influence of Carothers on Flory and his brilliant physico-chemical work. Like Carothers, Flory's study of polymers largely benefited from the "reciprocal approach" in which theoretical principles were developed by contact with industrial practice. While other members of Carothers's group worked out practical problems on fiber, Flory was allowed to remain a theorist. Surely Carothers must have found a kindred spirit in this midwestern newcomer who, too, was intro-

Figure 20. Carothers's DuPont associates at the dedication ceremony of the Carothers Research Laboratory, September 1946. Left to right, Julian W. Hill, Harry B. Dykstra, Gerard Berchet, James E. Kirby, Edgar W. Spanagal, Donald D. Coffman, Frank J. L. van Natta. Courtesy Hagley Museum and Library.

verted, theoretically-minded, and yet with some practical sense. Flory and Gerard Berchet were the last two select co-workers in Carothers's group before it finally disbanded.

Carothers's suicide in 1937 was "one of the most profoundly shocking events of my life," Flory remembered. "It just pulled the rug out from under my hopes and aspirations and plans—to the extent that I had any."[6] After losing his guiding spirit, Flory no longer found cause to stay at DuPont, although in his subsequent career he was not averse to moving back and forth between academia and industry.[7] In 1937 he left DuPont for the University of Cincinnati, where he began introducing macromolecular chemistry in chemistry courses he taught. But he soon perceived that industry could offer better research opportunities than his university post, which appeared increasingly precarious with the advent of war.[8]

Flory returned to industry, first at the Esso Laboratories of Standard Oil (1940–42), and then at the Goodyear Tire and Rubber Company (1943–48), where he was assigned to direct basic research on polymers.

In early 1948 Flory was invited by Peter Debye to deliver the George Fischer Baker Lectures at Cornell University, a distinguished lectureship immediately followed by his appointment as professor at Cornell. His Baker lectures were published five years later, as *Principles of Polymer Chemistry*, and were to play a definitive pedagogical role in this growing field. Having gone through twelve printings, the book served as the "bible" for generations of polymer scientists throughout the world, and is still in wide use today.[9] Cornell's chemistry department (from which the colloid chemist Bancroft had withdrawn after his tragic automobile accident in 1937) with Debye and Flory became an active research center of polymer physical chemistry in the late 1940s and early 1950s.

In 1938, the same year Flory left DuPont, Kraemer, a leader of the colloid chemistry group in DuPont's fundamental research program, also left the company to take a position at the Biochemical Research Foundation of the Franklin Institute in Newark, Delaware. A former student of the Svedberg school, he was a colloidalist, but no doubt became a full-fledged macromolecularist during his association with Carothers at DuPont. In 1941 he embarked on teaching "Colloid Chemistry" at the University of Delaware, a pioneering colloid course that incorporated to a considerable extent the new chemistry of macromolecules.[10]

Kraemer's case illustrates the transition in the colloid chemist's notion of colloids by adopting the macromolecular theory. In a lecture, "The Colloidal Behavior of Organic Macromolecular Materials," given at Western Reserve University in Cleveland, Ohio in the early 1940s, Kraemer explained, keeping Carothers's work in mind:

After years of uncertainty regarding the structure and constitution of these materials [such as the proteins, gums, starch, cellulose, rubber, and synthetic plastics], it finally became quite generally accepted about ten years ago (to a considerable extent owing to the synthesis of many similar materials by well-defined chemical reactions) that these materials are actually exceedingly large molecules, in the ordinary sense of the organic chemist.[11]

The triumph of macromolecular theory forced a change in colloid chemists' traditional definition and in their scope of colloids. Although colloid chemistry would continue to exist as a discipline, it no longer enjoyed the strong claim made by colloidalists, such as Wolfgang Ostwald and Wilder D. Bancroft, in the previous decades. It was symbolic that two decades after his death, Ostwald's *Kolloid-Zeitschrift* would have its title changed into *Kolloid-Zeitschrift und Zeitschrift für Polymere* (see Appendix

Table 7). Kraemer represented a new generation of colloid chemists who explored the colloid chemistry of large molecules.

Marvel's interest in polymers was stimulated through close contact with Carothers, both as an intimate friend and as a consultant for DuPont in 1928. Like Carothers, the professor at the University of Illinois was trained as an organic chemist. Heavily influenced by Carothers's pioneering work at DuPont, Marvel abandoned his study of classic organic compounds and from 1933 onward came to devote his research to the organic chemistry of polymers. His areas of investigation ranged from sulfur dioxide addition polymers to the mechanism of vinyl polymerization, and from copolymerization reactions to the development of synthetic rubber. He taught synthetic polymer chemistry as part of his organic chemistry courses and directed theses of his advanced students in this subject. While at Illinois, Marvel trained about a hundred and fifty doctoral students as organic polymer chemists and published some two hundred and fifty papers in the field.[12]

Thus, by the early 1940s, Carothers's successors had brought the study of macromolecules into an academic setting. American polymer chemistry, which first emerged from a basic research program in industry, gradually spread as an academic discipline in its universities by the late 1940s. DuPont's venture into fundamental research was rewarded by the discoveries of neoprene and nylon. Even beyond the perspective of a single company, industrial research turned out to play a seminal role in the birth of a new science in America. Carothers's work — which Herman Mark in 1976 likened to "a volcanic eruption the reverberations of which are still being felt"[13] — did not constitute a mere carbon copy of existing academic chemistry. Rather, the industrialized style of polymer science grew by itself and achieved autonomy, in turn powerfully influencing university scientists. In this respect, Carothers's case reversed the traditional relationship between science and industry, in which the latter only followed university science. His scientific activity at DuPont showed contemporaries that industry might now take the initiatives traditionally expected of pure science. It demonstrated as well that macromolecular chemistry could attain maturity by a fruitful interaction between pure science and industrial practice.

In 1940 the American chemical community welcomed a powerful organizer of the field from Europe. Herman F. Mark would bring with him to the New World both European industrial and academic experience. In many ways, Mark's new relationship with DuPont shaped his professional career in America. "A man of almost pathological optimism," as one biographer put it, Mark had stayed in Vienna until the late 1930s, without taking too seriously the darkening political situation caused by the threat of Nazi Germany.[14] When Hitler's troops invaded Austria in March 1938, Mark was immediately arrested and interrogated at the Gestapo prison about his relations with I. G. Farben and Engelbert Dollfuss — Mark's wartime comrade and close friend who became chancellor of Austria in 1930 but was assassinated during the abortive Nazi revolt four years later. Mark was released a few days later without his passport and "with a stern warning not to have contact with anyone Jewish."[15] Disguised as mountaineers, with picks, ropes, skis, and a Nazi flag on the car, he and his family succeeded in April 1938 in a dramatic escape by driving across the border into Switzerland. Continuing via France and England, they fled in September to Canada, where he became research manager at the Canadian International Paper Company in Hawkesbury, Ontario.

The Canadian company had a team of young researchers, but Mark found that "none of them knew much about polymer science."[16] He introduced the latest European version of polymer chemistry to them (such as the distribution of molecular weights and new methods of the characterization of cellulose), and endeavored to modernize the laboratory facility. The firm was the main supplier of cellulose pulp for DuPont, which was then one of the largest producers of rayon in the United States.[17] As a consequence, Mark got into a lively exchange with DuPont chemists who frequently came to Hawkesbury for technical discussion.[18]

After the loss of Carothers, DuPont had eagerly sought out the best polymer chemists, especially among European refugees. Although the company was unwilling to recruit Jewish chemists as full-time employees — perhaps because of an anti-Semitic bias among a few leaders in the Chemical Department, as David Hounshell and John K. Smith have pointed out — it made efforts to help secure academic positions for the rejectees while retaining them as consultants.[19] DuPont staff now showed a growing interest in Mark, who had also had "a very good rapport" with Carothers. Mark had corresponded with Carothers since 1929; after meeting with Carothers at the 1935 Faraday Society symposium, he invited the latter to see his Institute at the University of Vienna.[20] Instrumental for scouting out Mark was William F. Zimmerli, the former European representative of DuPont, who had also arranged a private London meeting for Carothers and Staudinger in 1935, as mentioned earlier. While in Europe, he had called Mark, Max Bergmann, and Kurt H. Meyer to the attention of the DuPont management.[21] Now a member of the Board of Directors of the Polytechnic Institute of Brooklyn, New York, Zimmerli was able to use his influence to persuade Brooklyn's president to invite Mark. By early 1940 Mark was happy to move to the United States, where he was appointed to a joint position as DuPont consultant and adjunct professor at Brooklyn Polytechnic, with the financial backing of DuPont.

The same year Mark started publication of an influential monograph series *High Polymers,* a title he preferred to Staudinger's coinage "macromolecules." Its first volume, edited with George S. Whitby, a British-born rubber chemist who would join the faculty of the University of Akron in 1942, was the *Collected Papers of Wallace Hume Carothers on High Polymeric Substances.* This "classical" work, the editors were convinced,

will always remain an essential part of the foundation on which the high polymeric chemistry of the future will be erected. . . . [T]here could be no better start for this series than to publish, as the first volume in it, a collection of the papers embodying Carothers' studies of high polymers and closely related matters.[22]

A considerable number of young polymer chemists read this book as a foundation for their studies. Mark later recalled that the Carothers volume was a big hit; it "sold very, very well."[23] It was Mark, especially through this handy collected work, who made American students and academics widely aware of the work of Carothers as a trailblazer in macromolecular chemistry. To this volume, three other volumes were added in the series by 1942, including works by Mark and his friend Meyer.[24] Today, the *High Polymers* series has amounted to over forty volumes.

Building on a foundation laid by Carothers and his successors, Mark served as an influential organizer and teacher in establishing the discipline of polymer chemistry in America. There was good reason for "Brooklyn Poly" to nurture Mark's field of activity. The Polytechnic Institute operated a national laboratory under the Shellac Bureau for testing and controlling incoming shipments of shellac, a natural resin, from India and Indonesia to the docks of Brooklyn. On assuming his position at Brooklyn, Mark was assigned to this work, and in 1941 he became director of the Bureau, which he expanded to include research on synthetic coatings and polymers, involving his students in the effort. At the same time, Mark was eager to teach his students a new introductory course, "General Polymer Chemistry," and to organize a series of weekly symposia and intensive summer courses in the field, involving outside scholars and industrial researchers, among them Kraemer of the Franklin Institute and Emit Ott, a Swiss-born physicist who was then research director for Hercules Powder Company in Wilmington, Delaware.[25]

In 1946, Mark's school was established as a newly independent "Polymer Research Institute," where he initiated a graduate program leading to M.S. and Ph.D. degrees with a major in polymer chemistry. In contrast to Marvel's Illinois program, which stressed the synthetic organic chemistry of polymers, Mark's new educational program covered all areas of polymer science: organic chemistry, physical chemistry, biochemistry, and industrial applications. As Mark stated, "it was more or less a duplication of what I had attempted to accomplish at the University of

Vienna."[26] This time, he was able to design a high level of teaching on a far larger scale and with more funds. A number of newly educated chemists worked in and helped to teach this field at the institute under Mark's directorship. Specialized courses then offered included "Polymerization Kinetics" by Turner Alfrey (Mark's 1943 Ph.D.); "Solution Properties of High Polymers" by Frederick Roland Eirich (Mark's former assistant at Vienna); "Organic Polymer Chemistry" by Charles Gilbert Overberger (Marvel's 1944 Ph.D. and an organic chemist Mark hired in 1946); and "Chemistry of Proteins," by Douglas McLaren (a specialist in the physical chemistry of proteins).[27] Mark's Brooklyn school soon became a mecca for advanced students and polymer researchers.

World War II accelerated polymer research in American universities and industries. In 1942 DuPont's entire production of nylon was allocated by the War Production Board for vital military uses such as parachutes, flak vests, and military tires. The government synthetic rubber program, begun in 1942 in the face of Japan's occupation of the Pacific area, mobilized a considerable number of leading academic and industrial scientists, including Marvel at Illinois, Flory and Debye at Cornell, William D. Harkins and Morris Kharasch at Chicago, Izaak M. Koltoff at Minnesota, Whitby at Akron, William O. Baker at Bell Telephone Laboratories, and Frank Mayo at U.S. Rubber, to name a few. The government helped DuPont's new neoprene plant at Louisville, Kentucky to operate for expanded production. But the program's major focus was production of GR-S (named after "Government Rubber-Styrene"), a styrene-butadiene copolymer rubber considered superior to neoprene for tire manufacture. Scientists and engineers involved in the program greatly improved its quality and production method; in 1945 the GR-S production reached 730,000 tons, more than six times that of Germany's annual peak production (1943) of Buna S rubber, a GR-S equivalent. Fifty-one plants were built for production of synthetic rubber, at a total cost of $700 million. Due to the postwar threat by the Soviet Union and the advent of the Korean War, the program was kept alive until the government withdrew from the synthetic rubber business in 1955.[28]

The literature on the history of the synthetic rubber program, often written by supporters of the program, tends to stress that this unprecedented government-sponsored program was responsible for the rapid growth of polymer science in postwar America.[29] But would this growth not have been possible without governmental mobilization? As we have seen in the case of Debye and Flory, this program certainly enhanced the advance of the fundamental study of polymers, notably in polymer physical chemistry, as well as the improvement of analytical techniques for macromolecules. The program also activated polymer science research at a number of U.S. corporations, which had no small effect in the course of postwar industrial research. On the one hand, what the rubber research program did in effect was to help research by providing scientists with financial support, a common meeting place for researchers of various backgrounds, and the urgent motivation to study polymers. On the other hand, as Peter Morris has rightly suggested, "Flory would have carried out his pathbreaking research in the absence of the rubber research program."[30] Flory and Marvel had started studying polymers long before the war. Their achievements may well be seen as an extension of their prewar experiences. The program did promote training of graduate students as polymer chemists at universities such as Illinois, which received a large portion of governmental funds. But it should also be noted that Mark and his Brooklyn Polytechnic Institute (which produced nearly twice as many Ph.D.s in polymer-related subjects as Illinois in the early 1950s) never directly took part in the mainstream of the rubber program as did many other scientists.[31] Such counter-examples are not few in number. Evidently, the growth of American polymer chemistry in the 1940s and 1950s cannot be fully explained without considering the flow of prewar development, in which Carothers's tradition and Mark's complementary role as an organizer were prominent.

Shortly after the war, as polymer study expanded, publication in polymer chemistry reached the point that the *Journal of the American Chemical Society* — a journal which had carried papers on polymers, including those of Carothers's group — could hardly accept the deluge in polymer manuscripts. There was another problem in that physical chemists and physicists published their polymer articles in various physics-related journals such as the *Journal of Chemical Physics,* the *American Journal of Physics,* and the *Journal of Applied Physics.* For example, Debye's landmark paper on light scattering appeared in 1943 in the *Journal of Applied Physics.* Mark thus felt it necessary to create a new specialty journal. After learning that the American Chemical Society was unwilling to be involved, Mark persuaded Interscience Publishers in New York to start an independent journal. The first English-language periodical devoted to the field of polymers, the *Journal of Polymer Science,* was inaugurated under Mark's editorship in March 1946 (see Appendix Table 7).[32] Its first volume carried fifty-seven papers, and the journal was soon to become the leading vehicle for the growing number of polymer scientists in postwar America. The ever increasing flux of papers on various aspects of polymers would later lead the editors to split the journal into three sections: "Polymer Chemistry," "Polymer Physics," and "Polymer Letters."[33]

Meanwhile, the "High Polymer Forum" was organized in 1946, a polymer symposium sponsored by the American Chemical Society's several divisions: Organic Chemistry; Physical and Inorganic Chemistry; Paint, Varnish and Plastics; Rubber Chemistry; Colloid Chemistry; and Cellulose Chemistry. In the next three years, 162 papers were presented at the forum, drawing an average of over three hundred attendants. This high level of activity led the society to form the Division of High Polymer Chemistry (later renamed the Division of Polymer Chemistry) in the fall of 1950, with Marvel as the first chairman and Mark as secretary-treasurer. Within a decade, the division would grow to over a thousand members, and by the mid-1960s it had reached twenty-five hundred.[34] As chairman of the management committee of the Gordon Research Conference, Mark was also instrumental in developing this event into a series of impact-making symposia on polymers, in which many academic and industrial chemists met informally to discuss up-to-date topics.[35]

During the 1950s the polymer industry grew to be one of the major industries in the United States. DuPont no longer enjoyed supreme dominance in industrial research on polymers, as it had done in the prewar period. Active polymer-related programs were set up at such competitors as Allied, Dow, Hercules, Monsanto, Phillips Petroleum, Polaroid, Rohm and Haas, Shell, and Standard Oil (New Jersey), not to mention major rubber companies such as Goodrich, Goodyear, and U.S. Rubber.[36] By the 1960s, besides Illinois, Brooklyn, and Cornell, a number of American universities had installed programs on polymer chemistry, University of Akron, Case Western Reserve University, University of Massachusetts, and North Carolina State among them.[37]

But Mark's institute dominated polymer education during the 1950s and 1960s, "the golden age of Brooklyn Poly," as a protégé of Mark's remarked.[38] With Mark on top and Overberger, Eirich, Harry Gregor, Herbert Morawetz, Gerald Oster, Robert Ullman, and Murray Goodman as its active faculty members, Brooklyn attracted students and postdoctoral scholars not only from the United States but also from overseas, such as Finland, France, Great Britain, India, Israel, Italy, Japan, the Soviet Union, and Sweden. During his tenure at Brooklyn, Mark published some two hundred and fifty papers as well as half a dozen books on polymer chemistry, and trained about four hundred students who earned their master's or Ph.D. degrees in the field.[39] After his retirement in 1964 as Dean of the Faculty of the Polytechnic Institute of Brooklyn, he was even more active and busy in editing the *Journal of Polymer Science*, the *Journal of Applied Polymer Chemistry* (founded in 1959), and the *Encyclopedia of Polymer Science and Technology* (which was started in 1965 and finished in 1972); organizing scientific meetings; and making lecture tours around the world. The two decades after the war saw the United States tower over polymer chemistry and the polymer industry, a growth that echoes Mark's immense contributions as a discipline builder.

Staudinger in Postwar Germany

During World War II the German chemical industry, particularly I. G. Farben, did continue to develop and produce synthetic rubber (such as Buna S), fibers (such as Perlon), and a variety of plastics — technological developments sufficient to impress such postwar visitors as Marvel and Mark.[40] Staudinger, however, felt keenly the stagnation of macromolecular chemistry in his country. Allied bombing in November 1944 had almost completely destroyed Staudinger's Chemical Institute at Freiburg. Out of the ruins, the German professor had to start all over again. Like other Germans, Staudinger had been uninformed about wartime development of polymer research outside Germany. In its aftermath, however, he soon learned that great progress had been made, especially in the United States, which made him feel uneasy. Exploring the possibility of annexing his Institute to a larger organization like the Max Planck Society (Max Planck Gesellschaft, the former Kaiser Wilhelm Gesellschaft), he wrote to the Society's new president Otto Hahn, "I cannot calmly see the essential results of my lifework move to the United States and other places."[41]

Staudinger wasted no time in reorganizing his science in postwar Germany. Resuming his journal was among his urgent concerns. As early as 1943, thanks to a proposal from the publishing house J. A. Barth in Leipzig to the Reich Union of German Journals, he had managed to change the whole title of the *Journal für praktische Chemie,* which he edited, into the *Journal für makromolekulare Chemie.* However, as a result of wartime conditions in Germany and resulting poor distribution, the journal ended after publication of only two volumes, which were issued in August 1943 and December 1944. After the war's end, Staudinger found it virtually impossible to revive publication of his journal in Leipzig, since Leipzig was occupied by the Russians, and their ordinances laid down that the editor must be a resident of their occupied zone. He then sought to negotiate with publishers in West Germany and Switzerland. Finally, in April 1947, he was able to enter into a contract with the Wepf Press in Basel to publish a new journal, *Die makromolekulare Chemie,* with the approval of the French military government which was occupying Freiburg. The first issue appeared in September 1947, one year and a half after the establishment of Mark's American journal (see Appendix Table 7). This was effectively the first German periodical devoted exclusively to macromolecular chemistry. Staudinger reigned as editor for the ensuing two decades.[42]

Apparently, the zenith of Staudinger's creative research activity had passed before the end of the war. He was now in his mid-sixties. Falling behind in physico-chemical issues, his research interest in postwar years turned more and more to biological aspects of macromolecules. Biology, dealing with living, truly organic matter, remained the organic chemist's final area of interest, and he was happy to return to where he had started his scientific study as a Gymnasium student fifty years ago. Since the mid-1920s, his thoughts on the biological implications of such macromolecules as proteins had been encouraged by his wife Magda, a plant physiologist who helped to develop some new concepts relating macromolecules to physiological and philosophical questions.[43] Biochemists were slower than physical chemists and physicists to respond to the macromolecular theory. Physical chemistry continued to influence the development of the study of proteins, enzymes, and biological phenomena. Yet most biochemists were rather indifferent to the achievements in polymer chemistry. As Eugene Guth observed, "Books on biochemistry at the end of the 1930s do not even mention Staudinger's name!"[44] In 1946 Staudinger published a monograph, *Makromolekulare Chemie und Biologie*, in which he attempted to explain the life processes of the living cell from the point of view of the organic macromolecular structurist. Although his own work on bio-macromolecules touched only on the fringe of the subject, it challenged biochemists and biologists to scrutinize the structure of proteins more intensively in tackling the enigma of life.[45] Symbolically, Staudinger received the Nobel Prize in chemistry "for his discoveries in the field of macromolecular chemistry" in 1953, the same year in which the Watson-Crick double helix theory of the DNA giant molecule emerged as a landmark of the new science, molecular biology.

Staudinger had been nominated for the Nobel Prize at least four times before the war (1931, 1932, 1934, and 1935).[46] When the prize finally came into his hands, he was seventy-two, the second oldest by that time of fifty-six recipients in chemistry (next to Otto P. H. Diels who won the 1950 Prize at seventy-four). Featuring the headline, "Better Late than Never," a *Newsweek* article quoted an American chemist as saying: "This is the final accolade before the curtain goes down. I regret that; he would have enjoyed it earlier."[47] Whether the lapse was due to the war interruption or Staudinger's polemical nature or the belated recognition of macromolecular chemistry,[48] Staudinger was wholeheartedly delighted to win the most coveted prize and to be officially recognized as the founder of macromolecular chemistry. "It was really wonderful that he lived to see the success of his labors; all too often the founder of a new science does not live to see its successful acceptance," said Magda Staudinger.[49]

In his autobiography, Staudinger listed over six hundred papers he and/or his associates had published on macromolecular chemistry covering forty years of his career. Of these, Staudinger authored or coauthored about three hundred and fifty papers, seven times as many as Carothers.[50] He published these treatises with a total of approximately one hundred co-workers (see Appendix Table 2). He also published six

Figure 21. Staudinger delivering his Nobel lecture, Stockholm, December 1953. Courtesy Deutsches Museum.

textbooks in the field. Although he trained only seven doctoral students after the war, the total number who studied macromolecules between 1920 and 1954 amounted to seventy-five (see Appendix Table 1). While most of them went to industry, a few became noted academicians who taught macromolecular chemistry in universities. They were Rudolph Signer (University of Bern), Werner Kern (University of Mainz), and Hans Batzer (Stuttgart Technische Hochschule).[51] Günther V. Schulz and Elfriede Husemann, though they did not earn doctorates from

Staudinger, served for many years as his chief collaborators at Freiburg, and both passed their *Habilitations* under Staudinger.

In 1951, when Staudinger retired as professor at Freiburg, his laboratory was transformed into the State Research Institute for Macromolecular Chemistry (Staatliches Forschungsinstitut für makromolekulare Chemie). He served as its honorary director for the next five years before Husemann took over the directorship in 1956. The "Makromolekulare Kolloquium," which Staudinger had began in 1934 from a small gathering of polymer researchers at Freiburg, grew to be an internationally known annual meeting of polymer chemists. Today, one may well sense the depth of the "Freiburg Mafia" from the large size of the colloquium, whose participants include numerous offspring of the Staudinger school.

The University of Mainz grew to be another mecca of polymer research and education in postwar Germany. In 1946 Schulz and Kern joined its faculty, the former as professor of physical chemistry and the latter as associate professor of organic chemistry. Schulz served as director of Mainz's Institute of Physical Chemistry, and Kern as director of the Institute of Organic Chemistry after 1954. The duo made Mainz an internationally recognized center of macromolecular chemistry, drawing many students not only from Germany but also from other countries, especially Japan. Kern, for example, trained at his institute over a hundred Ph.D.s in the field, including nine *Habilitation* recipients, and published over two hundred and fifty papers on macromolecules.[52] After Kern became active co-editor of Staudinger's *Die Makromolekulare Chemie* in 1962, Mainz became the editorial center of the journal. In 1971 the German Chemical Society presented both Kern and Schulz the Hermann Staudinger Prize for their contributions. In 1984 Mainz further succeeded in luring the Max Planck Institute for Polymer Research (Max Planck Institut für Polymerforschung) onto its campus, so that Staudinger's long-cherished dream finally came true.

Polymer Chemistry and Industry

Rapid industrialization undoubtedly helped legitimate polymer chemistry as a discipline. Staudinger's 1953 Nobel Prize not only was a recognition of his contributions to pure chemistry and their biological implications, but also owed much to the explosive growth of the postwar polymer industry. In fact, Arne Fredga of Uppsala University in his presentation speech to Staudinger on behalf of the Nobel Committee emphasized:

Professor Staudinger has not been involved directly in the technical and industrial development but without his energetic and bold pioneering work this development would scarcely have been conceivable. . . . In recognition of your services to the natural science and the material culture . . . the Royal Swedish Academy has resolved to award you the Nobel Prize for this year.[53]

Fredga was right to associate industrial development with Staudinger's work. The macromolecular theory laid the theoretical basis for industrial research on polymers. Staudinger himself indirectly contributed to the industrial development by consulting with chemical companies such as I. G. Farben and by sending industry a large number of his students who worked on plastics and synthetic rubber based on his macromolecular views.

Nevertheless, the impact of Carothers's work on industry was far more direct and visible. His study at DuPont was at once considered internationally to be a prototype for science-based industrial research on polymers. Foreign industrial chemists working on fibers in the 1930s and 1940s testified to the immediate effects that Carothers's macromolecular synthesis had on the course of their research. For example, Paul Schlack of I. G. Farben admitted that he was inspired by Carothers's 1931 paper, "Polymerization," and his subsequent fiber patents (and not Staudinger's work) when he synthesized "nylon 6" in 1938, a nylon formed from caprolactam, a carboxylamine that had a ring structure. Given the trade name "Perlon," the fiber was produced at the I. G. plants during World War II; its manufacture was adopted after the war by Japan, Italy, the Netherlands, and other countries that could not obtain DuPont's patent of nylon 66.[54] The English industrial chemists John Rex Whinfield and James Tennent Dickson, working in the laboratories of the Calico Printers' Association, synthesized in 1941 a polyester fiber of high quality, later called "Terylene." Almost simultaneously with the British development, DuPont discovered the same fiber, named "Dacron." This concurrence was by no means an accident, for both works were based on Carothers's studies on condensation polymers in the 1930s. As Whinfield wrote, their investigations were "a logical extension" of Carothers's series of papers, "Studies on Polymerization and Ring Formation."[55]

The impact of nylon was especially immediate in both academia and industry in Japan, a country where there was a long tradition in the fiber and textile industry. In the mid-1930s Japan was a major exporter of silk to the United States and the world's greatest producer of rayon. Correspondingly, early Japanese studies of polymers centered on fibers rather than rubber, a trend which contrasted with that of Germany and the United States.[56]

With this background, Japan's polymer chemistry sprang from a combined influence of Germany and the United States. Due in part to the friendly political and cultural ties between Japan and Germany in the interwar period, a large number of Japanese scientists went to Germany to carry out their doctoral and postdoctoral work. Macromolecular theory was introduced to Japan by those individuals who had studied in Germany in the 1920s and 1930s. While Staudinger had four students from Japan (Taizo Yamashita, Ryuzaburo Nodzu, Michizo Asano, and Eiji Ochiai), more Japanese students studied under his academic opponents. For example, Kurt Hess trained four active Japanese cellulose chemists (Tsukumo Tomonari, Ichiro Sakurada, Motoi Wadano, and Hiroshi Sobue), and Wolfgang Ostwald had over fifteen students (including Ryukichi Tanaka, Sakae Tsuda, Sakurada, Kisou Kanamaru, and Muneo Takei). Hess was known at the time by Japanese chemists as a German leader in cellulose chemistry. Wolfgang Ostwald had a prominent reputation among Japanese as the founder of colloid chemistry.[57]

Ironically, it was not Staudinger's adherents but students of the latter group who played a major role in the formation of macromolecular chemistry in Japan. Staudinger's four students from Japan, who studied with him in the 1920s and early 1930s though none received their doctorates under him, did not play a principal role in the growth of Japan's polymer chemistry during the late 1930s and the 1940s. In this regard, there was a parallel between Staudinger's American students (Herman A. Bruson and Avery A. Ashdown) and the Japanese students. Their failure to anticipate the building of a new discipline might be ascribed in part to their inability to form a research school with a coherent program in academia. However, their motives for studying with Staudinger must also be considered. The foreign students chose to come to Staudinger largely because of his early reputation in the field of low molecular weight organic compounds such as ketenes and insecticides. Three of the Japanese students — Yamashita, Asano, and Ochiai — were in fact pharmaceutical chemists; Asano and Ochiai later became professors of pharmacology at the University of Tokyo, and Yamashita worked for a pharmaceutical company. Ochiai was not even aware that Staudinger had given up his early study of ketene until he arrived at Freiburg University in 1930. Ashdown's interest too lay in normal organic reactions, and he had little intention of teaching polymer chemistry at MIT, where he obtained a post after his return to America. Although these students worked on polymers with Staudinger and found some interest in the new field of macromolecules, they did not venture to turn their careers toward this still polemical field. There was an often-told rumor in Japan that Nodzu and Ochiai went all the way to Germany only to measure viscosity. Perhaps their work at the Staudinger school constituted small assemblies of their master's grandiose program.[58]

Among the leading Japanese polymer chemists was Sakurada, who studied with Wolfgang Ostwald at Leipzig from 1928 to 1929 and then with Hess at the Kaiser Wilhelm Institute for Chemistry in Berlin-Dahlem between 1929 and 1931. While in Germany, Sakurada was a harsh critic of Staudinger's work, but after assuming a position at Kyoto University in 1931 he changed his mind and became the most active propagator of the macromolecular view of polymers in Japan's chemical community. By that time he and other Japanese students sensed that the macromolecular theory was clearly gaining significant acceptance in Europe and America.[59]

Japan's boycott of Australian wool in 1936, which signaled the beginning of its autarky program, resulted in the establishment of the Japanese Institute for Chemical Fibers Research (Kagaku Sen'i Kenkyusho) at Kyoto University. Here Sakurada and his students, including Seizo Okamura, began to study the synthesis of "artificial wool" from cellulose acetate.[60] Even though he was aware of some of Carothers's ongoing research through his publications and patents, Sakurada predicted that complete synthetic fibers, which would challenge existing natural and semisynthetic fibers, would not be feasible. Thus, he focused his investigative efforts on modified natural fibers like cellulose acetate.[61] DuPont's announcement of the invention of nylon in late 1938 dashed Sakurada's optimism. Many industrialists were now concerned that this new synthetic fiber would deal a fatal blow to the Japanese silk and rayon industry. A newspaper article reported that the United States "intended to shut out silk within ten years, which would be a matter of grave concern to forty million Japanese farmers."[62] The rumor further stated that the name "nylon" was taken from the reverse spelling of Japan's Ministry of Agriculture and Forestry (*Nolyn*), which was in charge of the silk and cellulose industry, in order to upset government officials. To calm the public down, the Nolyn Minister announced — with little evidence — that nylon would not be a threat.[63]

By January 1939 U.S. representatives of two Japanese companies, Mitsui Bussan and Kanegafuchi Boseki, had independently managed to obtain a few milligrams of nylon, and immediately sent pieces to Sakurada, to Kohei Hoshino at Toyo Rayon Company, and to Toshio Hoshino (a former student of Heinrich Wieland) at the Tokyo Institute of Technology. Their analyses quickly showed that nylon was a surprisingly excellent fiber.[64] In February, an emergency symposium on nylon was held in Osaka — Japan's center of the fiber and textile business — where Sakurada and other fiber chemists reported on the properties and possible starting materials of nylon. They also praised Carothers's scientific work as well as DuPont's well-planned program, while criticizing the backwardness of Japan's industrial research.[65] In late 1940 a second symposium was held in Osaka, in which nineteen chemists and industrialists gave lectures on the national policy on fibers and the latest development of fiber research. In the opening remarks, Gen'itsu Kita, Sakurada's mentor at

Kyoto, emphasized that "today, we must further contribute to national prosperity by strengthening more and more the status of the fiber industry which our country has established."[66] In November 1940, the University of Osaka set up the first program on polymer chemistry and opened the Institute for Fiber Science (Sen'i Kagaku Kenkyusho) under the directorship of Yukichi Go, who had intensively studied cellulose structure under Karl Freudenberg, Reginald O. Herzog, and Kurt H. Meyer between 1929 and 1934. At the Institute, Go directed X-ray studies of fibers, no doubt aimed at realizing what he had seen at Herzog's Kaiser Wilhelm Institute for Fiber Chemistry in Berlin.[67] Kyoto University quickly rivaled Osaka. In April 1941, the former opened the Department of Fiber Chemistry in which Sakurada, Okamura, Masao Horio, and Kiyohisa Fujino embarked on teaching fiber and polymer chemistry.[68] The "nylon shock" also led Japanese government, industry, and academia together to create the Japanese Research Association of Synthetic Fibers (Nihon Gosei Sen'i Kyokai) in the following year to promote research on and development of synthetic fibers. Under its umbrella, Sakurada's group studied polyvinyl alcohol, leading to the invention of the "Vinylon" fiber, and Hoshino's group at the Tokyo Institute of Technology, including a young polymer chemist, Yoshio Iwakura, worked on nylon and polyurethane fibers.[69]

In the midst of war, the activities of the Research Association of Synthetic Fibers were extended to synthetic rubber and plastics. In early 1943 it was transformed into the Association of Polymer Chemistry (Kobunshi Kagaku Kyokai), the world's first independent scholarly society in this field. It published the first issue of its journal, *Kobunshi Kagaku* (Polymer Chemistry) in October 1944, a year and a half earlier than the appearance of Mark's *Journal of Polymer Science,* and three years before that of Staudinger's *Die makromolekulare Chemie* (see Appendix Table 7).[70] In 1951, the Association was reorganized as the Society of Polymer Science (Kobunshi Gakkai), whose first four presidents—Katsumoto Atsuki at the University of Tokyo (served 1951–58), Toshio Hoshino (1958–62), Sakurada (1962–68), and Hiroshi Sobue (1968–72) at the University of Tokyo—were products of prewar German education. The society played an indispensable role in the phenomenal growth of the polymer community in postwar Japan.[71] The rapid growth of polymer chemistry as a discipline can thus be seen as a representative example of vital interactions between science, industry, and society.

The year 1953, in which Staudinger won the Nobel Prize, also marked the beginning of a technological breakthrough in Europe. In the fall of that year, Karl Ziegler, Director of the Max Planck Institute for Coal Research (Max Planck Institut für Kohlenforschung) at Mülheim, discovered catalysts for producing at atmospheric pressure crystalline polyethylene whose molecular chains were linear.[72] This discovery was soon followed by Giulio Natta's preparation of crystalline "isotactic" polypropylene with a Ziegler catalyst. The Italian chemist Natta had first learned about macromolecules from Staudinger when he visited Freiburg in 1932. As a professor at Milan Polytechnic, he worked during the postwar years in close collaboration with the large Italian chemical company Montecatini, which started commercial production of Ziegler-Natta polypropylene in 1957.[73] What is important, however, is that Natta's work on the syntheses of stereoregular polymers, for the first time, enabled chemists to control the three-dimensional shape of synthetic macromolecules, and even to synthesize natural rubber whose macromolecules possess a very specific spatial configuration (an all *cis* polymer of isoprene). By the end of the 1950s, industrial production of "synthetic natural rubber" had began at such American companies as Goodrich-Gulf, Phillips Petroleum, and Shell Chemical. By that time, the polymer industry had become more and more based upon petroleum in place of coal as raw material. Ziegler and Natta shared the 1963 Nobel Prize in chemistry for their discoveries. As illustrated by these developments, the third quarter of the twentieth century saw an explosive growth of the science-based polymer industry, developing a flood of new and better synthetic fibers, rubber, and plastics that brought about a material revolution. Just as science changed the polymer industry, polymer chemistry owed its wide recognition to the dramatic growth of industry.

End of an Epoch

Staudinger lived his last ten years actively but less polemically. He could afford to look back on the early period of stormy controversy with more charity and nostalgic sentiment. He was able to recall that the polemics were not fruitless. On the contrary, in retrospect, they did stimulate his thought and contribute to the strengthening of his spirit and study. He knew that many of his old opponents had adopted the macromolecular theory which he had first proposed three decades earlier. The passage of time no doubt healed the old antagonisms that had embraced a certain amount of ill-natured rivalry. After all, he now was a celebrity in the world's community of polymer science.

In April 1957, Japan's Society of Polymer Science invited Staudinger and his wife to Japan, where they received an enthusiastic welcome. Former students of Hess and Ostwald, as well as his own, and many other academicians and industrialists worked together to provide the couple cordial hospitality throughout their one-month tour in Japan: lectures in Tokyo and Osaka; parties; visits to textile companies; sightseeing in Kyoto, Nara, and Hiroshima; enjoying Japanese gardens and full-blown cherry blossoms; watching *kabuki*; and having an audience with the prime minister, Shinsuke Kishi, and the Emperor Hirohito, who as a biologist showed great interest in the biological significance of macromolecular chemistry.[74]

Mark thought that he became "a good friend" of Staudinger after the mid-1950s.[75] His *Journal of Polymer Science* dedicated its March 1956 issue to Staudinger on the occasion of his seventy-fifth birthday. In the preface, Mark, with Herman A. Bruson (Staudinger's former American student), wrote a generous tribute to Staudinger, calling him "the Founder of Macromolecular Chemistry." Their words were filled with praises:

> We have both witnessed numerous, unforgettable occasions in the 1920's and 1930's when history of chemistry was made in the eloquent clashes . . . Holding firm to his main ideas and introducing modifications whenever the facts demanded them, Staudinger emerged from these battles as the grand old man of macromolecular chemistry, the Nobel prize winner, the honorary doctor of many famous institutions of higher learning, as the fatherly friend of his pupils, and as the benevolent counsel of his colleagues.
>
> High ideals, creative imagination, and hard work were never more splendidly and more deservedly rewarded than in the case of the man to whom we dedicate this issue. To him and to his wife, who faithfully shared with him the dark and sunny days, go the greetings and felicitations of all polymer chemists in the world.[76]

In September 1958, when Staudinger visited the United States, Mark persuaded him to visit the Polytechnic Institute of Brooklyn to give a lecture. He introduced Staudinger as the originator of the macromolecular concept. Mark's well-disposed attitude surprised Staudinger. Staudinger, for his part, lectured in German on the development of macromolecular chemistry, in which he reviewed the work of his old opponents, such as Paul Karrer, Rudolf Pummerer, Max Bergmann, but consciously skipped the part of his controversy with Mark.[77] On his return to Germany, Staudinger told Helmut Ringsdorf, his last student, perhaps with a tinge of sarcasm, that: "Now Mark believes in macromolecules!"[78] Apparently, Staudinger's bias toward this old rival had disappeared. It is remarkable that, two years later, Ringsdorf chose Brooklyn Poly to study macromolecular synthesis under Mark's successor, Charles Overberger.[79]

Staudinger's U.S. tour in 1958 culminated in a pilgrimage to DuPont in Wilmington. At the Experimental Station where Carothers had ended his short professional career twenty years before, Staudinger offered a lecture, "Polyester Studies and Discussions with Carothers on Macromolecular Chemistry," which was dedicated to this "great researcher who regrettably died prematurely." He spoke poignantly to the DuPont audience about personal reminiscences of his first and last impressive meeting with this young American in England. "We researchers, who are engaged in the field of macromolecular chemistry, must keep the best possible memory of Mr. Carothers, whose work opened up this new and important field," concluded Staudinger.[80] It was unusual for Staudinger to pay such a special tribute with a highly respectful tone in memory of a single investigator in his science.[81]

On his return to Freiburg, Staudinger started writing *Arbeitserinnerungen*, which traced the development of his thought and research. Herein, he included his debate with Kurt H. Meyer and Mark as well as colloidalists and supporters of the aggregate theory, but talked about himself and personal feelings only insofar as they were needed to understand the development of his work. This somewhat impersonal autobiography, which appeared in 1961, was the last expression of his version of scientific strife.[82]

In the late summer of 1965, Staudinger's health was deteriorating rapidly. Magda brought all the flowers from their garden to decorate his hospital bed. In the midst of reciting Johann Ludwig Uhland's poem, "*Maientau* (May Dew)," Staudinger fell into a coma due to a sudden cerebral thrombosis. Near midnight of September 8, he passed away peacefully.[83]

Staudinger's funeral was simple but touching to the attendees. Hans Staudinger, his younger brother, wrote:

> Magda . . . arranged the funeral in such a way that it was more a last visit with my brother. The family gathered in a "Weinstube" under the huge dark shadow of the Freiburger Münster. We were sitting at the table where Hermann and his friends had been sitting, listening as we did to the older choir of the bells. So we were with him, his ideas and his problems and work. . . .
>
> The Pastor was the only one who spoke. The representatives from Bund, Staat, and the Stadt Freiburg, whose honorary citizen Hermann was, laid down their wreaths. Scholars and friends from many countries joined the silence. There were no eulogies . . . as Hermann willed it. 21 family members, all close to him, put the flowers Hermann loved so in the coffin, which went down with Bach's Cantata ascendant. . . .
>
> When the family and guests went together to the chapel the sound of trickling rain on the fir trees accompanied them; suddenly all stopped—a carpet of flowers was shining through the mist covering the stairs to the little temple—the flowers gleaming through the haze bade him the farewell from his own personal world.[84]

Hermann Staudinger closed his eighty-four-year life at the pinnacle of the prosperity of macromolecular chemistry and the polymer industry. His death signaled the end of a great epoch in the history of chemistry — an epoch of chaos, controversy, struggles, and the birth and growth of a new science.

Appendix

The tables in this appendix identify Staudinger's and Carothers's students and associates during their professional careers. Table 7 lists journals devoted to macromolecular chemistry.

Table 1. Staudinger's Doctoral Students in Macromolecular Chemistry, 1920–54

Name	Thesis completed	Name	Thesis completed
Eidgenössische Technische Hochschule, 1920–27			
E. Suter	1920	E. Geiger	1926
F. Felix	1923	E. Huber	1926
J. Fritschi	1923	E. W. Reuss	1926
M. Lüthy	1923	S. Wenrli	1926
A. Rheiner	1923	H. Harder	1927
H. A. Bruson	1925	H. W. Johner	1927
W. Widmer	1925	R. Signer*	1927
M. Brunner	1926	E. Urech	1927
K. Frey	1926		

continued

Table 1. *Continued*

Name	Thesis completed	Name	Thesis completed

Universität Freiburg im Breisgau, 1928–54

Name	Thesis completed	Name	Thesis completed
D. Russidis	1928	H. Schmidt	1938
W. Starck	1928	J. Schneiders	1938
H. Thron	1928	A. W. Sohn	1938
H. F. Bondy	1929	H. Warth	1938
W. Heuer	1929	K. F. Daemisch	1939
W. Feisst	1930	B. Lantzsch	1939
A. Schwalbach	1930	O. Nuss	1939
O. Schweitzer	1930	F. Zapf	1939
W. Schaal	1931	K. Eder	1940
W. Kern*	1932	F. Finck	1940
E. O. Leupold	1932	Hj. Staudinger*	1940
H. Lohmann	1932	O. Heick	1941
E. Trommsdorf	1932	H. Jörder	1941
A. Steinhofer	1933	E. Roos	1941
E. Dreher	1934	F. Berndt	1942
H. Eilers	1934	W. Döhle	1942
H. Schwalenstöcker	1934	W. Keller	1942
H. von Becker	1935	G. Lorentz	1943
H. Frey	1935	H. Hellfriz	1944
H.-P. Mojen	1935	P. Herrbach	1944
B. Ritzenthaler	1935	H. Schnell	1944
K. Rössler	1935	H. Sattel	1945
F. Staiger	1935	H. Batzer*	1946
M. Sorkin	1936	K.-H. In den Birken	1949
A. E. Werner	1936	M. Häberle	1952
G. Daumiller	1937	W. Hahn	1952
I. Jurisch	1937	T. Eicher	1953
F. Reinecke	1937	G. Niessen	1954
K. Fischer	1937	K. Wagner	1954

*Indicates those who became university teachers. All the rest went to industry. Sources: Staudinger, 1970, pp. 243–79; Magda Staudinger to the author, May 7, 1987.

Table 2. Staudinger's Co-authors of Papers on Macromolecular Chemistry,
1920–55

M. Asano	J. Hengstenberg	K. Rössler*
A. A. Ashdown	J. J. Herrera	D. Russidis*
R. C. Bauer	P. Herrbach*	E. Sauter
H. von Becker	W. Heuer*	W. Schaal*
G. Berger	E. Huber*	A. Schwalbach*
F. Berndt	O. Huntwyler	G. Schiemann
G. Bier	E. Husemann	W. Schilt
H. F. Bondy*	K.-H. In den Birken*	H. Schmidt*
F. Breusch	H. Jöder	J. Schneiders*
M. Brunner*	H. W. Johner*	H. Schnell*
H. A. Bruson*	H. Joseph	H. Scholz
K. F. Daemisch*	I. Jurisch*	G. V. Schulz
G. Daumiller*	W. Kern*	H. Schwalenstöcker*
W. Döhle	H. W. Klever	O. Schweitzer*
E. Dreher*	H. W. Kohlenschütter	J. R. Senior
T. Eicher*	B. Kupfer	R. Signer*
H. Eiler*	B. Lantzsch	A. W. Sohn*
H. af Ekenstam	L. Lautenschläger	M. Sorkin*
W. Feisst*	E. O. Leupold*	F. Staiger*
K. Feuerstein	H. Lohmann*	W. Starck*
K. Fischer*	G. Lorentz*	Hj. Staudinger*
T. Fleitmann	M. Lüthy*	M. Staudinger
E. Franz	H. Machemer	A. Steinhofer*
H. Fredenberger	G. Mie	H. Stock
K. Frey*	R. Mohr	E. Trommsdorf*
J. Fritschi*	H.-P. Mojen*	E. Urech*
W. Frost	H. Moser	K. Wagner*
P. Garbsch	G. Niessen*	H. Warth*
E. Geiger	R. Nodzu	S. Wehrli*
H. Haas	O. Nuss	K. W. Werner
M. Häberle*	E. Ochiai	G. Widmer
W. Hahn*	F. Reinecke*	W. Widmer*
O. Heick*	A. Rheiner*	W. Widerscheim
H. Hellfriz*	B. Ritzenthaler*	E. Zapf*

*Indicates Staudinger's doctoral students. Sources: Staudinger, 1969–76; 1970, pp. 243–79.

Table 3. Organization of the Chemical Department, E. I. du Pont de Nemours and Company, July 1930

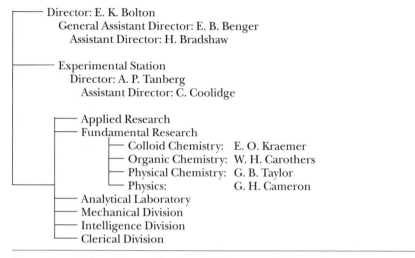

Director: E. K. Bolton
  General Assistant Director: E. B. Benger
  Assistant Director: H. Bradshaw

Experimental Station
  Director: A. P. Tanberg
  Assistant Director: C. Coolidge

Applied Research
Fundamental Research
  Colloid Chemistry: E. O. Kraemer
  Organic Chemistry: W. H. Carothers
  Physical Chemistry: G. B. Taylor
  Physics: G. H. Cameron
Analytical Laboratory
Mechanical Division
Intelligence Division
Clerical Division

Source: Organizational Chart of the Chemical Department, DuPont Company, July 11, 1930, HML.

Table 4. Carothers's Co-workers in the Fundamental Research Group in Organic
Chemistry at the DuPont Company, 1928–1937

Name	University (date of Ph.D.)	Period in Carothers's group
J. A. Arvin	Illinois (1928)	1928
G. J. Berchet	Colorado (1930)	1929–37
D. D. Coffman	Illinois (1930)	1931–35
A. M. Collins	Columbia (1920)	1930–34
M. G. Conner*		1934
M. E. Cupery	Illinois (1930)	1931–33
G. L. Dorough	Johns Hopkins (1929)	1929–32
H. B. Dykstra	Ohio State (1927)	1932
P. J. Flory	Ohio State (1934)	1934–37
J. W. Hill	MIT (1928)	1929–35
R. A. Jacobson	Illinois (1925)	1928–31
G. A. Jones*		1929–30
J. E. Kirby	Iowa State (1929)	1929–34
O. R. Kreimeier*		1931–32
S. B. Kuykendall	Ohio State (1935)	1930
W. L. McEwen	Harvard (1928)	1929–35
W. J. Merrill	Ohio State (1930)	1931–32, 1935
W. R. Peterson	Illinois (1927)	1934–36
G. W. Rigby	MIT (1930)	1934–35
E. W. Spanagel	McGill (1933)	1933–35
H. W. Starkweather	Harvard (1925)	1931–34
W. F. Talbot	Iowa (1929)	1930
W. H. Taylor*		1928
F. J. van Natta	Michigan (1928)	1928–35
F. C. Wagner	Johns Hopkins (1929)	1935

*Indicates non-Ph.D. chemists. Sources: "Experimental Station — Technical Staff,"
DuPont Company, October 23, 1928–May 14, 1937, HML; *American Men of Science*, 7th ed.
(1944), 11th ed. (1967).

Table 5. Number of Carothers's Co-workers, 1928–37

Date	Ph.D. workers	Non-Ph.D. workers	Total
1928	3	1	4
1929	7	1	8
1930	9	3	12
1931	11–12	1	12–13
1932	11–12	0–1	11–13
1933	10	0	10
1934	10–12	0–1	10–13
1935	9–10	0	9–10
Fall 1935–Spring 1936	3	0	3
Summer 1936	0	0	0
Fall 1936	1	0	1
1937	2	0	2

Source: "Experimental Station — Technical Staff," DuPont Company, October 23, 1928–May 14, 1937, HML.

Table 6. Carothers's Co-authors of Papers on Macromolecular Chemistry, 1929–36

J. A. Arvin*	J. W. Hill*
G. J. Berchet*	R. A. Jacobson*
D. D. Coffman*	J. E. Kirby*
A. M. Collins*	J. A. Nieuland
M. E. Cupery*	E. W. Spanagel*
G. L. Dorough*	F. J. van Natta*
H. B. Dykstra*	I. Williams

*Indicates Carothers's co-workers at DuPont. Source: Carothers, 1940, pp. 424–27.

Table 7. Journals Devoted to Macromolecular Chemistry Established by 1970

Journal	Year established	Country[a]
Journal für Makromolekulare Chemie[b]	1943	Germany
Kobunshi Kagaku[c]	1944	Japan
Polymer Bulletin[d]	1945	U.S.A.
Journal of Polymer Science[e]	1946	U.S.A.
Die makromolekulare Chemie[f]	1947	W. Germany
Kobunshi	1952	Japan
Journal of Applied Polymer Science	1959	U.S.A.
Vysokomolekularniye Soedineniya	1959	U.S.S.R.
Polymer	1960	Britain
Polymer Engineering and Science	1961	U.S.A.
Kolloid-Zeitschrift & Zeitschrift für Polymere[g]	1962	W. Germany
European Polymer Journal	1965	Britain
Journal of Macromolecular Chemistry[h]	1966	U.S.A.
Journal of Macromolecular Science	1967	U.S.A.
Macromolecules	1968	U.S.A.
Polymer Journal	1970	Japan

Journals devoted to specific subjects (such as rubber, cellulose, plastics, and fibers) are not included. [a]Name of country at the time of establishment. [b]Discontinued in 1944. [c]Renamed *Kobunshi Ronbunshu* in 1974. [d]Merged into *Journal of Polymer Science* in 1946. The German periodical of the same title (established in 1978) has no relation to this one. [e]The first issue was titled *Journal of Polymer Research*. [f]Renamed *Macromolecular Chemistry and Physics* in 1994. [g]Formerly titled *Kolloid-Zeitschrift*. [h]Merged into *Journal of Macromolecular Science* in 1967.

Abbreviations

Serial Publications

Amer. J. Sci. — *American Journal of Science*
Ann. Sci. — *Annals of Science*
Ann. Rev. Phys. Chem. — *Annual Review of Physical Chemistry*
Annalen — *Annalen der Chemie und Pharmacie*
Ber. — *Berichte der deutschen chemischen Gesellschaft*
Biog. Mem. FRS — *Biographical Memoirs of Fellows of the Royal Society of London*
Biog. Mem. NAS — *Biographical Memoirs of the National Academy of Sciences, U.S.A.*
BJHS — *British Journal for the History of Science*
Bull. Soc. chim. — *Bulletin de la Société chimique de France*
Chem. Abstracts — *Chemical Abstracts*
Chem. Ber. — *Chemische Berichte*
Chem. Eng. News — *Chemical and Engineering News*
Chem. Ind. — *Chemistry and Industry*
Chem. Reviews — *Chemical Reviews*
Chemiker-Ztg. — *Chemiker-Zeitung*
Colloid Symp. Monogr. — *Colloid Symposium Monograph*
Compt. rend. — *Comptes rendus hebdomadaires des Séances de l'Académie des Sciences*
Compt. rend. Trav. Lab. Carlsberg — *Comptes rendus des Travaux du Laboratoire Carlsberg, serie Chimique*
DSB — *Dictionary of Scientific Biography*
Gummi-Ztg. — *Gummi-Zeitung*
Helv. Chim. Acta — *Helvetica Chimica Acta*
His. Stud. Phys. Sci. — *Historical Studies in the Physical Sciences; Historical Studies in the Physical and Biological Sciences*
Ind. Eng. Chem. — *Industrial and Engineering Chemistry.*
India-Rubber J. — *India-Rubber Journal*
J. Amer. Chem. Soc. — *Journal of the American Chemical Society*
J. Appl. Phys. — *Journal of Applied Physics*
J. Appl. Poly. Sci. — *Journal of Applied Polymer Science*
J. Chem. Educ. — *Journal of Chemical Education*
J. Chem. Phys. — *Journal of Chemical Physics*
J. Chem. Soc. — *Journal of the Chemical Society, London*
J. Ind. Eng. Chem. — *Journal of Industrial and Engineering Chemistry*

J. Macromol. Sci. — Journal of Macromolecular Science
J. Phys. Chem. — Journal of Physical Chemistry
J. Poly. Sci. — Journal of Polymer Science
J. prakt. Chem. — Journal für praktische Chemie
J. Russ. Phys. Chem. Soc. — Journal of the Physical and Chemical Society of Russia
J. Soc. Chem. Ind. — Journal of the Society of Chemical Industry
Kolloid-Z. — Kolloid-Zeitschrift.
Liebigs Ann. Chem. — Justus Liebigs Annalen der Chemie.
Macromol. Chem. Phys. — Macromolecular Chemistry and Physics
Makromol. Chem. — Die Makromolekulare Chemie
Monatsh. — Monatshefte für Chemie und verwandte Teile anderer Wissenschaften
Naturwiss. — Die Naturwissenschaften
NTM — Schriftenreihe für Geschichte der Naturwissenschaften, Technik und Medizin
Pflüger's Arch. Physiol. — Pflüger's Archiv für die gesamte Physiologie des Menschen und der Tiere
Phil. Mag. — Philosophical Magazine
Phil. Trans. — Philosophical Transactions of the Royal Society of London
Proc. Roy. Soc. — Proceedings of the Royal Society of London
Quart. J. Sci. Arts — Quarterly Journal of Science and the Arts
Rubber Chem. Tech. — Rubber Chemistry and Technology
Sitzungsber. Preuss. Akad. Wiss. Berlin — Sitzungsberichte der Preussischen Akademie der Wissenschaften zu Berlin
Tech. Cult. — Technology and Culture
Trans. Amer. Electrochem. Soc. — Transactions of the American Electrochemical Society
Trans. Amer. Inst. Chem. Eng. — Transactions of the American Institute of Chemical Engineering
Trans. Faraday Soc. — Transactions of the Faraday Society
Trends Biochem. Sci. — Trends in Biochemical Science
Z. anal. Chem. — Zeitschrift für analytische Chemie
Z. angew. Chem. — Zeitschrift für angewandte Chemie
Z. Chem. Ind. Kolloide — Zeitschrift für Chemie und Industrie der Kolloide
Z. Elektrochem. — Zeitschrift für Elektrochemie
Z. Physik — Zeitschrift für Physik
Z. physik. Chem. — Zeitschrift für physikalische Chemie, Stöchiometrie und Verwandtschaftlehre
Z. physiol. Chem. — Hoppe-Seyler's Zeitschrift für physiologische Chemie

Papers, Archives, Libraries, and Institutions

AAAP — Avery Allen Ashdown Papers, Institute Archives and Special Collections, Massachusetts Institute of Technology, Cambridge, Massachusetts

ACS — American Chemical Society, Washington, D.C.

AGMPG — Archiv zur Geschichte der Max-Planck-Gesellschaft, Berlin-Dahlem

CHF — Chemical Heritage Foundation, Philadelphia, Pennsylvania

CSMP — Carl Shipp Marvel Papers, Chemical Heritage Foundation, Philadelphia, Pennsylvania

DCER — Department of Chemical Engineering Records, Institute Archives and Special Collections, Massachusetts Institute of Technology, Cambridge, Massachusetts

EFP — Emil Fischer Papers, Bancroft Library, University of California, Berkeley, California

FH — Firmenarchiv der Hoechst AG, Frankfurt-Hoechst

HABP — Herman Alexander Bruson Papers, Chemical Heritage Foundation, Philadelphia, Pennsylvania

HML — Hagley Museum and Library, Wilmington, Delaware

HSP — Hermann Staudinger Papers (Archiv, Nachlaß 88), Deutsches Museum, Munich. The material numbers cited in the notes are based on the new catalogue: Stephan Diller, Wilhem Füßl, and Rudolf Heinrich, eds., *Katalog des wissenschaftlichen Nachlasses von Hermann Staudinger* (München: Deutsches Museum, 1995).

HWC — Haber-Willstätter Correspondence, Leo Baeck Institute, New York, New York

HWSP — Hans Wilhelm Staudinger Papers, University Library, State University of New York, Albany, New York

JBCP — James Bryant Conant Papers, Nathan M. Pusey Library, Harvard University, Cambridge, Massachusetts

JLP — Jacques Loeb Papers, Library of Congress, Washington, D.C.

KEAP — Kenneth E. Appel Papers, Medical Library, Pennsylvania Hospital, Philadelphia, Pennsylvania

LHBP — Leo Hendrik Baekeland Papers, National Museum of American History, Washington, D.C.

MBP — Max Bergmann Papers, American Philosophical Society, Philadelphia, Pennsylvania

MHFC — Martin Henry Fischer Collection, Thomas Library, Wittenberg University, Springfield, Ohio

PJFP — Paul John Flory Papers, Chemical Heritage Foundation, Philadelphia, Pennsylvania

PP — Presidents' Papers, Joseph Regenstein Library, University of Chicago, Chicago, Illinois

RAP — Roger Adams Papers, University of Illinois, Urbana-Champaign, Illinois

SF — Staatarchiv Freiburg, Freiburg

WHCF — Wallace Hume Carothers File, University of Illinois, Urbana-Champaign, Illinois

Notes

Introduction

1. Leicester, 1956; Partington, 1964; Farber, 1969.
2. Ihde, 1964, pp. 711–19; Findlay, 1965, Chapter 8; Brock, 1992, pp. 647–53; also Hudson, 1992, pp. 255–56.
3. Mark, 1993, p. 119.
4. Ridgway, 1977, p. 343.
5. E.g., Carothers, 1931; Flory, 1953a, pp. 3–28; McGrew, 1958; Staudinger, 1961, 1970; Mark, 1967, 1973, 1976b, 1984b, 1987, 1993; Ulrich, ed., 1978; Mark, 1981; Marvel, 1981; Stahl, ed., 1981; Carra et al., eds., 1982.
6. Seymour, ed., 1982; Seymour and Cheng, eds., 1986; Seymour and Kirshenbaum, eds., 1986; Seymour, ed., 1989; Seymour and Mark, eds., 1990; Seymour and Porter, eds., 1993.
7. Morawetz, 1985; Fruton, 1972.
8. Olby, 1974, Chapters 1, 2.
9. McMillan, 1979.
10. Priesner, 1980.
11. Morris, 1986, 1989. See also Mossman and Morris, eds., 1994.
12. Hounshell and Smith, 1988, especially Chapter 12. The company is now officially spelled "DuPont" rather than "Du Pont." I adopt the DuPont spelling in the present book, while I leave the Du Pont spelling in references and quotations.
13. Meikle, 1995.
14. Hermes, 1996.
15. Flory, 1982; Flory's emphasis.
16. Staudinger and Fritschi, 1922, especially p. 788.
17. Earlier, Staudinger had called the field *die Chemie der hochmolekularen organischen Stoffe* (the chemistry of high molecular organic substances). But he came to use increasingly "*makromolekulare Chemie*" after the mid-1930s. E.g., Staudinger, 1929a, 1936c.
18. E.g., Carothers, 1932, p. 4471.
19. Adams, 1939. John R. Johnson's obituary of Carothers (1940) made no mention of "macromolecular chemistry" or "polymer chemistry" either.
20. Meyer and Mark, 1937; Mark, 1938a; Burk, Mark, and Whitby, 1940, pp. v, vii. This does not mean that Mark never used the term "macromolecular chemis-

try." He later used the term more freely in his articles, especially in those which described the history of the field.

21. Flory, 1953a; Billmeyer, 1957. Although less common, "macromolecular chemistry" or "macromolecular science" is also occasionally used in English-speaking countries, as exemplified by the title of an American journal, the *Journal of Macromolecular Science*, established in 1967. See Appendix Table 7.

22. Magda Staudinger, "Macromolecule and Polymer: Comments on the Publication by IUPAC Macromolecular Division, Commission on Macromolecular Nomenclature," unpublished manuscript (in possession of Magda Staudinger), 1995. She sent her comments to the editorial board of *Macromolecular Chemistry and Physics*, asking for their response. The published version of the glossary clearly reflects the effect of her protest: Commission on Macromolecular Nomenclature, 1996.

23. This demarcation has been developed in Furukawa, 1982. H. Zandvoort (1988) has adopted a similar scheme, calling the two approaches the "chemical" and the "physical" approach. John W. Servos (who prefers George Scatchard's division of colloid chemists into "unionists" and "isolationists") has construed my division as that of colloid chemists and identified The Svedberg, a Swedish physical and colloid chemist, as an organic-structurist. However, it is not my intention to divide *colloid chemists*—those who shared a disciplinary identity as colloid chemists—in such a way. As we shall see in the first three chapters, whether "unionists" or "isolationists," colloid chemists who worked in the 1910s and 1920s—including Richard Zsigmondy, Wolfgang Ostwald, Herbert Freundlich, Wilder Bancroft, and Svedberg—more or less all belonged to the physicalist tradition according to my classification. They, including Svedberg, generally showed little concern with the elucidation of precise molecular structure in their colloid studies. By contrast, Emil Fischer, Samuel Pickles, Staudinger, and Carothers were among typical organic-structuralists who studied polymeric substances, not on the basis of physical or colloid-chemical point of views, but on that of organic-structural chemistry. Even though Svedberg's ultracentrifugal study of proteins turned out to support the work of organic-structural chemists like Staudinger, he was far from an organic-structuralist. See Servos, 1990, p. 386; Scatchard, 1973, p. 103. Of course my demarcation between organic-structural and physicalist traditions is a relative one. None of Staudinger's contemporaries would have claimed to be 100 percent within the "organic-structural" tradition, nor 100 percent within the "physicalist" tradition. There were organic-structural chemists who were affected by the physicalist approach, and vice versa. While researchers in the polymer field often openly expressed their positions one way or another, however, it is clear that there was divergence of opinion and approach within each tradition, as we will see in this study.

24. Marcel Florkin, for instance, has called the era in which the colloid doctrine of Wolfgang Ostwald flourished "the dark age of biocolloidology." Florkin, 1972, pp. 279–84.

Chapter 1: Background, 1800–1920

1. On Liebig's school, see Morrell, 1972; Fruton, 1990, pp. 16–71.

2. Quoted in Benfey, 1964, p. 110.

3. The German translation from the Swedish original (1832) is in Berzelius, 1833, p. 64; cf. 1832, p. 44.

4. Watts, 1863–68, vol. 4 (1866), s.v. "Polymerism."
5. Gmelin, 1848–60, vol. 7: *Organic Chemistry* 1 (1852), pp. 67–69.
6. Graham, 1861, p. 183.
7. Ibid., p. 220.
8. Ibid., p. 221.
9. Ibid., p. 223. Graham also related colloids to life processes, as he thought that the colloid possessed "ENERGIA," which might be the primary source of the force in the phenomena of vitality. Ibid., pp. 183–84.
10. On the Karlsruhe Congress, see De Milt, 1948.
11. Kekulé, 1858; Couper, 1858a, b.
12. J. A. Johnson, 1990, p. 33.
13. On the history of the synthetic dye industry, see Beer, 1959; Travis, 1993.
14. See Meyer-Thurow, 1982; Homburg, 1992.
15. J. A. Johnson, 1985, p. 510.
16. On the Baeyer research school, see Fruton, 1990, pp. 118–62, 327–74.
17. Willstätter, 1965, p. 113.
18. Ibid., p. 118; J. A. Johnson, 1985, p. 510.
19. J. A. Johnson, 1990, pp. 33–34.
20. Quoted in Fruton, 1990, p. 162.
21. Kekulé, 1878, p. 212.
22. Pflüger, 1875.
23. Kekulé, 1878, p. 212.
24. Berthelot, 1866a, pp. 294–96; 1866b.
25. For example, Arnold F. Holleman's textbook defined polymerization as follows: "The union of two or more molecules of a substance to form a body from which the original compound can be regenerated is called 'polymerization.'" Holleman, 1915, p. 139.
26. Kleeberg, 1891.
27. Baekeland, 1909b, p. 150.
28. Raoult, 1882, 1887.
29. Brown and Morris, 1888; 1889, p. 473.
30. Gladstone and Hibbert, 1889, pp. 39, 42.
31. Nastukoff, 1900.
32. Van't Hoff, 1887, 1888.
33. Rodewald and Kattein, 1900, pp. 588ff.
34. Perrin, 1916, p. 11, n. 1. For Perrin's work, see Nye, 1972.
35. Wilhelm Ostwald, 1926–27, vol. 3 (1927), pp. 111–12.
36. On Fischer's career, see Bergmann, 1930; Feldman, 1973; Moy, 1989.
37. E. Fischer, 1913b, pp. 3288–89; 1914, p. 1201.
38. E. Fischer, 1907b, pp. 1757–58.
39. E. Fischer, 1913b, p. 3288; 1914, pp. 1200–1201. Zinoffsky's estimation was made by elemental analysis on the assumption that the molecule contained a single iron atom. Zinoffsky, 1886.
40. E. Fischer, 1907a, p. 916.
41. E. Fischer, 1916; 1923, pp. 24ff.
42. E. Fischer, 1913b, p. 3288; also 1914, p. 1200.
43. See Zsigmondy, 1909.
44. On Ostwald's view of *allgemeine Chemie*, see Servos, 1990, Chapter 1. On Wilhelm Ostwald, see Donnan, 1951; Hiebert and Körber, 1978.
45. Willstätter, 1965, pp. 94–95.
46. Wilhelm Ostwald, 1926–27, vol. 2 (1927), pp. 48–49.

47. Donnan, 1951, p. 9.

48. Wilhelm Ostwald, 1926–27, vol. 3 (1927), pp. 111–12.

49. Servos, 1990, p. 51; J. A. Johnson, 1985, pp. 502–3.

50. J. A. Johnson, 1990, p. 30.

51. On Donnan's career, see Freeth, 1957. On McBain, see Hauser, 1955, p. 7. On Bancroft, see Findlay, 1953; Servos, 1990, pp. 299–324. On Freundlich, see Donnan, 1942; Hauser, 1955, pp. 5–7; Ross, 1978.

52. Wolfgang Ostwald, 1909, 1915, 1920. English translations are respectively, 1919, 1917, 1924. For his career, see Lottermoser, 1943; Hauser, 1955, pp. 2–3; Körber, 1975; Sühnel, 1989; Elliott, 1989a.

53. Wolfgang Ostwald, 1917, p. 76.

54. Ibid., p. 134.

55. Ibid., p. 180.

56. Ibid., pp. 218–19.

57. Ibid., p. 34.

58. Ibid., p. 6. See also Thomas, 1925, p. 324.

59. Weimarn, 1907, p. 83; Wolfgang Ostwald, 1919, pp. 58–61. On Weimarn, see Hauser, 1955, p. 5.

60. Wolfgang Ostwald, 1917, p. 76.

61. Wolfgang Ostwald, 1919, pp. 2, 111.

62. E.g., Wolfgang Ostwald's "Forward" to Buzágh, 1937, pp. v–vii. The term "sister science" appeared in Wolfgang Ostwald, 1917, p. 136.

63. Wolfgang Ostwald, 1919, pp. 140–45, on p. 145. See also Buzágh, 1937, p. v; Hauser, 1955, pp. 2–3.

64. Freundlich, 1926, pp. 2–4.

65. "The particles of colloidal solutions are larger than the molecules, and must frequently be regarded as aggregates of molecules," said Zsigmondy in his 1925 Nobel lecture: Zsigmondy, 1925, p. 57.

66. Martin H. Fischer, "Translator's Preface to the First Edition," dated June 1915, in Wolfgang Ostwald, 1919, pp. ix–x, on p. ix.

67. The aggregate theory was also sometimes called "association theory" or "micelle theory."

68. Werner, 1891; also Kauffman, 1967.

69. Werner, 1902, especially pp. 268ff. Werner distinguished the molecular compounds from "valence compounds" (*Valenzverbindungen*) or ordinary compounds which were constituted by Kekulé's primary valence bonds. See also Russell, 1971, pp. 213–23.

70. Thiele, 1899, p. 89.

71. Although Thiele himself did not develop the aggregate theory, he suspected that perhaps in such compounds as polystyrene the molecules of styrol were bound together by partial valences. Thiele, 1899, p. 92.

72. Harries, 1905, 1919; Pummerer and Burkard, 1922; Pummerer, Nielsen, and Gündel, 1927; Pummerer and Koch, 1924; Pummerer and Gündel, 1928; Hess and Wittelsbach, 1920; Hess, 1924, 1928; Karrer, 1920, 1925; Pringsheim, 1925, 1926; Bergmann, 1925, 1926; Abderhalden, 1924.

73. Fischer's students are listed in Fruton, 1990, pp. 375–403.

74. Dalton, 1836; Faraday, 1826; Liebig, 1835; C. G. Williams, 1860; Bouchardat, 1875; Tilden, 1884. For a good account of the early history of rubber studies, see Törnqvist, 1968. See also Whitby and Katz, 1933, pp. 1204–5; Fisher, 1944, pp. 3–4.

75. Williams proposed the isoprene formula $C_{10}H_8$, because he thought that carbon was divalent and had an atomic weight of 6, an idea which was not unusual at the time. C. G. Williams, 1860. The formula was later altered to C_5H_8. Isoprene does not exist in nature. Williams and others obtained and isolated isoprene by dry distillation of rubber. But the dry distillation yielded larger quantities of a dimer of isoprene ($C_{10}H_{16}$) and the like than isoprene itself. Thus Carl Otto Weber suggested in 1900 that rubber was composed of chain molecules of the dimer ($C_{10}H_{16}$). The colloidalist Weber believed that rubber colloid was an aggregate of crystalline molecules held together by certain weak forces. In many ways, Harries refined Weber's views. Weber, 1900a,b.

76. On Harries, see Whitby, 1989b.

77. Crawford, Heilbron, and Ullrich, 1987, pp. 192, 196, 200, 208, 305.

78. Harries, 1901.

79. Harries, 1905, p. 1196; 1910, p. 852; cf. 1904, p. 2709. Harries in 1913 also called the partial valence "colloid secondary valence" (*Kolloid-Nebenvalenz*). Harries, 1913a, p. 735.

80. Harries later proposed a larger ring of five isoprene units, and eventually of seven isoprene units. Harries, 1914a, 1919. On the development of Harries's rubber formula, see Whitby, 1921; Wada, 1987.

81. Hess and Wittelsbach, 1920; Hess, 1920. Formula is taken from Purves, 1954, p. 42.

82. Schroeter, 1916, p. 2697.

83. Nägeli, 1877, pp. 424ff.

84. See Hauser, 1939, p. 104.

85. Kraft and Stern, 1894; Kraft and Strutz, 1896; Zsigmondy and Bachmann, 1912; Wolfgang Ostwald, 1919, p. 143.

86. McBain and Salmon, 1920; also McBain, 1929. The term "micelle" had occasionally been used by other scientists before McBain's time.

87. Bergmann, 1926, p. 2973.

88. Ibid., p. 2981.

89. Nishikawa and Ono, 1913. Nishikawa's papers were collected in Nishikawa Commemorative Committee, ed., 1982.

90. On this Institute, see Engle, 1984, pp. 243–45; Brocke, 1990, pp. 241–42; Löser, 1991. For the X-ray research there, see Herzog, 1923; Polanyi, 1962; Mark, 1962; 1981, pp. 527–29.

91. Katz, 1925. On Katz, see Whitby, 1989c.

92. E.g., Herzog and Janke, 1921; Herzog, 1925.

93. Wolfgang Ostwald, 1917, p. 181.

94. Ibid., pp. 179–219.

95. Ibid., p. 179.

96. Elliott, n.d.; Brocke, 1990, pp. 246–47. Fischer's study of tannins included E. Fischer, 1913a, b; 1914.

97. See Planck, ed., 1936, vol. 1, AI–13; vol. 2, BI–13. On Bergmann, see Harington, 1945; Neuberger, 1945; Clarke, 1945; Helferich, 1969; Fruton, 1970.

98. On the history of the rubber industry, see Schidrowitz and Dawson, eds., 1952. On Hancock, see Whitby, 1989a.

99. Quoted in Morawetz, 1985, p. 54.

100. See Törnqvist, 1968, p. 28.

101. See Dawson, 1952. *Gummi Zeitung* was also founded in Dresden, Germany, in 1886.

102. Weber, 1902. See also Dawson, 1952, p. 381. On Weber, see Marckwald and Frank, 1906.
103. Weber, 1906. See also note 79.
104. Wolfgang Ostwald, 1917, pp. 211–12.
105. Törnqvist, 1968, pp. 28–29; Dinsmore, 1951, p. 119.
106. Dunstan, 1906, p. 274.
107. On synthetic rubber research at Bayer and I. G. Farben, see Morris, 1982, 1994.
108. Hofmann, 1936.
109. Ibid., p. 693.
110. Törnqvist, 1968, p. 46.
111. Tilden, 1908, p. 322.
112. Hofmann, 1912.
113. Harries, 1910, 1911.
114. See Törnqvist, 1968, pp. 47–58; Whitby and Katz, 1933, p. 1207.
115. Perkin, 1912, p. 617.
116. Harries, 1913b, p. 217. See also Törnqvist, 1968, pp. 49–58; Naunton, 1952, pp. 101ff.
117. Emil Fischer to Carl Duisburg, August 26, 1915, EFP.
118. Burgdorf, 1926; Whitby and Katz, 1933, p. 1209; Hofmann, 1936, p. 700; Törnqvist, 1968, pp. 66–69.
119. On the history of the cellulose industry, see Haynes, 1953.
120. Ibid., pp. 105–24.
121. Kenyon, 1951, pp. 143–53, on p. 145.
122. Haynes, 1953, p. 348.
123. Ibid., pp. 154ff.
124. Wolfgang Ostwald, 1917, p. 209.
125. L. E. Wise, 1960, p. 4. W. O. Kenyon called the decade of 1920–30 "the renaissance of cellulose chemistry." Kenyon, 1951, p. 147.
126. Beside fibers and plastics, there were other uses of cellulose. In 1908 Eastman Kodak produced cellulose triacetate photographic films. Cellophane was made from viscose. Shortly after World War I, "lacquer," based on low-viscosity nitrocellulose, or pyroxylin, became a popular coating material, with the increasing demand for cars.
127. See, e.g., Mossman, 1994.
128. On the history of Celluloid, see Friedel, 1983.
129. Baekeland, 1914a, pp. 91, 93.
130. Friedel, 1983, pp. 103ff. Baekeland, 1909c. On Baekeland, see Kettering, 1947; Houwink, 1964; Morris, 1986, pp. 39–41.
131. See Baekeland, 1909a, b; "Condensation Product of Phenol and Formaldehyde and Method of Making the Same," U.S. Patent, 942,699, filed December 4, 1907, patented December 7, 1909.
132. Baekeland, 1913, p. 509.
133. Baekeland, 1909b, pp. 173–74.
134. Baekeland, 1913, pp. 506, 509.
135. Baekeland, 1914b, p. 172.
136. Baekeland, 1913, p. 507.
137. Ibid., p. 510; idem., 1914b, on. p. 172.
138. Kienle and Ferguson, 1929. General Electric started research on this type of resins in the early 1910s. See Haynes, 1949, p. 187. On the general history of plastics in America, see DuBois, 1972; Meikle, 1995.

139. On institutions associated with the rubber industry, see Drakeley, 1952, pp. 364–71. On the early divisional history of American Chemical Society, see, e.g., American Chemical Society, ed. 1951.

Chapter 2: Staudinger and the Macromolecule

1. Nodzu, 1954, p. 376.

2. Staudinger had an elder brother, Wilhelm; an elder sister, Louise; and a younger brother, Hans Wilhelm. The genealogy of the Staudinger family is outlined in Ernst Wolrad Staudinger, "Stammbaum der Familie Staudinger," 1972, HSP, E I 1. Literature on Staudinger's life and work includes Quarles, 1951; Staudinger, 1961, 1970; Yarsley, 1967; Whitby, 1967; Hopff, 1969; Olby, 1976; Krüll, 1978; Furukawa, 1982; Morris, 1986; M. Staudinger, 1987; Batzer and Ringsdorf, 1987; James, 1993. Staudinger's papers were brought together in Staudinger, 1969–76.

3. F. Staudinger, 1907. Franz published more than fifteen books on culture, society, ethics, politics, and philosophy.

4. On Hans, see Rutkoff and Scott, 1981; Hans Staudinger, "Lebenslauf von Staatssekretär a.D. Dr. Hans Staudinger," n.d., HWSP, Box 1, Folder 2.

5. Staudinger, 1970, p. 1.

6. At Munich, Staudinger also took courses in mineralogy, physics, microscopy, botany, geology, and philosophy. See Remane and Weise, 1993.

7. Staudinger, 1920b, p. 3. This article was drawn from a lecture he gave at a meeting of the Naturforschenden Gesellschaft Zürich, January 12, 1920.

8. Staudinger, 1903.

9. Vorländer and Staudinger, 1903a, b; Staudinger, 1905, 1906.

10. See Krüll, 1976.

11. See Staudinger, 1970, pp. 11–17. Ketene is a class of organic compounds with the general formula, $R_2=C=C=O$, in which "R" symbols represent organic radicals or hydrogen atoms.

12. Staudinger, 1912.

13. See Staudinger, 1970, pp. 18–23, 36–38.

14. Einstein taught at the ETH from 1912 to 1914.

15. Staudinger, 1961, p. 4.

16. Interview with Signer by Koeppel, 1986, transcript, p. 6; interview with Signer by the author, 1989. For the Chemistry Department at the ETH, see Eidgenössische Technische Hochschule, ed., 1955, pp. 443–69.

17. Quarles, 1951, p. 120.

18. Yarsley, 1967, pp. 250–51.

19. Whitby, 1967, p. xx.

20. Interview with Signer by Koeppel, 1986, transcript, p. 10.

21. Ružička was awarded with Adolf Butenandt the 1939 Nobel Prize in chemistry for his work on polymethylenes and higher terpenes. Reichstein received the 1950 Nobel Prize in physiology and medicine, jointly with Edward C. Kendall and Philip S. Hench, for hormone research. On Ružička, see Prelog and Jeger, 1980.

22. Ibid., p. 413.

23. Other scientist-signers included Emil Fischer, Ernst Haeckel, Felix Klein, Philipp Lenard, Walter Nernst, Wilhelm Ostwald, Max Planck, and Wilhelm Röntgen. For the manifesto and the list of signers, see Nicolai, 1918, pp. ix–xiv.

24. Apparently, Dorothea became an active socialist under the influence of

Franz Staudinger. Hermann Staudinger was a sympathizer of social democracy but probably not a member of SDP. Magda Staudinger to the author, July 14, 1987; interview with Magda Staudinger by the author, 1989.

25. See Ragaz, 1982, pp. 26, 185, 206, 207, 224, 225, 238, 315, 352; Mattmüller, 1968. On Ragaz's activities, see Paul Bock, "Introduction," in Ragaz, 1984, pp. xi–xxii.

26. Although firsthand evidence has not been uncovered, Germany's General Council in Zürich reported to the Nazi Gestapo in 1933 that Staudinger supported Nicolai's anti-war activities. See Chapter 4. Staudinger possessed the 1919 edition of Nicolai's *Die Biologie des Krieges* as well as books on socialism and social democracy. These are now stored in the Staudinger Papers at Deutsches Museum: HSP, F IV 7, 8, 10, 31, 33, 34.

27. Staudinger, 1917b, p. 196.

28. Interview with Magda Staudinger by the author, 1989; Magda Staudinger to the author, September 24, 1987; Hermann Staudinger, "Zur Beurteilung Amerikas," March 1917, HSP, B II 3. See also Sachsse, 1984; M. Staudinger, 1987, p. 13.

29. On gas warfare, see L. F. Haber, 1986.

30. Rolland, 1952. This lengthy diary, written by Rolland during the war, was published posthumously following his will. In the diary, Staudinger's name was misspelled "Standinger." See p. 1404; cf. Strube, 1987.

31. Rolland, 1952, pp. 1400–1401.

32. Ibid., p. 1401.

33. Ibid., p. 1404.

34. Ibid., p. 1405.

35. Staudinger, 1919a, p. 512.

36. Fritz Haber to Hermann Staudinger, October 18, 1919, HSP, D II 11.4a; Hermann Staudinger to Fritz Haber, October 19, 1919, HSP, D II 11.4b.

37. Fritz Haber to Hermann Staudinger, October 23, 1919, HSP, D II 11.4c.

38. Hermann Staudinger to Fritz Haber, November 17, 191a, HSP, D II 11.4d. The dispute between Haber and Staudinger has been discussed in Sachsse, 1984, p. 976; Furukawa, 1993, pp. 4–6; 1993–94, p. 4; Stoltzenberg, 1994, pp. 313–20.

39. Later, Staudinger wrote to his mentor, Carl Engler, explaining his quarrel with Haber. Sympathetic to Staudinger, Engler attempted, in vain, to calm Haber's anger. See Hermann Staudinger to Carl Engler, November 17, 1920, HSP, D II 11.12; Hermann Staudinger to Volkmar Kohlschütter, January 24, 1922, HSP, D II 11.22.

40. Staudinger, 1970, p. 77.

41. Vorländer, 1894.

42. Engler, 1897.

43. Kronstein, 1902.

44. Cf. Morawetz, 1985, pp. 24–25, 90.

45. Staudinger and Klever, 1911. Staudinger filed two patents reflecting this method to prepare isoprene. Hermann Staudinger, "Verfahren zur Darstellung von Isopren aus Terpenkohlen wasserstoffen," DRP 257640, September 4, 1910; "Verfahren zur Darstellung von Isopren," DRP 264923, August 15, 1911.

46. Staudinger, 1917, p. 77.

47. Ibid., p. 45.

48. Hermann Staudinger, "Historische Uebersicht über frühere angebliche und wirkliche Kautschuk-Synthesen," ca. 1912, HSP, B II 59.

49. Staudinger, 1917a; 1969–76, vol. 1, p. 39.

50. Staudinger and Klever, 1911, p. 2215.

51. Staudinger, 1917a; 1969–76, vol. 1, pp. 24–25.

52. On Pickles, see "Dr. Samuel S. Pickles," 1951; "Pickles, Samuel Shrowder," 1953; Löser, 1983, pp. 53–54; Morris, 1986, pp. 42–43; Wada, 1987, p. 21.

53. Pickles, 1910.

54. Ibid., pp. 1088–90.

55. Harries, 1911.

56. Harries, 1914a; cf. 1914b. See also Törnqvist, 1968, pp. 40–41.

57. Wada, 1987, pp. 57–59.

58. Whitby, 1967, p. xix.

59. In 1912, Pickles moved to George Spencer, Moulton and Company, Ltd., where he stayed until his retirement in 1950. Cf. Pickles, 1951.

60. Staudinger, 1917a; 1969–76, vol. 1, pp. 25, 27.

61. Ibid., p. 25.

62. Staudinger, 1919b.

63. Staudinger, 1920a.

64. Ibid., pp. 1073–74. Here, Staudinger criticized Georg Schroeter's interpretation of ketene polymerization as a molecular aggregation.

65. Ibid., p. 1074.

66. Ibid., p. 1082.

67. Ibid., p. 1083.

68. Ibid., p. 1081.

69. Staudinger, 1970, p. 83.

70. Mark, 1976b, p. 180.

71. Staudinger and Fritschi, 1922, especially p. 788. Staudinger considered the hydrogenation of natural rubber to be the following reaction, in which the saturation of double bonds occurred without changing the size of the large molecule:

$$
\cdots \cdot CH_2\text{-}\underset{\underset{CH_3}{|}}{C}\text{=}CH\text{-}CH_2\text{-}CH_2\text{-}\underset{\underset{CH_3}{|}}{C}\text{=}CH\text{-}CH_2 \cdot \cdots \cdot CH_2\text{-}\underset{\underset{CH_3}{|}}{C}\text{=}CH\text{-}CH_2 \cdot \cdots \cdot \overset{H_2}{\longrightarrow}
$$

$$
\cdots \cdot CH_2\text{-}\underset{\underset{CH_3}{|}}{CH}\text{-}CH_2\text{-}CH_2\text{-}CH_2\text{-}\underset{\underset{CH_3}{|}}{CH}\text{-}CH_2\text{-}CH_2\text{-} \cdots \cdot CH_2\text{-}\underset{\underset{CH_3}{|}}{CH}\text{-}CH_2\text{-}CH_2 \cdot \cdots
$$

See ibid., p. 790.

72. Staudinger, 1924, p. 1206.

73. According to Staudinger, the primary colloidal particles represented the macromolecules in most cases. But he admitted a few exceptions to this rule; for example, he thought that soap formed colloidal particles which consisted of small molecules. He suggested calling this group of colloids "*Pseudo-Kolloide,*" as distinguished from "*Eu-Kolloide*" in which colloidal particles were macromolecules. Staudinger, 1925; 1970, pp. 98–103.

74. Staudinger, 1926.

75. Staudinger, 1970, pp. 104–5.

76. Friedrich Oltmanns to Hermann Staudinger, July 25, 1925, HSP, D II 16.1. See also Priesner, 1987, pp. 151–54; Furukawa, 1993, pp. 6–8.

77. Willstätter was head of the Kaiser Wilhelm Institute for Chemistry, 1912–15, and professor at the University of Munich, 1915–24. He considered Wieland the

most versatile and outstanding organic chemist at that time and chose him as his successor for a chair once occupied by Liebig and Baeyer. See Witkop, 1992, p. 207.

78. Staudinger, 1970, p. 79.

79. Hans Z. Lecher to Hermann Staudinger, November 9, 1953, HSP, E III 12. Another student of Staudinger's, Rudolf Signer, related, "when he [Staudinger] gave his first general lecture in Freiburg, his famous colleague with a Nobel Prize, Wieland, afterwards clapped him on the shoulder and said, "Lieber Staudinger, organische Moleküle mit mehr als fünfzig C-Atomen gibt es nicht! [Dear Staudinger, organic molecules with more than fifty carbon atoms do not exist!]." Interview with Signer by Koeppel, 1986, transcript, p. 33.

80. Oltmanns to Staudinger, July 25, 1925 (n. 76 above).

81. Interview with Signer by Koeppel, 1986, transcript, p. 8; interview with Signer by the author, 1989.

82. Hermann Staudinger to Friedrich Oltmanns, September 14, 1925, HSP, D II 16.4.

83. According to Magda Staudinger, "Hermann Staudinger was very sorry about the error of translation of 'unheilvoll' into 'criminelle.' Surely he would have corrected it if he had seen the translation before printing. But I have no written proof for this fact." Magda Staudinger to the author, July 14, 1987.

84. Staudinger to Oltmanns, September 14, 1925 (n. 82 above).

85. Hermann Staudinger to Heinrich Wieland, September 14, 1925, HSP, D II 11.27. Staudinger, 1925a.

86. Heinrich Wieland to Hermann Staudinger, September 24, 1925, HSP, D II 11.28.

87. Friedrich Oltmanns to Hermann Staudinger, November 28, 1925, HSP, D II 16.6. The other two candidates were Hans L. Meerwein and Jakob Meisenheimer.

88. Staudinger and Dorothea had three daughters, Eva, Hilde, and Klara; and a son, Hansjürgen. Hansjürgen later studied chemistry under Hermann Staudinger at Freiburg (see Appendix Table 1), and between 1959 and 1979 he served as director of Physiological Chemistry and Biochemistry Institute at the University of Giessen. Dorothea died in 1964. Ernst Staudinger, "Stammbaum der Familie Staudinger" (n. 2 above); M. Staudinger, 1987, p. 12; interview with Magda Staudinger by the author, 1989.

89. Born in 1903, Magda studied biology at the University of Berlin and later served as an assistant in the Botanical Institute at the University of Riga. In the summer of 1927, she met Staudinger at the Biological Institute in Helgoland. Interview with Magda Staudinger by the author, 1989; Staudinger, 1970, pp. 237–40; M. Staudinger, 1987, pp. 17–18; Schwarzhaupt, 1983.

90. On Pummerer, see Oesper, 1951.

91. Pummerer and Burkard, 1922.

92. Harries, 1923, p. 1050.

93. Staudinger, 1924; 1970, pp. 203–5.

94. Katz, 1936, p. 77.

95. Staudinger, 1970, pp. 84–85.

96. Frey-Wyssling, 1964, p. 5. On the basis of X-ray crystallography, Emil Ott at the ETH supported the aggregate structure of polymers. E. Ott, 1926a, b.

97. Staudinger, 1970, pp. 84–85. Karrer received the Nobel Prize in chemistry in 1937, together with Walter Haworth, for research in the carotenoids, flavins, and vitamins A and B_2. On Karrer, see Miller, 1993.

98. Yarsley, 1967, p. 250.

99. See Appendix Table 1. Their dissertations are listed in the bibliography in Staudinger, 1970, Pt. B. See also Staudinger, 1955.

Outside the Freiburg school, Karl Freudenberg at the University of Heidelberg was among the few chemists who sided with Staudinger in the early 1920s. In 1921 Freudenberg, probably not yet familiar with Staudinger's work, published a paper suggesting the possibility of a long chain structure of cellulose, though his paper remained less influential. Freudenberg, 1921.

100. Mark, 1981, p. 529. A similar statement is in an interview with Mark by Bohning and Sturchio, 1986, transcript, pp. 12–13.

101. Haber's views on the aggregate theory were mentioned by Ryuzaburo Nodzu, Staudinger's student, in Nodzu, 1930, p. 52.

102. Interview with Mark by Bohning and Sturchio, 1986, transcript, pp. 12–13.

103. Bergmann, 1926; Ernst Waldschmidt-Leitz, 1926; Pringsheim, 1926.

104. Quoted from the printed version of the lecture: Staudinger, 1926, p. 3041.

105. Ibid., pp. 3041–43.

106. Reis, 1920; Weissenberg, 1925. Mark also knew about a 1920 lecture by Michael Polanyi, which commented on Herzog and Jancke's paper that measured X-ray diffraction spots of cellulose could be in agreement with either long chain molecules or small ring molecules. Polanyi, then unaware of Staudinger's 1920 paper on polymerization, recalled forty years later, "I was gleefully witnessing the chemists at cross-purposes with a conceptual reform when I should have been better occupied in definitely establishing the chain structure as the only one compatible with the known chemical and physical properties of cellulose. I failed to see the importance of the problem." Polanyi, 1962, p. 631; also 1921a, b; Mark, 1982, p. 179. On Polanyi, see Wigner and Hodgkin, 1977; Szabadváary, 1990.

107. Mark, 1926. Whereas Mark, in his memoir, stressed the neutrality of his position at this meeting, Staudinger recalled that Mark's talk had not sounded neutral but in favor of the aggregate theory. Cf. Mark, 1981, pp. 529–30; Staudinger, 1970, p. 85.

108. Nodzu, 1954, p. 376. Nodzu attended this meeting.

109. Mark, 1981, p. 530.

110. Interview with Signer by Koeppel, 1986, transcript, p. 13.

111. Staudinger, 1970, p. 85.

112. Mark, 1982, p. 2.

113. Katz, 1936, p. 77.

114. Sponsler and Dore, 1926, which was read in the American Colloid Symposium, held in June 1926 at the Massachusetts Institute of Technology. Its German translation appeared in *Cellulosechemie* 11 (1930): 186–97. Sponsler and Dore were probably unfamiliar with Mark's 1926 speech. Interview with Mark by Bohning and Sturchio, 1986, transcript, p. 14.

115. Staudinger et al., 1927, p. 448. Although this paper was co-authored by Staudinger, the X-ray measurements were largely carried out by the two physicists, Mie and Hengstenberg. See also Hengstenberg, 1927. Signer also studied the shape of macromolecules in solution by introducing an apparatus for flow birefringence, a simple device that measured optically the approximate length to breadth ratio of long-chain molecules. Interview with Signer by Koeppel, 1986, transcript, p. 12.

116. Katz, 1936, p. 79.

117. Staudinger, 1970, p. 85.

118. Pummerer, Nielsen, and Gündel, 1927. Cf. Pummerer and Gündel, 1928.

119. Staudinger, 1970, p. 205.

120. Hess, 1928.

121. Quoted in Ochiai, 1954, p. 226.

122. Ibid.

123. Quoted in Yarsley, 1967, p. 268.

124. On Mark's career, see Stahl, 1979; Mark, 1981; interview with Mark by the author, 1982; interview with Mark by Bohning and Sturchio, 1986; Stahl, 1981a–c; Mark, 1993.

125. On Meyer's career, see Mark, 1952; Picken, 1952; Van der Wyk, 1952; Jeanloz, 1956; Farrar, 1974.

126. I. G. Farbenindustrie A.G. (Gez. Bosch, Kurt H. Meyer) to the Ministerium der geistl. u. Unterrichts-Angelegenheiten, 1926; quoted in Elliott, 1989b.

127. I. G. Farbenindustrie A.G. (Gez. Bosch, Kurt K [*sic.*] Meyer) to Richard Willstätter, March 12, 1926; quoted in Elliott, 1989b. See also Fritz Haber to Richard Willstätter, October 21, 1926, HWC.

128. Interview with Mark by Bohning and Sturchio, 1986, transcript, pp. 22–27.

129. Mark, 1981, p. 530.

130. Herman F. Mark to Hermann Staudinger, January 5, 1927, HSP, D II 12.1.

131. E.g., Meyer and Mark, 1928a, b; Meyer, 1930.

132. Staudinger, 1929a,b.

133. Stahl, 1979, p. 84.

134. The debate between Staudinger and Meyer-Mark has been extensively documented in Priesner, 1980, pp. 65–216.

135. Hermann Staudinger to Hermann F. Mark, October 31, 1928, HSP, D II 12.13.

136. Herman F. Mark to Hermann Staudinger, November 2, 1928, HSP, D II 12.14.

137. Meyer, 1929. This article was followed by Staudinger's "Schlusswort" or final word: Staudinger, 1929b.

138. Meyer and Mark, 1930a. In the same year, they also published an almost identical book under the title, *Der Aufbau der hochpolymeren organischen Naturstoffe* (Meyer and Mark, 1930b).

139. Interview with Mark by Bohning and Sturchio, 1986, transcript, p. 26.

140. HSP, C 58. Staudinger also criticized Mark's 1932 treatise. HSP, B I 57.

141. Takei, 1948, p. 60. Takei studied colloid chemistry under Wolfgang Ostwald from 1940 to 1943.

142. Meyer and Mark's theory aimed at a general understanding of polymers as an alternative to Staudinger's macromolecular theory, whereas rubber chemists, cellulose chemists, fiber chemists, protein chemists, and starch chemists confined their investigative efforts to their own specialized fields, interpreting their substances somewhat differently. Mark later analogized, "for us it was all the same; all were long molecules carrying different groups. . . . it was as with astronomers. For centuries, Jupiter, Saturn, and Uranus; each was a special world. After Copernicus, it was all the same." Interview with Mark by Bohning and Sturchio, 1986, transcript, p. 24.

143. Wolfgang Ostwald, "Organisatorische Arbeiten," AGMPG, KWG Generalverwaltung KWI für Faserstoffchemie, Hauptarken, 10.4.1934–10.4.1935, alt II 32a, Band IV, 2118.

144. Staudinger, 1930.

145. On colloid chemists' views on viscosity, see Morawetz, 1985, pp. 105ff.

146. The results were published in Staudinger and Nodzu, 1930; Staudinger and Ochiai, 1932. See also Goto and Maruyama, 1983.

147. On Svedberg's ultracentrifuge, see Svedberg and Pedersen, 1940; Olby, 1974, pp. 11–21; J. W. Williams, 1979; Elzen, 1993; Kay, 1993, pp. 112–14; Ede, 1996. See also Chapter 3.

148. Pedersen, 1940; Elzen, 1993, p. 355.

149. See Olby, 1974, p. 14; Morawetz, 1985, p. 106. For the Emergency Community, see L. F. Haber, 1971, p. 368; Stoltzenberg, 1994, pp. 530–39.

Later, in 1933, Staudinger sent his *Dozent* Rudolph Signer as a Rockefeller fellow to Svedberg's laboratory at Uppsala. There, Signer worked on the ultracentrifuge, and then went to Manchester to learn the X-ray analysis at William L. Bragg's laboratory. Interview with Signer by Koeppel, 1986, transcript, pp. 13–16.

150. Go, 1957, p. 527; interview with Go by the author, 1985. Go visited Staudinger's laboratory in 1932.

151. Staudinger, 1930, pp. 28ff. See also Staudinger and Heuer, 1930. Staudinger's Law was expressed by the following equation:

$$\eta_{sp}/C = K_m \times M$$

where η_{sp} = specific viscosity; C = concentration; K_m = constant; and M = molecular weight.

152. See Staudinger's discussion remarks in *Kolloid-Z.* 53 (1930), p. 42.

153. Meyer, 1930.

154. Mark, 1930.

155. Herzog, 1930; Hess, 1930; Pummerer, 1930.

156. Sakurada, 1969, p. 49.

157. Hess, 1930, pp. 74–75.

158. Sakurada, 1969, p. 50.

159. "Colloquium im Kaiser Wilhelm-Institut für physikalische und Elektrochemie," 1929; Nodzu, 1930, p. 52.

160. Interview with Mark by Bohning and Sturchio, 1986, transcript, p. 18.

161. See Appendix Table 1. Their dissertations are listed in the bibliography in Staudinger, 1970, Pt. B.

162. Staudinger, 1932. Although the book represented an extensive scholarly effort, it lacked the comprehensiveness and lucid style of the Meyer-Mark textbook. Whereas four editions of the latter appeared, Staudinger's book ended with first edition, with the exception of the photographically reproduced version made in the United States during World War II (1943), and the 1960 reprint edition by Springer Verlag.

163. Mark and Bruson, 1956, p. 388.

164. Quarles, 1951, p. 120.

165. Emil Ott, quoted in "Nobel Prize to German, Hollander," 1953, p. 4761.

166. Staudinger, 1926, pp. 3021–22.

167. Staudinger, 1947, especially pp. 134ff; 1953, pp. 415–16.

168. Staudinger, 1935; 1937, p. 681; 1953, pp. 404ff.

169. Staudinger, 1953, pp. 404ff.

170. Staudinger, 1970, p. 92; see also 1953, p. 414.

171. Staudinger held a holistic conception of life as well as matter. In reference to the minimum size of organisms, he stated, "The known facts of macromolecular chemistry show further that an individual macromolecule is still not living, however large it is and however complex its structure. On the contrary, the term is relevant to a certain amount of substance comprising numerous macromolecules with the constituent small molecules combined together by strictly pre-

scribed order, an 'atomos' of living matter which is indivisible without losing its livingness." Thus, living nature could not be properly understood by reducing the ordered whole of the "atomos" to its small parts. Staudinger, 1953, pp. 416ff. Cf. Olby, 1979, p. 194. On physiological holism in the early twentieth century, see, e.g., Allen, 1975, pp. xix–xxiii, 103ff; Fruton, 1972, especially pp. 485–503; Florkin, 1972, especially pp. [1]–[6].

172. Bernhard Welte, quoted in Staudinger, 1970, pp. 240–41.

173. Staudinger, 1953, p. 414.

174. Staudinger, 1970, pp. 91–97.

175. Quoted in Yarsley, 1967, p. 263.

176. See Staudinger's patents, which are listed in the bibliography in Staudinger, 1970, pp. 243–77.

177. Staudinger, 1970, p. 98.

178. Georg Kränzlein to Hermann Staudinger, March 21, 1941, HSP, E II 12.

179. Staudinger, 1926.

180. Correspondence between Staudinger and Kränzlein, 1926–28 has been brought together in Staudinger and Kränzlein, 1966. See also A. Tanaka, 1992–93, 20: 243–58.

Chapter 3: Carothers and the Art of Macromolecular Synthesis

1. Stahl, 1981b, p. 73; interview with Mark by Bohning and Sturchio, 1986, transcript, p. 23.

2. Mark and Whitby, 1940, p. viii.

3. Roger Adams, Carothers's mentor, highly respected Carothers's gifted writing style and clear thinking. He stated that Carothers

expresses himself in writing as well as any scientific man I know regardless of the field; he is exact, concise, and has a pleasing style. If I am in doubt about any of my own manuscripts which are in preparation for publication, I prefer criticism from Carothers on the scientific reasoning involved as well as upon the mode of presentation to that of any other man. He has an unusually active, clear, keen mind and is a wide reader in subjects of general cultural interest as well as in science.

Roger Adams to Frederic Woodward, November 8, 1934, PP, 1925–1945, Box 101, Folder 3.

4. Biographical memoirs and studies of Carothers include Adams, 1939; J. R. Johnson, 1940, 1958; Hill, 1971, 1977; Schwartz, 1981; Morris, 1986, pp. 58–60; Hounshell and Smith, 1988, Chapter 12; Furukawa, 1994; Hermes, 1996. All Carothers's papers on polymer chemistry were brought together in Carothers, 1940.

5. Wallace H. Carothers to Wilko G. Machetanz, January 15, 1928, HML, Acc. 1850.

6. Ira H. Carothers to Roger Adams, December 10, 1937, RAP, Box 54.

7. Ira H. Carothers to Roger Adams, December 2, 1937, RAP, Box 54.

8. Ira Carothers to Adams, December 10, 1937 (n. 6 above).

9. Ira Carothers to Adams, December 2, 1937 (n. 7 above).

10. Cf. Hardy, 1974.

11. Ira Carothers to Adams, December 2, 1937 (n. 7 above).

12. Wallace H. Carothers to Pauline G. Beery, April 4, 1932, HML, Acc. 1784 C5.
13. Duncan, 1905.
14. Duncan, 1907. Quotations are from pp. 3, 9. On Duncan's activities as a popularizer of chemistry, see Rhees, 1987, pp. 22–34.
15. Ira Carothers to Adams, December 2, 1937 (n. 7 above); Arthur M. Pardee to Roger Adams, February 19, 1938, RAP, Box 54.
16. Ibid.
17. Arthur M. Pardee, "Contribution to the Biographical Memoir of Wallace Hume Carothers," February 19, 1938, RAP, Box 54.
18. Pardee, 1916.
19. Pardee, "Contribution to the Biographical Memoir of Wallace Hume Carothers" (n. 17 above).
20. Carothers to Beery, April 4, 1932 (n. 12 above).
21. Carroll, 1982, Chapter 4.
22. E.g., Kohler, 1975, pp. 442ff.
23. Lewis, 1916; Langmuir, 1919a,b, 1920.
24. Carothers, 1923.
25. Carothers, 1924.
26. Kohler, 1975, pp. 436–45.
27. Fry, 1928, p. 539; also Kohler, 1975, pp. 444–45.
28. John R. Johnson, " . . . This Uncommon Man," *Du Pont Magazine*, clipping, HML, Acc. 1842. Although Carothers's 1924 paper was "reviewed unfavorably by several referees," Johnson wrote in 1938, "Looking back on it about fifteen years I think that there can be no question of the significance of that paper." John R. Johnson to Roger Adams, November 3, 1938, RAP, Box 54.
29. In 1928 Carothers submitted a paper, "The Ethyl Anion and the Structure of the Grignard Reagents," which included an application of the octet theory, to the *Journal of the American Chemical Society*. Although criticized by the referee, this paper was approved for publication but never appeared in the Journal. Wallace H. Carothers, "The Ethyl Anion and the Structure of the Grignard Reagents," unpublished manuscript, 1928; "Comments of the Referee," October 1928, p. 2, HML, Acc. 1842.
30. For Roger Adams, see Leonard, 1969; Kohler, 1978; Corey, 1977; Tarbell and Tarbell, 1981.
31. In Jackson's opinion, Willstätter was then the only German organic chemist comparable to Emil Fischer. See Tarbell and Tarbell, 1981, pp. 27ff.
32. *Organic Syntheses: An Annual Publication of Satisfactory Methods for the Preparation of Organic Chemicals* (New York: John Wiley & Sons, 1921–). Carothers edited the thirteenth volume of this series in 1933.
33. See, e.g., Adams, 1954.
34. Tarbell and Tarbell, 1981, pp. 95–100, 221–28; Tarbell, Tarbell, and Joyce, 1980, p. 625.
35. Wallace H. Carothers to Wilko G. Machetanz, April 22, 1923, HML, Acc. 1850.
36. Carothers, 1924a. This thesis consisted of the following published papers: Carothers and Adams, 1923, 1924b, 1925.
37. Adams, 1939, p. 296.
38. Louis F. Fieser to Roger Adams, July 29, 1938, RAP, Box 54.
39. Carothers to Machetanz, January 15, 1928 (n. 5 above).
40. James B. Conant to Frederic Woodward, November 13, 1934, PP, 1925–1945, Box 11, Folder 3.

41. J. R. Johnson, 1958. Paul J. Flory also recalled that Carothers "was a very cultured person, very refined, and privately charming. Talking to a small group (no more than three people), he was an excellent conversationalist, but if the group became larger he shut up like a clam. The classroom was anathema [to] him." Interview with Flory by Overberger, 1982; partly transcribed in "Flory on Carothers and Du Pont," 1983. Julian W. Hill, another DuPont co-worker, said of Carothers: "He was very shy. After a drink or two, he was the funniest man alive, but it took a few people he knew well to warm him up." Quoted in Wallick, 1988.

42. James B. Conant to Roger Adams, August 19, 1938, RAP, Box 7.

43. G. Wise, 1985, p. 73.

44. Thackray et al., 1985, pp. 113, 345–46.

45. Ibid., p. 114.

46. For basic research at General Electric, see Reich, 1985; G. Wise, 1985; Furukawa, 1987, pp. 247–57.

47. Hounshell and Smith, 1988, Chapters 6, 8.

48. Charles M. A. Stine to the Executive Committee, "Pure Science Work," December 18, 1926, HML, Acc. 1784, Box 16. For the DuPont fundamental research program, see also Stine, 1936; L. G. Wise and N. G. Fisher, "History, Activities, and Accomplishments of Fundamental Research in the Chemical Department of the Du Pont Company, 1926–1939 Inclusive," August 14, 1940, HML, Acc. 1784; Elmer K. Bolton, "Du Pont Research," 1961, HML, Acc. 1689; Sturchio, 1981, especially pp. 144ff.; Hounshell and Smith, 1988.

49. Stine, 1932, p. 2034. This was a lecture delivered at the Johns Hopkins University, on April 29, 1932. For Stine's career, see "Biographical Sketch: Dr. C. M. A. Stine," March 11, 1954, HML, Acc. 1497, Box 7.

50. Stine, 1936, p. 131.

51. Ibid., p. 128.

52. Stine, "Pure Science Work" (n. 48 above).

53. Charles M. A. Stine to Roger Adams, December 2, 1938, RAP, Box 54.

54. Cole Coolidge to A. P. Tanberg, March 9, 1927, HML, Acc. 1784, Box 16.

55. Elmer O. Kraemer to Hamilton Bradshaw, March 9, 1927, HML, Acc. 1784, Box 16.

56. Bancroft suggested as possible subjects: "the general theory of rayon," "the theory of lithophone [a pigment]," and "the colloid chemistry of cellulose and its derivatives." Wilder D. Bancroft to Charles M. A. Stine, April 16, 1927, HML, Acc. 1784, Box 16. See also Charles M. A. Stine to Wilder D. Bancroft, April 21, 1927; Wilder D. Bancroft to Charles M. A. Stine, April 22, 1927, HML, Acc. 1784, Box 16.

57. E. Emmet Reid to Charles M. A. Stine, December 23, 1926, HML, Acc. 1784, Box 16.

58. "Notes on Meeting of the Directors of the Chemical Sections of E. I. Du Pont de Nemours & Co. and Subsidiaries at Wilmington, Del.," April 22, 1927, HML, Acc. 1662, Box 17.

59. Charles M. A. Stine to the Executive Committee, "Fundamental Research by the Du Pont Company," March 31, 1927, HML, Acc. 1784, Box 16.

60. It was estimated that expenditures for fundamental research in 1927 would amount to 200,000 U.S. dollars. Wise and Fisher, "History, Activities, and Accomplishments of Fundamental Research" (n. 48 above).

61. Organizational charts, 1928, HML, Acc. 1784, Box 18.

62. A. P. Tanberg to Henry Gilman, September 13, 1927, HML, Acc. 1784, Box 16.

63. Stine, "Fundamental Research by the Du Pont Company" (n. 59 above).

64. E.g., interview with Bolton by Chandler, Williams, and Wilkinson, 1961, transcript, p. 20.

65. Wallace H. Carothers to A. P. Tanberg, October 13, 1927, HML, Acc. 1784.

66. Charles M. A. Stine to Wallace H. Carothers, September 20, 1927, HML, Acc. 1784.

67. Wallace H. Carothers to Charles M. A. Stine, September 23, 1927, HML, Acc. 1784.

68. Charles M. A. Stine to Wallace H. Carothers, September 26, 1927, HML, Acc. 1784.

69. Wallace H. Carothers to Charles M. A. Stine, October 9, 1927, HML, Acc. 1784.

70. Carothers to Tanberg, October 13, 1927 (n. 65 above).

71. Hamilton Bradshaw to Wallace H. Carothers, October 17, 1927, HML, Acc. 1784.

72. Hamilton Bradshaw to James B. Conant, November 4, 1927, UAI.15.898.12, JBCP; Hermes, 1996, pp. 84–85.

73. Wallace H. Carothers to Hamilton Bradshaw, October 31, 1927, HML, Acc. 1784.

74. Carothers to Machetanz, January 15, 1928 (n. 5 above).

75. Wallace H. Carothers to Hamilton Bradshaw, November 9, 1927, HML, Acc. 1784. Probably, Pummerer's paper to which Carothers referred here is Pummerer, Nielsen, and Gündel, 1927. See Chapter 2.

76. Rogers, 1952, p. 45; Wallace, 1952, pp. 334, 336.

77. Thackray et al., 1985, pp. 351–52.

78. Kenyon, 1951, pp. 146, 148, 153.

79. Gray, 1927, p. 177.

80. Geer, 1927, p. 171.

81. Levene, 1927, p. 162.

82. Jacques Loeb to Thomas H. Morgan, February 17, 1920, JLP.

83. Conant, 1970, p. 67.

84. These meetings included the Chemical Society of France (1931), the Faraday Society in Manchester (1932) and in Cambridge (1935), and the Madrid meeting of the International Union of Pure and Applied Chemistry (1934). As a German chemist, Staudinger exceptionally was allowed to attend an International Solvay Conference, held in Brussels in 1925, probably due to his pacifism during the war. At the meeting, however, he did not give a paper but only commented on M. J. Duclaux's paper. Hermann Staudinger, "Remarques relatives à la communication de M. Duclaux," April, 1925, HSP, B I 32.

85. Staudinger's 1920 paper was summarized in *Chem. Abstracts* 14 (1920), Part 3: 3423–24; and in *J. Chem. Soc., Abstracts of Papers* 143 (1920), Part I: 517–18.

86. Rossiter, 1975; Hannaway, 1976; Dolby, 1977; Kohler, 1982; Servos, 1990.

87. Staudinger and Bruson, 1926a,b; Staudinger et al., 1929.

88. For Bruson's career, see, e.g., "Bruson, Dr. Herman A(lexander)," 1965; Seymour, 1980. His resume, list of publications, obituaries, and correspondence are stored in HABP, especially, Box 1, nos. 1, 3; Box 11, no. 28.

89. Staudinger et al., 1929; Staudinger and Ashdown, 1930.

90. For information on Ashdown's career, see "Ashdown, Prof. Avery A(llen)," 1965. Ashdown's "Curriculum Vitae," "Biographical Sketch," together with his diaries and personal correspondence, are stored in AAAP. It is not known whether Carothers met him at Cambridge in the years between 1925 and 1928. In

any event, Carothers's colleague at DuPont, Julian W. Hill (an MIT graduate who knew Ashdown personally), agreed that Ashdown did not play an important role in the introduction of Staudinger's theory in the United States. Indeed, Hill himself was not interested in polymer research until he joined Carothers's group in 1929. Interview with Hill by the author, 1982.

91. Haworth, 1928; Haworth and Hirst, 1930.
92. Whitby and Katz, 1928. Cf. Whitby, 1921, p. 315.
93. E. Fischer, 1914.
94. Wolfgang Ostwald, 1917, p. ix.
95. Oesper, 1945, p. 264.
96. Wolfgang Ostwald, 1915, p. xiii; 1917, p. xii.
97. The English translation (Wolfgang Ostwald, 1917) was reprinted periodically until 1925.
98. Martin H. Fischer, "Translator's Preface to the First Edition," dated June 1915, in Wolfgang Ostwald, 1919, pp. ix–x, on p. x. The correspondence between Ostwald and Fischer is stored in MHFC, Wittenberg University. See Elliott, 1993.
99. See, e.g., Holmes, 1954, p. 600.
100. See Ede, 1993a, pp. 94–107.
101. Ede, 1993b, p. 11; 1993a, pp. 145–47.
102. "Editor's Outlook," 1925. *Journal of Chemical Education* was then edited by Neil E. Gordon, founder of the journal and of the Gordon Research Conferences on Gibson Island.
103. Servos, 1990, pp. 53–99.
104. A. A. Noyes, 1905, p. 85.
105. Servos, 1990, p. 385.
106. On Bancroft, see Servos, 1990, Chapter 7; Ede, 1993a, pp. 242–69.
107. Bancroft, 1926, p. 110; also 1908, especially pp. 979–80.
108. See Servos, 1990, p. 305; Ede, 1993b, p. 146.
109. Bancroft, 1921.
110. Ibid., p. 2.
111. Ibid., p. 1.
112. Ibid., p. 187.
113. Sørensen, 1917.
114. Edsall, 1962, p. 18.
115. Julius Stieglitz to Max Mason, December 28, 1925, PP, Box 16, Folder 2. This plan of invitation, however, never materialized.
116. W. K. Lewis to Samuel Wesley Stratton, January 6, 1927, DCER, AC 122. In 1928 Hauser accepted a nonresident professorship and later a permanent position at MIT.
117. Quoted in Ihde, 1990, p. 479.
118. Ibid., p. 482. This is the comment of Aaron J. Ihde who later heard several of Svedberg's lectures.
119. Ibid.
120. Milligan, Williams, and Miller, 1951, p. 171.
121. Svedberg and Fåhraeus, 1926.
122. Bancroft, 1932 (3rd edition of Bancroft, 1921), p. 232.
123. Jacques Loeb to Svante Arrhenius, June 8, 1922, JLP.
124. Jacques Loeb to Svante Arrhenius, January 8, 1923, JLP.
125. Draft of Jacques Loeb's letter to Fritz Haber, May 26, 1923, JLP. This part was deleted in his letter actually sent to Haber.

126. Jacques Loeb to Svante Arrhenius, December 14, 1923, JLP.

127. Edsall, 1962, p. 18. On Loeb, see Pauly, 1987.

128. Herzog and Kobel, 1924.

129. Cohn and Conant, 1926.

130. See Conant, 1970, pp. 67–73; Hershberg, 1993, p. 61.

131. James B. Conant to Hermann Staudinger, November 17, 1953, HSP, E III. Ashdown recorded Conant's visit in his diary on July 6, 1925, AAAP. See also Hermann Staudinger to James B. Conant, July 10 and 29, 1925, UAI 15.898.12, Box 16.1, JBCP.

132. Staudinger, 1970, P. 84.

133. Conant to Staudinger, November 17, 1953 (n. 131 above).

134. Scatchard, 1973, p. 105.

135. Conant, 1928, p. 279.

136. Ibid., p. 54.

137. Ibid., p. 55.

138. Conant to Adams, August 19, 1938 (n. 42 above).

139. Wallace H. Carothers to A. P. Tanberg, November 20, 1927, HML, Acc. 1784.

140. Conant to Adams, August 19, 1938 (n. 42 above).

141. Interview with Hill by the author, 1982.

142. Wallace H. Carothers to John R. Johnson, February 14, 1928, HML, Acc. 1842. Johnson received his Ph.D. in 1922 under Roger Adams at Illinois. After working as Instructor of Organic Chemistry at Illinois (1924–27), he moved to Cornell University in 1928, where he taught until his retirement in 1965. The bulk of their correspondence between 1926 and 1937 is now stored in HML.

143. Carothers and Hill, 1932, in Carothers, 1940: 179–89, on p. 186.

144. Carothers, 1931, in Carothers, 1940: 81–140, on p. 129.

145. See Carothers and Dorough, 1930, in Carothers, 1940: 42–53, on p. 52.

146. Carothers to Bradshaw, November 9, 1927 (n. 75 above).

147. Wallace H. Carothers to Charles M. A. Stine, "Proposed Research on Condensed or Polymerized Substances," March 1, 1928, HML, Acc. 1784, Box 24.

148. Carothers to Johnson, February 14, 1928 (n. 142 above). In this letter Carothers, for whatever the reason, incorrectly wrote of "Fischer's record of 4200." That he knew it to be 4021 (and not 4200) is evident from his memorandum, "Proposed Research on Condensed or Polymerized Substances" (n. 147 above). As Johnson later deduced, "Wallace probably did not have the figure at hand when he wrote the letter to me and merely put it down from memory." John R. Johnson to C. H. Greenewalt, March 14, 1941, HML, Acc. 1842. Cf. E. Fischer, 1913b, p. 3288; 1914, p. 1200.

149. In his letter to Bradshaw of November 9, Carothers made no mention of the bifunctional reaction, but suggested a far more complicated route of step-by-step reactions to big molecules. Nor did he allude to the reaction in his letter of November 20, which reported his talk with James Conant on polymers. From these documents, it is apparent that he did not conceive the use of bifunctional reaction at the time. Carothers to Bradshaw, November 9, 1927; Wallace H. Carothers to A. P. Tanberg, November 20, 1927, HML, Acc 1784.

150. Wallace H. Carothers to R. T. Dufford, September 22, 1927; R. T. Dufford to Wallace H. Carothers, September 27, 1927; Wallace H. Carothers to R. T. Dufford, October 3, 1927, HML, Acc. 1896. Dufford, of the University of Missouri, published an article on Grignard compounds in electric fields, which caught

Carothers's attention. While in Illinois in 1925, Carothers had become interested in the Grignard reagent. Correspondence on this subject between Carothers and Henry Gilman, Professor of Iowa State College, in spring 1925, is found in HML, Acc. 1896.

151. Wallace H. Carothers, "Early History of Polyamide Fibers," February 19, 1936, HML, Acc. 1784.

152. Wallace H. Carothers, "Fundamental Research in Organic Chemistry at the Experimental Station: A Review," August 5, 1932, HML, Acc. 1784, Box 16.

153. Carothers to Tanberg, November 20, 1927; A. P. Tanberg to Wallace H. Carothers, January 13, 1928, HML, Acc. 1784.

154. Carothers, "Early History of Polyamide Fibers" (n. 151 above). In this memorandum, he wrote that the event happened in "the latter part of 1927." From the context, it is apparent that this was his second visit, not the first (September 1927). Cf. Stine to Carothers, September 20, 1927 (n. 66 above); Carothers to Stine, September 23, 1927 (n. 67 above).

155. W. Smith, 1901. For General Electric's work on resins including Glyptal, see Haynes, 1949, p. 187.

156. Hounshell and Smith, 1988, p. 144.

157. Roy H. Kienle, a chemist at General Electric, coined the generic word "alkyd" for these resins, as they were made from alcohols and acids. Kienle and Ferguson, 1929; Kienle and Hovey, 1929.

158. Carothers, "Early History of Polyamide Fibers" (n. 151 above); "Fundamental Research in Organic Chemistry" (n. 152 above).

159. J. R. Johnson, 1940, p. 102. A similar statement is found in Johnson to Adams, November 3, 1938 (n. 28 above).

160. Carothers to Johnson, February 14, 1928 (n. 142 above).

161. See Morawetz, 1985, p. 116.

162. Carothers, "Fundamental Research in Organic Chemistry" (n. 152 above).

163. Carothers to Stine, "Proposed Research on Condensed or Polymerized Substances" (n. 147 above).

164. In September 1927, Stine first approached Adams "with the idea of interesting you [Adams] in becoming one of our consultants." Charles M. A. Stine to Roger Adams, September 13, 1927, RAP, Box 9. Adams accepted the invitation on the condition that he and Marvel would pay alternate monthly visits to Wilmington. Charles M. A. Stine to Roger Adams, April 16, 1928, RAP, Box 9; interview with Adams by Mellecker, 1965, tape; interview with Marvel by Gorter and Price, 1983, transcript, p. 25.

165. Roger Adams to Wallace H. Carothers, February 18, 1928, HML, Acc. 1896.

166. Cf. J. W. Williams, 1979, pp. 77–91.

167. See Ihde, 1990, pp. 502–3; Hounshell and Smith, 1988, p. 227; Wise and Fisher, "History, Activities, and Accomplishment" (n. 48 above).

168. This thesis was published as Nichols, 1924.

169. James B. Nichols, "Biographical Sketch," August 25, 1978, HML, Acc. 1850. For DuPont's hiring Nicholas, see Elmer O. Kraemer to Hamilton Bradshaw, May 17, 1927, HML, Acc. 1784, ADD Box 19.

170. Wise and Fisher, "History, Activities, and Accomplishment" (n. 48 above).

171. Wallace H. Carothers to Roger Adams, December 26, 1928, RAP, Box 7.

172. Carothers, "Early History of Polyamide Fibers" (n. 151 above).

173. Carothers, 1929, in Carothers, 1940, pp. 4–17.

174. The condensation reactions included esterification, amide formation, ether formation, and anhydride formation.

175. Carothers, 1929, in Carothers, 1940, p. 7.

176. Ibid., pp. 6, 9.

177. Carothers, 1931, in Carothers, 1940, p. 124.

178. See Carothers, 1940, pp. 3–270.

179. Carothers, "Fundamental Research in Organic Chemistry" (n. 152 above).

180. Kienle, 1930, p. 593. For Kienle and his work, see "Physical Chemical Research Discussed by Kienle in Mattiello Memorial Lecture," 1949; Kauffman, 1990b, pp. 149–162.

181. "Editorial," 1969, p. 245. Cf. Kienle and Ferguson, 1929.

182. Carothers, 1931, in Carothers, 1940, pp. 121–23, 134.

183. Ibid., p. 125.

184. Ibid., p. 123.

185. Marvel, 1981, p. 536.

186. Conant, 1933, p. 78.

187. Carothers, 1931, in Carothers, 1940, pp. 82–83.

188. Carothers and Dorough, 1930, in Carothers, 1940, pp. 42–53, on pp. 52–53.

189. Carothers and Hill, 1932e, in Carothers, 1940, pp. 179–89, on p. 186.

190. Carothers and Van Natta, 1933, in Carothers, 1940, pp. 195–202, on p. 196.

191. As Carothers put it, the primary object of his study was "to synthesize giant molecules of known structure by strictly rational methods." Carothers and Hill, 1932e, in Carothers, 1940, p. 186. Concerning the addition polymerization of unsaturated compounds, Carothers stated in 1931: "So far as the formation of materials of high molecular weight is concerned, such reactions are much less clear-cut than bifunctional condensations, for the latter involve only the application of known reactions of typical functional groups, and the general structural plan of the product may be inferred directly from the structure of the starting materials. On the other hand, no clue to the intimate details of the mechanism of self-addition can be found in the reactions of the compound concerned with any compounds other than itself." Carothers, 1931, in Carothers, 1940, pp. 113–14.

192. Carothers, 1932, p. 4470.

193. Staudinger, 1932, pp. 11, 40, 148, 255.

194. Go's remark in March 1939; quoted in Soma, 1976, p. 17.

195. Ridgway, 1977, p. 342.

196. Interview with Flory by Overberger, 1982; "Flory on Carothers and Du Pont," p. 9.

197. Charles M. A. Stine to the Executive Committee, "Annual Report—1929," January 15, 1930, HML, Acc. 1784.

198. Cf., e.g., "Nylon," 1940, p. 58; Mosley, 1980, pp. 362–63.

199. Carothers, "Early History of Polyamide Fibers" (n. 151 above).

200. See, e.g., Carothers, 1931, in Carothers, 1940, pp. 88–89, 129–36.

201. Carothers, "Early History of Polyamide Fibers" (n. 151 above).

202. Ibid.; Washburn, 1929. Carothers attended the Swampscott meeting, held on September 13, 1928. Wallace H. Carothers to Roger Adams, September 23, 1928, RAP, Box 7. The program of this meeting is in *Ind. Eng. Chem., News Edition* 6 (1928): 1–9. See also Brønsted and Hevesy, 1922; Burch, 1929.

203. Carothers and Hill, 1932a, in 1940, pp. 154–56.

204. Carothers and Hill, 1932b–d; Carothers, 1940, pp. 156–79. The first super-polymer was prepared on April 16.

205. Hill initially noticed this phenomenon when he pulled threads of a molten superpolyester with a rod. See memorandum from J. W. Hill to W. H. Carothers, "Review of Work on Superpolymers," DuPont Company, February 6, 1931, HML, Acc. 903, Rutledge Scrapbook, vol. 598.

206. Julian W. Hill to Bettina Sargeant, November 17, 1960, HML, Acc. 1497, Box 24.

207. Carothers and Hill, 1932e, in Carothers, 1940, p. 187. Although Carothers rejected the prevailing concept of the aggregate force of small molecules, he admitted the existence of cohesive forces which exerted themselves between macromolecules. Carothers, 1931, in Carothers, 1940, p. 130. See also Wallace H. Carothers to E. B. Benger, July 29, 1930, HML, Acc. 1784 C5.

208. Staudinger and Signer, 1929, p. 208.

209. Wallace H. Carothers and Julian W. Hill, "Artificial Fibers from Synthetic Linear Condensation Superpolymers," manuscript, 1931, HML, Acc. 1784.

210. Carothers and Hill, 1932e, in Carothers, 1940, p. 188.

211. Ibid., pp. 187–88. Cf. Carothers, 1931, in Carothers, 1940, pp. 132–36.

212. Carothers and Hill, 1931. Cf. Carothers and Hill, 1932e, in Carothers, 1940, especially p. 180.

213. "Chemists Produce Synthetic 'Silk,' " 1931.

214. Carothers and Hill, "Artificial Fibers" (n. 209 above). Cf. Hooke, 1665, p. 7.

215. Carothers, "Early History of Polyamide Fibers" (n. 151 above).

216. Interview with Hill by the author, 1982.

217. Carothers, "Early History of Polyamide Fibers" (n. 151 above).

218. Wallace H. Carothers, "Linear Condensation Polymers," U.S. 2,071,250 patented February 16, 1937; "Fiber and Method of Producing It," U.S. 2,071,251, patented February 16, 1937. In these patents, Carothers wrote, "So far as I am aware, no synthetic material has hitherto been prepared which is capable of being formed into fibers showing appreciable strength and pliability, definite orientation along the fiber axis, and high elastic recovery in the manner characteristic of the present invention."

219. Carothers, "Fundamental Research in Organic Chemistry" (n. 152 above).

220. Bolton, 1942a, pp. 54ff. See also interview with Bolton by Chandler, Williams, and Wilkinson, 1961, transcript, pp. 20–21.

221. Bolton, 1942b, p. 1365.

222. Carothers, "Early History of Polyamide Fibers" (n. 151 above).

223. Ibid.

224. Ibid.; Hounshell and Smith, 1988, pp. 244–45.

225. Carothers, "Early History of Polyamide Fibers" (n. 151 above).

226. Bolton, 1942a, p. 55.

227. Gerard J. Berchet, "Adipate of Hexamethylene Diamine" (experimental record), February 28, 1935 (in possession of Gerard J. Berchet). The chemical structure of the polymer 66 is

$$\ldots-NH-R-NH-CO-R'-CO-NH-R-NH-CO-R'-CO-\ldots$$

where R is $-(CH_2)_6-$ and R′ is $-(CH_2)_4-$.

Cf. the structure of silk fibroin (cf. Carothers, 1929, in Carothers, 1940, p. 9):

. . . –NH–R–CO–NH–R′–CO–NH–R–CO–NH–R′–CO– . . .
where R is –CH– and R′ is –CH$_2$–.
$\quad\quad\quad$|
$\quad\quad$CH$_3$

228. On the discovery of neoprene, see Stine, 1934; Collins, 1973; J. K. Smith, 1985.

229. Collins, 1973, pp. 649–50; J. K. Smith, 1985, pp. 4–5. The preparation of monovinyl acetylene and divinyl acetylene from acetylene is shown as follows:

$$\underset{\text{acetylene}}{CH\equiv CH} \underset{NH_4Cl}{\overset{CuCl}{\longrightarrow}} \underset{\text{monovinyl acetylene}}{CH_2=CH-C\equiv CH} + \underset{\text{divinyl acetylene}}{CH_2=CH-C\equiv C-CH=CH_2}$$

230. Wallace H. Carothers to Elmer K. Bolton, January 20, 1930, HML, Acc. 1784 C5; Carothers, "Fundamental Research in Organic Chemistry" (n. 152 above).

231. Carothers, "Fundamental Research in Organic Chemistry" (n. 152 above).

232. Collins, 1973, p. G50. See also Wallace H. Carothers to E. K. Bolton, April 18, 1930, HML, Acc. 1784 C5; Wallace H. Carothers to E. K. Bolton, "History of Work on D.V.A., M.V.A. and C.D.," November 19, 1931, HML, Acc. 1850.

233. According to Paul J. Flory, Carothers was able to clarify the formation process of the rubbery material several hours after Collins found it in his laboratory. Flory told this story in a lecture (translated into Japanese as Flory, 1953b, p. 4) he gave at the Siga Plant of Toyo Rayon Company, Japan, on September 24, 1953. The original lecture notes are not to be found in PJFP; therefore he probably gave the talk without notes.

234. Carothers to Bolton, April 18, 1930; Wallace H. Carothers, "Reactions of M.V.A. and C.D. Suggested for Study (Excluding Polymerization)," July 22, 30, 1930, HML, Acc. 1784 C5. The formation process of neoprene is shown as follows:

$$\underset{\substack{\text{monovinyl} \\ \text{acetylene}}}{CH_2=CH-C\equiv CH} \overset{HCl}{\to} \underset{\text{chloroprene}}{CH_2=CH-\overset{\overset{\textstyle Cl}{|}}{C}=CH_2} \to \underset{\text{neoprene}}{(-CH_2-CH=\overset{\overset{\textstyle Cl}{|}}{C}-CH_2-)_n}$$

Cf. the structure of isoprene:

$$CH_2=CH-\overset{\overset{\textstyle CH_3}{|}}{C}=CH_2$$

235. Carothers, Williams, Collins, and Kirby, 1931.

236. J. K. Smith, 1985, pp. 48–55.

237. Whitby, 1940, p. 273. As early as 1933, Whitby and Katz had provided an extensive account of the properties and industrial significance of neoprene rubber. It read, "Recently a novel form of synthetic rubber has been prepared . . .

Chloroprene rubber resembles vulcanized natural rubber in elastic properties more than any previous synthetic rubber preparation, and, moreover, has advantages over natural rubber in certain respects." Whitby and Katz, 1933, p. 1204.

238. Carothers to Beery, April 4, 1932 (n. 12 above).

239. Carothers, 1940, pp. 273–398.

240. Carothers, "Fundamental Research in Organic Chemistry" (n. 152 above). Whitby judged that in addition to its industrial significance, "the work also makes a brilliant and important contribution to our knowledge of (1) the influence of substitution on the polymerization of conjugated systems and (2) the chemical behaviour of conjugated systems in which an ethylenic and an acetylenic bond are disposed in a conjugated relationship to one another." Whitby, 1940, p. 273.

Chapter 4: Triumph and Struggles of Two Giants

1. Charles L. Reese, "Report Visit of Dr. Chas. L. Reese to Europe June 23–August 1, 1928," August 14, 1928, HML, Acc. 1662, Box 17. German and Austrian guests included Max Bodenstein (Berlin), Fritz Haber (Berlin), W. Marckwald (Berlin), A. Stock (Technische Hochschule, Karlsruhe), F. Pregl (Graz), and R. Wegscheider (Vienna). Reese attended as an American delegate to this meeting of the International Union of Pure and Applied Chemistry.

2. Staudinger, 1931a,b.

3. Papers given at this meeting were published in *Trans. Faraday Soc.* 29 (1933), Pt. 1.

4. Staudinger, 1934b.

5. On the Kaiser Wilhelm Society during the Third Reich, see Albrecht and Hermann, 1990; Macrakis, 1993.

6. Obituaries on Bergmann include Harington, 1945; Neuberger, 1945; Clarke, 1945.

7. Bergmann responded to the question by Felix Haurowitz of Istanbul about the fate of German scientists in the United States: "The number of scientists from Europe who have found jobs here in America is enormous and all of them are highly satisfied to live in this country, notwithstanding the fact that nearly all of them have positions here somewhat inferior to the jobs they held in Europe. The living conditions in America are very satisfactory and the 'milieu' very stimulating." Max Bergmann to Felix Haurowitz, July 8, 1943, MBP, Box 6. On Bergmann's resignation, see voluminous correspondence and documents in AGMPG, 1 Abt. Rep. 0001, 538, "KWG, Generalverwaltung, Gesetz zur Wiederherstellung des Berufsbeamtentums — KWI Lederforschung, 1933–1944."

8. Ewald Reche, "Die Kaiser Wilhelm-Institute in Dahlem. Eine Brutstätte jüdischer Ausbeuter, Bedrücker und Marxisten!" 1933; Max Planck to Reichsminister des Innern, June 16, 1933; R. O. Herzog, "Aktennotiz. Besprechung am 17 Juni 1933," June 19, 1933, AGMPG, 1 Abt. Rep.0001A, 535, "KWG, Generalverwaltung, Gesetz zur Wiederherstellung des Berufsbeamtentums — KWI Faserstoffchemie, 1933–1944." This file includes correspondence and documents relating to Herzog's resignation.

9. Before Herzog's death, Harold Hibbert, a cellulose chemist, and Bergmann had attempted to secure a post at Eastman Kodak for him. Harold Hibbert to Max Bergmann, June 20, 1934; Max Bergmann to Harold Hibbert, June 26, 1934; Harold Hibbert to Max Bergmann, July 3, 1934; and Max Bergmann to Harold

Hibbert, July 9, 1934; Max Bergmann to Harold Hibbert, March 18, 1935; Max Bergmann to Karl P. Herzfeld, March 29, 1935, MBP, Box 6; Herzfeld, 1935.

10. Mark, 1952; Picken, 1952; Van der Wyk, 1952; Farrar, 1974.

11. For Mark's activity during this period, see Mark, 1981, pp. 531–32; Hunt, 1958; Stahl, 1981b, pp. 76–78; Mark, 1993, pp. 61–85.

12. Donnan, 1953, p. 511. For a history of the Faraday Society, see Sutton and Davies, 1996.

13. Melville, 1953, p. 565. Before 1935, the Society's general discussions on colloids included "Colloids and Their Viscosity" (1913), "Physics and Chemistry of Colloids" (1920), "Capillarity" (1921), "Colloid Science Applied to Biology" (1930), "The Colloid Aspects of Textile Materials" (1932), and "Colloidal Electrolytes" (1934). See Sutton and Davies, 1996, pp. 391–92.

14. Papers given at this meeting were published in *Trans. Faraday Soc.* 32 (1936), Pt. 1: 3-412.

15. Carothers, 1936.

16. Staudinger, 1936a.

17. Staudinger, discussion in *Trans. Faraday Soc.* 32 (1936), Pt. 1, p. 52. For this reason, Staudinger had called addition polymers "true polymerization products" (*echte Polymerisationprodukte*), as distinguished from "false polymerization products" (*unechte Polymerisationprodukte*), i.e., condensation polymers; cf. 1920a, pp. 1074–75.

18. Carothers, discussion in *Trans. Faraday Soc.* 32 (1936), Pt. 1, p. 53. See also Carothers, 1936, p. 39.

19. International Union of Pure and Applied Chemistry, 1952, pp. 258, 261.

20. Melville, 1953, p. 566.

21. Mark, 1982, p. 5.

22. Dostal and Mark, 1936, p. 54.

23. Mark, 1982, p. 5.

24. Hermann Staudinger, "Polyesterarbeiten und Gespräche mit Carothers über die makromolekulare Chemie," lecture at DuPont, Wilmington, Delaware, September 23, 1958, HSP, B I 144.

25. Magda Staudinger to the author, May 7 and January 11, 1988; Staudinger et al., 1927.

26. Staudinger, "Polyesterarbeiten und Gespräche mit Carothers über die makromolekulare Chemie" (n. 24 above).

27. Roger Adams to Frederic Woodward, November 8, 1934, PP, 1925–1945, Box 101, Folder 3. See a similar statement in Roger Adams to Carl Stephens, November 21, 1934, WHCF.

28. Interview with Bolton by Chandler, Williams, and Wilkinson, 1961, transcript, p. 20.

29. Fred E. Wright (Home Secretary, National Academy of Sciences) to Wallace H. Carothers, April 29, 1936; telegram from Charles A. Kraus to Wallace H. Carothers, April 29, 1936; Claude S. Hudson to Wallace H. Carothers, April 30, 1936, HML, Acc. 1896. Quotation from Hudson's letter. Roger Adams wrote, "he had practically a unanimous vote. . . . This is very unusual." Roger Adams to Ira H. Carothers, December 6, 1937, RAP, Box 54.

30. Fred E. Wright to Leo H. Baekeland, April 29, 1936; Jerome Alexander to Leo H. Baekeland, April 30, 1936, LHBP, I-97. The content of Wright's letter to Baekeland was precisely the same as the one to Carothers.

31. See P. L. Salzberg to C. H. Greenewalt, April 22, 1959, HML, Acc. 1784, Add. Box 18.

32. Wallace H. Carothers to Roger Adams, May 1, 1936, RAP, Box 7; Wallace H. Carothers to Claude S. Hudson, May 1, 1936, HML, Acc. 1896. Adams was instrumental in recommending Carothers's election, as he was a member of the Academy's Council from 1931 to 1934. See Cochrane, 1978, p. 642. Carothers's election had been considered since 1934. See Roger Adams to Wallace H. Carothers, July 25, 1934; Wallace H. Carothers to Roger Adams, August 21, 1934, RAP, Box 7; Adams to Woodward, November 8, 1934 (n. 27 above).

33. Interview with Berchet by the author, 1982. In reality Carothers was not nominated for the Nobel Prize in his lifetime. Nominees in chemistry until 1937 are listed in Crawford, Heilbron, and Ullrich, 1987.

34. Ernest E. Benger to Elmer K. Bolton, November 7, 1930, HML, Acc. 1784, Box 16.

35. L. G. Wise and N. G. Fisher, "History, Activities, and Accomplishments of Fundamental Research in the Chemical Department of the Du Pont Company, 1926–1939 Inclusive," August 14, 1940, HML, Acc. 1784.

36. Wallace H. Carothers, "Fundamental Research in Organic Chemistry at the Experimental Station: A Review," August 5, 1932, HML, Acc. 1784, Box 16.

37. Interview with Dykstra and Cupery by Strange, 1978, transcript.

38. Wise and Fisher, "History, Activities, and Accomplishments of Fundamental Research" (n. 35 above).

39. Wallace H. Carothers to Wilko G. Machetanz, May 10, 1933 (in possession of Wilko G. Machtanz; courtesy of John K. Smith).

40. On Conant's assumption of the Harvard presidency, see Hershberg, 1993, pp. 65–75.

41. Wallace H. Carothers to Roger Adams, May 10, 1933, RAP, box 7.

42. Wallace H. Carothers to Roger Adams, May 21, 1933, RAP, Box 7.

43. Roger Adams to Wallace H. Carothers, May 26, 1933, RAP, Box 7.

44. James B. Conant to Roger Adams, January 23, 1934, RAP, Box 64.

45. E. P. Kohler to Roger Adams, January 23, 1934, RAP, Box 64; Conant to Adams, January 23, 1934 (n. 44 above).

46. James B. Conant to Roger Adams, March 23, 1934, RAP, Box 7. The exact reasons for Adams's decline of the professorship are unclear. Karl Compton, President of MIT, also offered Adams a professorship almost simultaneously. Once again, Adams turned it down. Karl T. Compton to Roger Adams, March 14, 28, and April 6, RAP, Box 64.

47. See Tarbell and Tarbell, 1981, pp. 106–9.

48. Wallace H. Carothers to Roger Adams, August 21, 1934, RAP, Box 7.

49. Robert M. Hutchins to Wallace H. Carothers, October 17, 1934; Frederic Woodward (Vice President) to James B. Conant, November 7, 1934, PP, 1925–1945, Box 101, Folder 3.

50. Wallace H. Carothers to Robert M. Hutchins, November 4, 1934, PP, 1925–1945, Box 101, Folder 3.

51. Ibid.

52. James B. Conant to Frederic Woodward, November 13, 1934, PP, 1925–1945, Box 101, Folder 3.

53. Roger Adams to Frederic Woodward, November 8, 1934, PP, 1925–1945, Box 101, Folder 3.

54. Conant to Woodward, November 13, 1934 (n. 52 above).

55. Wallace H. Carothers to Robert M. Hutchins, November 20, 1934, PP, 1925–1945, Box 101, Folder 3.

56. Robert M. Hutchins to Wallace H. Carothers, November 20, 1934; Wal-

lace H. Carothers to Robert M. Hutchins, November 26, 1934, PP, 1925–1945, Box 101, Folder 3.

57. Wallace H. Carothers to John R. Johnson, January 6, 1936, HML, Acc. 1842.

58. Words of W. S. Carpenter, Jr., then president of the DuPont Company; quoted in Bolton, 1942a, p. 56. For a history of the development of nylon, see Hounshell and Smith, 1988, Chapter 13.

59. Wallace H. Carothers, "Early History of Polyamide Fibers," February 19, 1936, HML, Acc. 1784.

60. Roger Adams, "Wallace Hume Carothers, 1896–1937," draft for *Biog. Mem. NAS*, ca. 1938, RAP, Box 54.

61. Wallace H. Carothers to John R. Johnson, February 26, 1936, HML, Acc. 1842. Their private wedding took place at a Presbyterian Church in New York without giving notice to the majority of their friends and colleagues. Their parents and relatives did not attend the wedding. In this letter, written soon after the wedding, Carothers wrote to Johnson, "Certainly it calls for apologies to you for failing to confide in advance in any way. However as a matter of fact no one was confided in. We both thought that we would settle it ourselves without any fuss or bother (you know my pathological attitudes toward such matters)."

62. Carothers's coworker Gerard J. Berchet said, "coming from that strict background [midwestern Protestant family] — no experience with drinking — he just overused it. . . . He just drank too much and it got worse." Interview with Berchet by Strange, n.d.

63. Wallace H. Carothers to John R. Johnson, July 9, 1936, HML, Acc. 1842.

64. Ira H. Carothers to Roger Adams, December 2, 1937, RAP, Box 54. Appel was later professor of psychiatry at the University of Pennsylvania and president of the American Psychiatric Association. On Appel, see Blain, 1980; Brady, 1980; Braceland, 1980. Publications by and on Appel are stored in two boxes of KEAP materials. According to the Institute's annual report, 245 patients were admitted to the Institute during 1936; 30 of the total 95 male inpatients were diagnosed as "manic depressive psychoses," 16 "dementia praecox (schizophrenia)," and 1 "alcoholic psychosis." Pennsylvania Hospital, 1937, p. 61.

65. Adams, "Wallace Hume Carothers" (n. 60 above).

66. Carothers to Johnson, July 9, 1936 (n. 63 above).

67. Adams had traveled to Europe to attend the 12th International Conference of Chemistry, held August 16–22 in Lucerne and Zürich. In England Adams learned from Elmer Bolton (then visiting there) of Carothers's admission to the "insane asylum" and probably suggested that Carothers might join his tour. Roger Adams to Lucile Adams, July 1, 8, and 25, 1936, RAP, Box 4.

68. Roger Adams to Lucile Adams, August 2, 1936, RAP, Box 4.

69. Roger Adams to Lucile Adams, August 7 and 15, 1936, RAP, Box 4.

70. Wallace H. Carothers to John R. Johnson, September 21, 1936, HML, Acc. 1842.

71. Roger Adams to Ira H. Carothers, December 6, 1937, RAP, Box 54.

72. Carothers to Johnson, September 21, 1936 (n. 70 above).

73. Isobel Carothers Berolzheimer died of a heart attack as a consequence of rheumatic fever. Carothers took a leave of absence for ten days. Willard Sweetman to J. A. Horty, January 27, 1937, HML, Acc 1784 C5.

74. "Synthetic Fiber," U.S. Patent 2,130,948, applied for on April 9, 1937 and patented on September 20, 1938. His first patent on fiber-forming polymers, applied for on July 3, 1931, was issued on February 16, 1937 (U.S. Patent 2,072,250).

75. Quoted in Molly (Mary) E. Carothers to Roger Adams, November 23, 1937, RAP, Box 54. Mary described it as "the longest and dearest letter I ever had from him."

76. "Dr. W. H. Carothers, Chemist, Found Dead," 1937. The *Philadelphia Inquirer* ("Du Pont Scientist Ends Life by Poison," 1937) reported, "No suicide note was found in the room. . . . He had $52 in his pockets." The *Philadelphia Evening Bulletin* ("Dr. Carothers Suicide by Poison," 1937) also wrote, "Mrs. Carothers said she last saw her husband late Wednesday [April 28] when he left home for 'a short trip' to an undisclosed destination. Associates at the duPont Experimental Station said he suffered a nervous breakdown some months ago and had been unable to work for several weeks." See also "Scientist Ends Life by Poison in Hotel Room," 1937. According to Joseph Labovsky, Carothers's laboratory technician, his sudden suicide was not foreseen by many of his close friends. Interview with Labovsky by Bohning and the author, 1993.

Although relying largely on hindsight, colleagues, friends, and journalists have conjectured about direct motives of Carothers's suicide, ranging from his private life (e.g., a broken affair with an attractive married woman named Sylvia Moore, his friends' disapproval of his marriage with Helen, Helen's pregnancy that he was likely to have noticed in the few days before his death and fear that their child might inherit his mental illness, and Isobel's death), to his relationship with the DuPont Company (e.g., disappointment about takeover of the nylon project by other departments, boredom after accomplishment, fear of his inability to make another big discovery, and the conflict he felt between academic and industrial values), to the socio-economic structure (American capitalism at large). It seems misleading to rationalize his suicide by simply ascribing it to these negative happenings or aspects. Whatever the real motivation, it is important to understand Carothers's behavior and suicide in the context of his complex psychogenic illness, coupled with alcoholism. Furukawa, 1989, and Hermes, 1996, have documented and analyzed Carothers's illness and alcoholism that led to his suicide.

77. Earlier, Carothers had suggested the names "Dualin," "Fintex," "Linex," and "Superin." The naming committee decided on "nylon" after modifying the finalist "Norun." On the derivation of the term, see Charles H. Rutledge, "The Name Nylon and Some of Its Adventures," Product Information Group, Textile Fibers Department, DuPont Company, June 20, 1966, HML, Acc. 903, *Rutledge Scrapbook*, vol. 598; "The Naming of Nylon," DuPont Company, HML, Acc. 1497, Box 17.

78. "Ladies' Hose and History," 1938. See also "New Fiber Called Superior to Silk for Making Stockings," 1938; "New Silk Made on Chemical Base Rivals Quality of Natural Product," 1938; "Coal and Castor Oil Challenge Silkworm," 1938.

79. Charles M. A. Stine, "What Laboratories of Industry Are Doing for the World of Tomorrow: Chemicals and Textiles," an address delivered before the New York Herald Tribune Eighth Annual Forum on Current Problems, October 27, 1938, HML, Acc. 903, *Rutledge Scrapbook*, vol. 598. Nylon was officially defined as "a man-made protein-like chemical product (polyamide) which may be formed into fibers, bristling filaments, sheets and other forms which are characterized when drawn by extreme toughness, elasticity, and strength." See, e.g., "Some Facts about Nylon," Public Relations Department, Du Pont Company, November 1939, HML, Acc. 903, *Rutledge Scrapbook*, vol. 598. Stine's claim that nylon was "the first man-made organic textile fiber" is perhaps debatable, since the invention of such synthetic fibers as "PeCe Faser" (I. G. Farben's polyvinylchloride fiber), "Vinyon" (C & C Chemicals' polyvinylchloride fiber), and "Per-

lon U" (I. G. Farben's polyurethane fiber), had already been announced by 1938. But that was how DuPont advertised nylon and how people saw it. Nylon was certainly unrivaled in its properties, applicability, output, and social impact, enough to overshadow other existing synthetic fibers.

80. "Nylon," 1940, p. 58.

81. Hounshell and Smith, 1988, p. 248. In this respect, Bolton's steering role as a research manager was indispensable to the birth of nylon. He received the Chemical Industry Medal in 1941, the Perkin Medal in 1945, and the Willard Gibbs Medal in 1954, all for his accomplishments as a successful leader of industrial research. On Bolton, see "Dr. Elmer Keiser Bolton," June 1, 1959, HML, Acc. 1689; Roger Adams, "Elmer K. Bolton," (speech at the presentation of the Willard Gibbs Medal to Bolton), May 21, 1954, RAP, Box 54.

82. Wise and Fisher, "History, Activities, and Accomplishments of Fundamental Research" (n. 35 above).

83. M. E. Carothers to Adams, November 23, 1937 (n. 75 above).

84. Adams, "Wallace Hume Carothers," draft for *Biog. Mem. NAS;* Elmer K. Bolton to Roger Adams, October 14, 1938, RAP, Box 54; Julian Hill to Elmer K. Bolton, October 11, 1938, RAP, Box 54.

85. Adams, 1939.

86. William Edward Hanford, a notable DuPont chemist, related: "I would say he [Carothers] was a scientist, but he was not able to make judgments as to how that science should be used to the greatest advantage of the business. He paid more attention to the chemistry than [he] did to business, just like he thought that 5-10 was a better polymer than 6-6." Interview with Hanford by the author, 1993.

87. Interview with Hill by the author, 1982. Gerard J. Berchet, another co-worker of Carothers, agreed with Hill on this point. Interview with Berchet by the author, 1982.

88. Furukawa, 1987, 1994.

89. Furukawa, 1993, pp. 9–14; 1993–1994, pp. 5–6.

90. Magda Staudinger to the author, November 1, 1988.

91. Baden Minister des Kultus und Unterrichts to Rektorat der Universität Freiburg, September 17, 1935, HSP, D II 17.18.

92. Heidegger, 1991. This interview was conducted on September 23, 1966, but was not published, by request, until after his death.

93. See Heidegger, 1933; Schneeberger, 1962; Farias, 1987; Rockmore, 1992.

94. H. Ott, 1988, pp. 201–13.

95. Badisches Landeskriminalpolizeiamt, Geheimes Staatspolizeiamt to Minister des Kultus, des Unterrichts, und der Justiz, Abteilung Kultus und Unterrichts (Nr. 14316), October 4, 1933; Minister des Kultus, des Unterrichts, und der Justiz, Abteilung Kultus und Unterrichts (Nr. A 27902), October 11, 1933, SF, A5, Nr. 180.

96. Martin Heidegger to Eugen Fehrle, February 10, 1934, SF, A5, Nr. 180.

97. Interview with Magda Staudinger by the author, 1989.

98. H. Ott, 1988, p. 207.

99. Hermann Staudinger, "Die Bedeutung der Chemie für das deutsche Volk," *Völkische Zeitung,* February 25, 1934, p. 7, HSP, A IV 171. The article was reprinted under the title, "Die Bedeutung der Chemie für die Existenzmöglichkeit des deutschen Volkes," in *Chemiker-Ztg.,* 59 (1935): 201–2, HSP, A IV 175 and B II 65.

100. Quoted in H. Ott, 1988, p. 207.

101. Ibid., pp. 208–9.

102. Interview with Magda Staudinger by the author, 1987; Magda Staudinger to the author, July 14, 1987.

103. Hermann Staudinger, "Einfluss des Nationalsozialismus auf die Entwicklung der makromolekularen Chemie," May 22, 1945, HSP, B II 68.

104. Interview with Magda Staudinger by the author, 1989.

105. Friedrich Metz, May 31, 1945, HSP, D II 17.87. This memo was attached to Staudinger's answer to a questionnaire, entitled "Military Government of Germany," May 30, 1945, HSP, D II 17.86.

106. Der preußische Minister für Wirtschaft und Arbeit to Hans Staudinger, August 30, 1933; a record of custody, Staatspolizei, Polizeibehörde, Hamburg, July 22, 1933, HWSP, Box 1, File 5; Rutkoff and Scott, 1981, pp. 149–150. His biographical material is in HWSP, Box 1, Folder 2 which includes the transcript of his interview with Riemer, 1978. On Else, see Judith Sallet, "Special Ceremony Honors Dr. Else Staudinger," news release, The American Council for Emigrés in the Professions, May 22, 1964, HWSP, Box 1, Folder 19; Toni Stolper, "Address to a Memorial Gathering on May 5, 1966 at Carnegie Peace Foundation, New York," May 5, 1966, HWSP, Box 1, Folder 22; Toni Stolper, "The American Committee for Emigré Scholars, Writers and Artists — Now, The American Council for Emigrés in the Professions," April 29, 1964, HWSP, Box 1, Folder 19.

107. Staudinger, "Einfluss des Nationalsozialismus auf die Entwicklung der makromolekularen Chemie" (n. 103 above); see also "Military Government of Germany" (n. 105 above).

108. Staudinger, "Einfluss des Nationalsozialismus auf die Entwicklung der makromolekularen Chemie" (n. 103 above).

109. Sakurada, 1969, p. 51.

110. Hermann Staudinger to Hans Wislicenus, June 22, 1934, HSP, D II 17.13c. In this letter, Staudinger also criticized Hess's student Sakurada.

111. Hahn, 1975, p. 54; Engel, 1984, p. 228.

112. Wolfgang Ostwald to Martin Fischer, December 28, 1933, MHFC; quoted in Elliott, 1993.

113. Wolfgang Ostwald to Martin Fischer, November 28, 1921, MHFC; quoted in Elliott, 1993.

114. Max Planck to F. M. Trautz, April 10, 1935; Wolfgang Ostwald, "Lebenslauf," ca. 1934, AGMPG, A 1, alt II 32a, Band IV, 2118, "KWG Generalverwaltung KWI für Faserstoffchemie Hauptakten, 10.4.1934–10.4.1935."

115. Quotation from Staudinger, "Einfluss des Nationalsozialismus auf die Entwicklung der makromolekularen Chemie" (n. 103 above). Staudinger's *Organische Kolloidchemie* (1940), as its title indicated, intended to show that the study of colloids belonged to the territory of organic chemistry, and not to traditional colloid chemistry. Wolfgang Ostwald, 1940; Liesegang, 1940. Cf. Liesegang, 1933. See also Staudinger, 1970, p. 102.

116. Ibid. For correspondence between Staudinger and Ostwald, see HSP, D II 13.1–13.23. See also Priesner, 1980, pp. 219–62.

117. Staudinger, "Einfluss des Nationalsozialismus auf die Entwicklung der makromolekularen Chemie" (n. 103 above).

118. Staudinger, 1940b.

119. See M. Staudinger, 1982.

120. Takei, 1948, p. 60.

121. Wolfgang Ostwald died of liver cancer near Dresden. On his death see, Steinkopff, Steinkopff, and Lottermoser, 1943, front pages; Takei, 1947, 1948;

Hermann Stadlinger (editor of *Chemiker-Zeitung*) to Hermann Staudinger, December 20, 1943, HSP, D II 17.65.

122. E.g., Hermann Staudinger to Reichsministerium für Wissenschaft, Erziehung und Volksbildung, October 26, 1937, HSP D II 17.26; Rektorat der Universität Freiburg to Reichsministerium für Wissenschaft, Erziehung und Volksbildung, January 26, 1938, HSP, D II 17.28; Reichsministerium für Wissenschaft, Erziehung und Volksbildung to Rektrat der Universität Freiburg, May 12, 1938, HSP, D II 17.36 and November 2, 1938, HSP, D II 17.38.

123. Quotation is from Hermann Staudinger to Bauer (Regierngsrat, Ministerium des Kultus und Unterrichts), December 4, 1938, HSP, D II 17.39. See also Badisches Ministerium des Kultus und Unterrichts to Reichsministerium für Wissenschaft, Erziehung und Volksbildung, January 26, 1940, HSP, D II 17.50; Reichsministerium für Wissenschaft, Erziehung und Volksbildung to Badisches Ministerium des Kultus und Unterrichts to Reichsministerium, April 27, 1940, HSP, D II 17.51.

124. Staudinger, "Einfluss des Nationalsozialismus auf die Entwicklung der makromolekularen Chemie" (n. 103 above).

125. E.g., Staudinger, 1934c, 1942.

126. For Aryan physics, see Beyerchen, 1977.

127. Wolfgang Ostwald to M. Fischer, December 28, 1933 (n. 112 above); quoted in Elliott, 1993.

128. Wolfgang Ostwald to Martin Fischer, July 1, 1935, MHFC; quoted in Elliott, 1993.

129. Macrakis, 1993, p. 89. On the Four Year Plan, see ibid., pp. 102–5.

130. E.g., Engel, 1984, p. 411.

131. Takei, 1948, p. 63.

132. The number of publications (those authored or co-authored by Staudinger) drawn from bibliography in Staudinger, 1970, pp. 243–79. For his doctoral students, see Appendix Table 1.

133. Hermann Staudinger, "Die Gründung eines Kaiser-Wilhelm-Instituts für Holz- und Cellulose-Forschung," March 4, 1937; Hermann Staudinger to Max Planck, March 4, 1937, AGMPG, 1 Abt. Rep. 0001A, 967, "KWG zur Gründung von Forschungsinstituten und zur Förderung anderer wissenschaftlicher Unternehmungen, Errichtung eines KWI für Holz- und Zelluloseforschung, Korrespondenz mit H. Staudinger, 4.3.1937–10.1.1938."

134. Peter Debye to Max Planck, May 26, 1937, AGMPG, 1 Abt. Rep.0001A, 967.

135. Hermann Staudinger, "Zur Errichtung eines Kaiser Wilhelm-Institutes für makromolekulare Chemie in Freiburg/Brsg.," September 15, 1937, HSP, D II 17.25; Hermann Staudinger to Carl Bosch, October 27, 1937, AGMPG, 1 Abt. Rep.0001A, 967.

136. Kurt Hess, "Betrifft Neugründung eines Institutes für Holz- und Zellstoff-Forschung," October 14, 1937, AGMPG, 1 Abt. Rep.0001A, 1146, "KWG, Generalverwaltung/KWI für Chemie Hauptakten, Korrespondenz mit Prof. K. Hess, 15.1.1937–16.11.1944."

137. Ernst Telschow to Kurt Hess, November 25, 1937, AGMPG, 1 Abt. Rep.0001A, 1146; Ernst Telschow to Hermann Staudinger, January 10, 1938, AGMPG, 1 Abt. Rep.0001A, 967; Reichsministerium für Wissenschaft, Erziehung und Volksbilding to Präsidenten der industrie- und Handelskammer in Freiburg/i.B., January 14, 1938, HSP, D II 17.27; Reichsministerium für Wissenschaft, Erziehung und Volksbildung to Badisches Ministerium des Kultus und Unter-

richts, January 14, 1938, ibid. Whatever the rhetoric, the reason for the Kaiser Wilhelm Society's refusal of Staudinger's second proposal contradicted that of his first proposal (i.e., the Society's unwillingness to consider mere expansion of the university institute).

138. This institute had nothing to do with the Kaiser Wilhelm Society. However, Hess officially used various names for his institute (such as Forschungsinstitut Hess im Hause Kaiser Wilhelm-Institut für Chemie, Institut für Kautschukforschung, and Institut für Rohstofforschung). The Kaiser Wilhelm Society warned him about the misuse, but Hess reacted bitterly. See W. Seelmann-Eggebert to Walter Forstmann, June 14, 1944; Franz Arndt to Kurt Hess, July 15, 1944; Kurt Hess to Franz Arndt, July 19, 1944; Franz Arndt to Kurt Hess, August 1, 1944; Otto Hahn to Franz Arndt, August 18, 1944; Franz Arndt to Kurt Hess, August 21, 1944, AGMPG, 1 Abt. Rep.0001A, 1146. On Hess's career, see "Professor Dr. Kurt Hess, 70 Jahre Alt," collected reprints of articles celebrating Kurt Hess's 70th birthday by his former students, 1958 (in possession of Christiane Hess, daughter of Kurt Hess); Ronge, 1972.

139. Magda Staudinger to the author, September 24, 1987.

140. Konrad worked on polymers with Staudinger's assistant Rudolf Signer in the 1920s at Freiburg and developed Buna synthetic rubber in the 1930s at I. G. Farben. See A. Tanaka, 1992–1993, 19, pp. 256–57.

141. Hermann Staudinger, "Bericht über den Einfluss des Nationalsozialismus auf die Unterrichtstätigkeit des chemischen Institutes, Freiburg," July 6, 1945, HSP, B II 43.

142. Hermann Staudinger, "Anlage 1, zu Punkt D, I des Fragebogens ['Military Government of Germany']," May 29, 1945, HSP, D II 17.86; Hermann Staudinger to E. Tscheulin, November 15, 1939, HSP, D II 17.47; Hermann Staudinger, "Satzung des Forschungsinstituts für makromolekulare Chemie im Verbindung mit dem chemischen Universitätlaboratorium Freiburg i. Breisgau," November 1939, ibid.; Minister des Kultus und Unterrichts to Hermann Staudinger, January 11, 1940, HSP, D II 17.48; E. Tscheulin, "Satzung der Förderungsgemeinschaft der Forschungsabteilung für makromolekulare Chemie in Verbindung am chemischen Universitätslaboratorium Freiburg/Breisgau," January 16, 1940, HSP, D II 17.49; Hermann Staudinger, "Bericht über die Forschungsarbeiten auf dem Gebiet der makromolekularen Chemie in der Forschungsabteilung des chemischen Universitätlaboratorium Freiburg i/Br.," September 3, 1940, HSP, D II 17.52 and September 17, 1940, HSP, D II 17.53; "Bericht für die erste Verwaltungsratsitzung der Förderungsgemeinschaft der Forschungsabteilung für makromolekulare Chemie am 17 September 1940 in Freiburg/Br.," September 23, 1940, HSP, D II 17.54.

143. Kurt Hess to Max Planck, May 26, 1937, AGMPG, 1 Abt. Rep.0001A, 1146.

144. Staudinger and Kränzlein, 1972. Other correspondence between them is stored in FH. On Kränzlein, see Simon, 1980. On Staudinger's relationship with Kern, see Iwabuchi, 1985, especially p. 475. Magda Staudinger to the author, May 7, 1987; Staudinger, "Bericht über den Einfluss des Nationalsozialismus auf die Unterrichtstätigkeit des chemischen Institutes, Freiburg" (n. 141 above).

145. Georg Kränzlein to Hermann Staudinger, March 21, 1941, HSP, E II 14.

146. Staudinger, 1936b, p. 1185.

147. Georg Kränzlein to Hermann Staudinger, June 3, 1936, HSP, D II 15.12.

148. Although he was not known to have made anti-Semitic remarks against Meyer, Staudinger continued to feel antipathetic to him. When the second edition of Meyer's book *Hochpolymere Chemie* appeared in 1940, he wrote, "According

to Meyer, of the 13 noteworthy recent papers cited, only two are from German universities. So, this book can be quoted by foreign countries at once as evidence for the decline of German science." HSP, C 58. Kern stayed with I. G. Farben until the end of the war and remained loyal to Staudinger throughout. He reported to Staudinger about his suspicion that Meyer continued to steal valuable data from the Ludwigshafen laboratory of I. G. Farben. Werner Kern to Hermann Staudinger, March 21, 1941, HSP, E II 5.

Chapter 5: Restoration of the Physicalist Approach

1. Flory, 1974, p 25. See also idem., 1973. For a discussion of disciplinary tensions between physics and chemistry, see Nye, 1993.

2. Cofman, 1927, p. 1.

3. As a group member in DuPont's fundamental research program, Cofman joined the company in early 1927 but left their employ in 1929. See L. G. Wise and N. G. Fisher, "History, Activities, and Accomplishments of Fundamental Research in the Chemical Department of the Du Pont Company, 1926–1939 Inclusive," August 14, 1940, HML, Acc. 1784.

4. Flory, 1974, pp. 24–25.

5. Staudinger, a discussion remark, in *Trans. Faraday Soc.* 29 (1933), pt. 2, pp. 43–44, on p. 44. See also Staudinger, 1933, pp. 26ff.; discussion in *Trans. Faraday Soc.* 32 (1936), Pt. 1, pp. 311–13.

6. Go, 1957, p. 202. Go attended this meeting.

7. Mark, discussion in *Trans. Faraday Soc.* 32 (1936), Pt. 1, p. 312. See also Mark, discussion in ibid., 29 (1933) Pt. 1, pp. 40–43. In 1934 Eugen Guth and Mark developed the random-chain model of large molecules and argued the entropic origin of the rubber elastic force. Guth and Mark, 1934. See also Mark, 1940, pp. 288–93; 1981, pp. 531ff.; Pritykin, 1981.

8. W. Kuhn, 1934.

9. This was what Staudinger told Mark and Kuhn at the 1934 Madrid meeting. Go, 1957, p. 202.

10. To put the viscosity formula in a general way:

$$\eta_{sp}/C = \text{constant} \times M^a$$

where η_{sp} = specific viscosity, C = concentration, M = (weight average) molecular weight, and a was a characteristic value for a given type of macromolecule. According to Staudinger, a should always be 1. However, these authors suggested that if the exponent a was numerically large, then the molecular chain was less folded; if it was small, the chain was highly folded and did not show strong influence on viscosity. They assumed different values of a. Kuhn thought that it should be between 0.6 and 0.9; Mark between 2/3 and 3/2; Houwink 0.6; and Sakurada between 0 and 2.0. W. Kuhn, 1934; Mark, 1938b, p. 103; Houwink, 1940; Sakurada, 1940 and 1972. On the other hand, Maurice L. Huggins, an American physical chemist at Eastman Kodak, suggested that Kuhn's concept of coiled chain molecules be in harmony with Staudinger's original formula (a=1). Huggins, 1938. It is now accepted that a lies between 0.5 and 0.8. See Flory, 1949.

11. Morawetz, 1985, p. 108.

12. Flory, 1936, p. 1877.

13. Flory, 1953a, pp. 23–24.

14. Stockmayer and Zimm, 1984.

15. Meyer, Susich, and Valkó, 1932; Guth and Mark, 1934; W. Kuhn, 1936; Guth and James, 1941; James and Guth, 1943.

16. Mark, 1981, p. 532.

17. Guth, 1979, p. 2. Guth's claim stemmed largely from a sense of priority to Kuhn's work in the study of rubber elasticity, as the former complained that the latter's comparable paper, published in the same year, was misunderstood and overrated by the later chemical literature. Cf. Guth and Mark, 1934; W. Kuhn, 1934. A. J. Staverman (1975, p. 46) in Leiden evaluated the 1934 Guth-Mark paper as "one of the great strides forward in the history of science, comparable to the conception of the kinetic theory of gases," since it introduced for the first time the dynamic conception of the macromolecule. In 1937 Guth moved to the United States where he taught at Notre Dame University, Indiana, and continued to advance the theory of rubber elasticity.

18. Interview with Mark by Bohning and Sturchio, 1986, transcript, pp. 32–34, quotation from p. 33.

19. See Chapter 2, n. 99. On Freudenberg, see Meinel, 1990a.

20. W. Kuhn, 1936.

21. On Kuhn, see Hermann, 1973; H. Kuhn, 1964, 1984; Morris, 1986, pp. 67–69.

22. Günther V. Schulz, "Ein erfülltes Forscherleben: Lebens- und Arbeitsinnerungen eines deutschen Wissenschaftlers in den Wirren des zwanzigsten Jahrhunderts," unpublished autobiography, October 1995, p. 36 (in possession of Günther V. Schulz; courtesy of Hiroshi Inagaki).

23. Ibid., pp. 37–38; idem., 1935.

24. E.g., Schulz, 1936a, b.

25. See Chayut, 1993, 1994.

26. E.g., A. W. Kenney and E. O. Kraemer reviewed in detail Carothers's manuscript, "Artificial Fibers from Synthetic Linear Condensation Superpolymers," from their physico-chemical standpoint. In the published version, Carothers acknowledged their criticism and advice. See A. W. Kenney and E. O. Kraemer to Julian W. Hill, August 26, 1931, HML, Acc. 903, Rutledge Scrapbook, vol. 598; Carothers and Hill, 1932e, p. 1587.

27. Roger Adams to Frederic Woodward, November 8, 1934, PP, 1925–1945, Box 101, Folder 3.

28. For Flory, see Stockmayer, 1974; Mark, 1976a; Ridgway, 1977; Flory, 1978; Scheraga, 1980; Seltzer, 1985: 27–30; Morris, 1986; Chayut, 1993; Kovac, 1993. Flory's scientific papers were brought together in Flory, 1985b.

29. Flory, 1934.

30. Interview with Flory by Overberger, 1982; "Flory on Carothers and Du Pont," 1983, p. 10.

31. Flory, 1985a.

32. Carothers, 1936, p. 49; Flory, 1936.

33. Flory, 1939.

34. Flory, 1937.

35. Flory, 1941a–c; 1942a,b; 1949.

36. Stockmayer, 1974, p. 724.

37. On Debye see Baker, 1977.

38. On the synthetic rubber program, see Morris, 1989.

39. See Stockmayer and Zimm, 1984, pp. 11–13.

40. Gehman and Field, 1937.

41. Debye, 1944.
42. See Stockmayer and Zimm, 1984.
43. Debye and Bueche, 1948; Flory, 1949.

Chapter 6: The Legacy of Staudinger and Carothers

1. Carothers, 1932, p. 4471.
2. Staudinger, 1953, p. 397.
3. Natta, 1963, p. 261.
4. Anderson, Bartron, and Collette, 1980, p. 807.
5. General discussions on and case studies of research schools include Morrell, 1972; Hannaway, 1976; Geison, 1978, 1981; Fruton, 1990; essays in Geison and Holmes, ed., 1993. Here I have borrowed, with a slight change, Geison's definition (1981, p. 23) of the research school.
6. Interview with Flory by Overberger, 1982; "Flory on Carothers and Du Pont," 1983, p. 10.
7. Ibid.
8. Flory, 1978, p. 69.
9. Flory, 1953a.
10. See Carraher, 1982, p. 177.
11. Kraemer, 1943a, pp. 73–74. See also Kraemer, 1943b; Burk and Grummit, eds., 1943, p. ix.
12. Marvel taught at Illinois for over forty years until he moved to the University of Arizona in 1961. Beginning in 1925 he trained 176 Ph.D.s and 145 postdoctoral students—the majority of them in the field of polymer chemistry. He served as a consultant to DuPont for sixty years until his death in 1988. Forty-six of his Ph.D.s joined DuPont. He published over 500 papers on organic and polymer chemistry, of which about 350 were published while he was at Illinois, and the rest at Arizona. Of his 500 papers, 400 were on polymer chemistry, of which 250 were published during his Illinois years. See "Publications of C. S. Marvel," CSMP, Box 2; Roger Adams, "Carl S. Marvel," paper read at Perkin Award Dinner, February 11, 1965, RAP, Box 54; Mark, 1984a, pp. 1579–1606. For Marvel, see Mulvaney, 1976; Marvel, 1978, 1981; "Carl Shipp Marvel—An Autobiography, 1894–," unpublished paper prepared in 1983 for *Kobunshi*, CSMP, Box 1; Ohtsu, 1983; interview with Marvel by Gortler and Price, 1983; Anderson and Lipscomb, 1984; Morris, 1986, pp. 61–63; Kauffman, 1990a; Tarbell and Tarbell, 1991.
13. Mark, 1976b, p. 184.
14. Hunt, 1958, p. 68.
15. Mark, 1993, p. 85; also Hunt, 1958, p. 69.
16. Mark, 1993, p. 87.
17. The paper company's other customers for pulp included Celanese, Eastman Kodak, Beaunit, and Aviso. Ibid., p. 88.
18. Mark, 1993, pp. 88–89; 1976b, p. 532.
19. Hounshell and Smith, 1988, p. 296.
20. Interview with Mark by the author, 1982; interview with Mark by Bohning and Sturchio, 1986, transcript, pp. 23, 36.
21. On Zimmerli's concern with Mark, see Hamilton Bradshaw, Memorandum for File, June 2, 1933, HML, Acc. 1784; with Bergmann, see Elmer K. Bolton, "Jewish Chemists in Germany," to J. E. Crane, September 29, 1933, HML, "Crane Files," Acc. 500, Series II, Part II, Box 1035; and with Meyer, see Jasper E. Crane to

Elmer K. Bolton, May 25, 1938, ibid.; Elmer K. Bolton to J. E. Crane, June 3, 1938, ibid. These letters reveal that Elmer K. Bolton was loath to offer positions to Bergmann and Meyer as early as 1933 when "the agitation against men of Jewish origin in Germany is resulting in forced resignation in many of the universities." F. W. Pickard, then a vice president of DuPont, also had shown interest in Michael Polanyi and Herbert Freundlich, asking Bolton to examine "an opportunity to secure the services of one or two highly competent and experienced men for specialized work." F. W. Pickard to E. K. Bolton, April 25, 1933, HML, Acc. 1662, Box 17, Folder c-22d.

22. Mark and Whitby, 1940, p. ix.

23. Interview with Mark by Bohning and Sturchio, 1986, transcript, pp. 81–82.

24. Vol. 2: Mark, 1940; Vol. 3: Mark and Raff, 1941; and Vol. 4: Meyer, 1942.

25. The X-ray physicist Ott was a former critic of Staudinger's while they were colleagues at the ETH in Zürich in the 1920s. He moved to the United States in 1927. Ott, who was also on the board of Rutgers University, had wanted to offer Mark a professorship at Rutgers, before the latter decided to come to Brooklyn. See Twiss, ed., 1945, p. 74; interview with Mark by Sturchio and Bohning, 1986, transcript, p. 48.

26. Mark, 1993, p. 115.

27. Carraher, 1981, p. 131; 1982.

28. See Morris, 1989, pp. 83–85; 1994, pp. 54–69.

29. E.g., Morton, 1982, pp. 225–38.

30. Morris, 1989, p. 132.

31. During the war, Mark's group studied the permeability and impact strength of thin films, emulsion polymerization, and phase transitions in polymer blends, sponsored by the Office of Scientific Research and Development. See Mark, 1993, p. 95.

32. Shortly before the establishment of the *Journal of Polymer Science,* Mark had launched the publication of *Polymer Bulletin,* but this small journal contained mostly work carried out at Brooklyn and appeared for only two years. The first issue of the new journal was titled *Journal of Polymer Research;* it renamed *Journal of Polymer Science* after the second issue. In 1961 the Interscience Publishing Company was annexed to John Wiley & Sons in New York. On the creation of the *Journal of Polymer Science,* see Mark, 1993, p. 127; 1988b; interview with Mark by Sturchio and Bohning, 1986, transcript, pp. 56, 74.

33. Ibid., pp. 82–83. See also Proskauer, 1988.

34. The first three chairmanships of the Division were held by DuPont-related chemists: Carl S. Marvel (1951); William E. Hanford, Marvel's student and a group leader of nylon development at DuPont, 1936–1942 (1952); and Paul J. Flory (1953). Raymond M. Fuoss of Yale University served as its fourth chairman (1954), and Herman Mark the fifth (1955). Ulrich, 1978, especially p. 3; Skolnik and Reese, eds., 1976, pp. 367–73.

35. Interview with Mark by Sturchio and Bohning, 1986, transcript, pp. 75–77.

36. See DuBois, 1972; Hochheiser, 1986, Chapters 5, 7; Boyer, 1986a,b.

37. Carraher, 1982, pp. 181–82. On polymer education at the University of Akron, see Morton, 1989.

38. Ogata, 1982, p. 864.

39. From the early 1920s to the mid-1980s, Mark published a total of over six hundred papers. Herman F. Mark, "Publications of H. F. Mark," CHF.

40. Marvel visited Germany in 1945, and Mark in 1946. See Mark, 1981, p. 533; Marvel, "An Autobiography — 1894–" (n. 12 above).

41. Hermann Staudinger to Otto Hahn, September 27, 1949, HSP, D II 17.150a. Hahn's response was negative. Otto Hahn to Hermann Staudinger, October 4, 1949, HSP, D II 17.150a.

42. M. Staudinger, 1982; Höcker, 1994. After 1952, the journal was published by the Dr. A. Hüthig Press in Heidelberg. In 1994, the journal was renamed *Macromolecular Chemistry and Physics*.

43. Staudinger, 1970, pp. 237–38.

44. Guth, 1979, p. 2.

45. Staudinger, 1947. See also Staudinger, 1953; 1970, pp. 237–42; Edsall, 1962; Olby, 1970, 1979; Fruton, 1976.

46. Crawford, Heilbron, and Ullrich, 1987, pp. 258, 260, 270, 274, 329.

47. "Better Late Than Never," 1953, p. 87.

48. Nodzu suspected that Staudinger's polemical nature delayed his Nobel Prize. Nodzu, 1954, p. 376.

49. Quoted in Yarsley, 1967, p. 268. Magda wrote a long report on the 1953 Nobel Prize Ceremony in Stockholm: Magda Staudinger, "Nobelpreis für Chemie 1953: Ein Bericht," HWSP, Box 1, Folder 32.

50. Staudinger, 1970, pp. 243–79.

51. Batzer served as part-time professor at the Technische Hochschule in Stuttgart, while working for the Ciba-Geigy Company, Basel. Staudinger's son Hansjürgen received a doctorate under his father in 1940 and later became a professor at the University of Giessen. His field, however, turned to medical science, and he rarely taught macromolecular chemistry.

52. Iwabuchi, 1985.

53. Fredga, 1953, pp. 395–96.

54. For Perlon, See Achilladelis, 1970.

55. Whinfield, 1946, p. 931. This superb polyester fiber was one which Carothers's group had simply overlooked in their systematic search for fibers during the 1930s. As Julian Hill explained this oversight to me with a smile: "We were not smart enough." Interview with Hill by the author, 1982.

56. For a useful survey on the history of Japanese polymer chemistry and industry, see Chemical Society of Japan, ed., 1978, pp. 853–921.

57. See Wolfgang Ostwald, "Lebenslauf," ca. 1934, AGMPG, A 1, alt II 32a, Band IV, 2118, KWG Generalverwaltung KWI für Faserstoffchemie Hauptakten, 10.4.1934–10.4.1935; R. Tanaka, 1930; Takei, 1948, p. 63; Wadano, 1952; Sakurada, 1969, p. 39. Besides students of Hess and Ostwald, there were several Japanese chemists who studied for a shorter or longer period with Reginald O. Herzog at Berlin, Herbert Freundlich at Berlin, Carl Freudenberg at Heidelberg, and Kurt Meyer at Geneva. See Go, 1957, pp. 527–30, 656–59; Tamamushi, 1978.

58. See Nodzu, 1954, 1957; Ochiai, 1953, 1954; Yoshio Kobayashi, "Translator's Preface' (in Japanese), in Staudinger, 1966, pp. iii–vi, on p. v; Goto and Maruyama, 1983. Nodzu did some work on rubber at Kyoto University.

59. According to Sakurada, he began to sense the decline of the aggregate theory at the 1930 Frankfurt meeting of the German Colloid Society. After his return to Japan, he introduced Hess's work on cellulose favorably for a brief period, but soon shifted to the side of the macromolecular view. Sakurada, 1931; 1969, pp. 36–52; 1983. On Sakurada, see also Okamura, 1995. My information on Sakurada is also based on my interviews with Okamura, 1994, and with Yutaka Sakurada (Sakurada's son), 1994.

60. See Sakurada, 1969, pp. 75–83; Iwakura, 1995, p. 53.

61. In a lecture given at the 41st annual meeting of the Society of Chemical Industry, held in April 1938, Sakurada made a remark that underestimated the possibility of synthetic fibers. Sakurada, 1938, p. 480.

62. "What Does Nylon Teach Us?" 1939.

63. Various episodes on nylon's impact on Japan were collected in *Nairon*, 1939, pp. 378–82.

64. Sakurada, 1969, p. 92; T. Hoshino, 1952; K. Hoshino, 1977. Shortly before Carothers's death, Shigeru Sakai at the New York office of Mitsui Bussan had learned about DuPont's new fiber from a newspaper article. He then telephoned Carothers asking for a sample, a request which the latter refused. See Kohei Hoshino's remarks in Sakurada et al., 1965, p. 1046.

65. The lectures and discussions at the symposium were published in *Nairon*, 1939. Speakers' references to Carothers and DuPont are on pp. 351–77.

66. The lectures were published in Japanese Union of Fiber Research, ed., 1940; quotation on p. 1.

67. Interview with Go by the author, 1985.

68. Sakurada, 1969, pp. 115–17.

69. Ibid., pp. 108–12; Iwakura, 1974, pp. 6–13; 1995, p. 53; Sakurada et al., 1965.

70. As early as 1942 the Association launched the periodical *Gosei Sen'i Kenkyu* (Synthetic Fiber Studies), two volumes of which were published until it was merged into *Kobunshi Kagaku* in 1944. See Iwakura, 1995. *Kobunshi Kagaku* was renamed *Kobunshi Ronbunshu* (Polymer Papers) in 1974.

71. As of 1997 the Society had over 13,000 members and published four journals: *Polymer Journal* (English), *Kobunshi Ronbunshu, Kaigai Kobunshi Kenkyu*, and *Kobunshi*.

72. On Ziegler, see McMillan, 1979; Morris, 1986, pp. 78–80; Meinel, 1990b.

73. Natta (Milan Polytechnic, 1939–79), along with his student Piero Pino (Milan Polytechnic, 1946–55, University of Pisa, 1955–68, and the ETH Zürich, 1968–89) were instrumental in the birth of Italian polymer chemistry. On Natta, see McMillan, 1979; Carra, Parisi, Pasquon, and Pino, eds., 1982; Morris, 1986, pp. 81–83; Cerruti, 1990.

74. Hermann Staudinger, "Bericht über die Japanreise im Frühjahr 1957," March 23, 1957, HSP, B I 139; "Die makromolekulare Chemie, ein neues Gebiet der organische Chemie," lecture at the University of Tokyo, April 5, 1957, HSP, B I 138; "Über die Konstitution der Cellulose," lecture in Osaka, April 12, 1957, HSP, B I 140; *Kobunshi* 6 (1957): 77–89. On his meeting with the Emperor, see Staudinger, 1970, p. 104.

75. Hunt, 1958 p. 62; interview with Mark by the author, 1982.

76. Mark and Bruson, 1956, p. 388. Mark wrote a similar preface to the English translation of Staudinger's *Arbeitserinnerungen*: Staudinger, 1970, pp. vii–ix.

77. Staudinger, 1958.

78. Quoted in Mark, 1993, p. 122.

79. Ringsdorf started his doctoral research under Staudinger just before the latter's retirement. He stayed at Brooklyn from 1960 to 1962, before he taught at Kern's Institute for Organic Chemistry of the University of Mainz.

80. Hermann Staudinger, "Polyesterarbeiten und Gespräche mit Carothers über die makromolekulare Chemie," lecture given at DuPont, Wilmington, September 23, 1958, HSP, B I 144.

81. Staudinger was a great admirer of Carothers. Magda later related, "My husband did like Carothers very much." Magda Staudinger to the author, May 7, 1987.

82. Staudinger, 1961.

83. Ochiai, 1966, p. 409.

84. Hans Staudinger, New Year's letter to his friends, February 8, 1966, HWSP, Box 2, Folder 7.

Bibliography

Published Sources, Dissertations, and Conference Presentations

Abderhalden, Emil. 1924. "Das Eiweiss als seine Zusammenfassung assoziierter, Anhydride enthaltenden Elementarkomplexe." *Naturwiss.* 12: 716–720.

Achilladelis, Basil G. 1970. "A Study in Technological History: Part I. The Manufacture of 'Perlon' (Nylon 6) and Caprolactam by IG Farbenindustrie." *Chem. Ind.* (December): 1549–1554.

Adams, Roger. 1939. "Biographical Memoir of Wallace Hume Carothers, 1896–1937." *Biog. Mem. NAS* 20: 293–309.

———. 1954. "Universities and Industry in Science." *Ind. Eng. Chem.* 46: 506–510.

Albrecht, Helmuth, and Armin Hermann. 1990. "Die Kaiser-Wilhelm-Gesellschaft im Dritten Reich (1933–1945)." In Vierhaus and Brocke, eds., 1990. Pp. 354–406.

Alfrey, Turner. 1986. *Selected Papers of Turner Alfrey.* Edited by Raymond F. Boyer and Herman F. Mark. New York and Basel: Marcel Dekker.

Allen, Garland E. 1975. *Life Science in the Twentieth Century.* New York, London, Sydney, and Toronto: John Wiley and Sons.

American Chemical Society, ed. 1951. *Chemistry . . . Key to Better Living: Diamond Jubilee Volume.* Washington, D.C.: American Chemical Society.

Anderson, B. C., L. R. Bartron, and J. W. Collette. 1980. "Trends in Polymer Development." *Science* 208: 807–812.

Anderson, Burton C., and Robert D. Lipscomb. 1984. "Carl Shipp Marvel: 'Speed' at 90." *Macromolecules* 17: 1641–1643.

Anschütz, Richard, 1929. *August Kekulé.* Vol. 2: *Abhandlungen, Berichte, Kritiken, Artikel, Reden.* Berlin: Verlag Chemie.

"Ashdown, Prof. Avery A(llen)." 1965. *American Men of Science,* 11th edition. New York: Bowker.

Baekeland, Leo H. 1909a. "Bakelite, A New Composition of Matter: Its Synthesis, Constitution, and Uses." *Scientific American* Supplement 1768: 322–323.

———. 1909b. "The Synthesis, Constitution, and Uses of Bakelite." *J. Ind. Eng. Chem.* 1: 149–175.

———. 1909c. "The Use of Bakelite for Electrical and Electrochemical Purposes." *Trans. Amer. Electrochem. Soc.* 15: 593–612.

———. 1913. "The Chemical Constitution of Resinous Phenolic Condensation Products." *J. Ind. Eng. Chem.* 5: 506–511.

———. 1914a. "The Invention of Celluloid." *J. Ind. Eng. Chem.* 6: 90–95.

———. 1914b. "Synthetic Resins." *J. Ind. Eng. Chem.* 6: 167–173.

Baker, William O. 1977. "Peter Joseph Wilhelm Debye." In Milligan, ed., 1977. Pp. 154–199.

Bancroft, Wilder D. 1908. "The Future in Chemistry." *Science* 27: 978–980.

———. 1921. *Applied Colloid Chemistry: General Theory.* New York: McGraw-Hill.

———. 1926. "Physical Chemistry." In *A Half-Century of Chemistry in America, 1876–1926,* edited by Charles A. Browne, 89–110. Philadelphia: American Chemical Society, 1926.

Batzer, Hans, and Helmut Ringsdorf. 1987. "Staudinger, Hermann." *Badische Biographien,* Neue Folge, vol. 2.

Beer, John J. 1959. *The Emergence of the German Dye Industry.* Urbana: University of Illinois Press.

Benfey, O. Theodor. 1964. *From Vital Force to Structural Formulas.* Boston: Houghton Mifflin. Reprint, Philadelphia: Chemical Heritage Foundation, 1992.

Bergmann, Max. 1925. "Über den hochmolekularen Zustand der Proteine und die Synthese proteinänlicher Piperazin-Abkömmlinge." *Naturwiss.* 13: 1045–1050.

———. 1926. "Allgemeine Strukturchemie der komplexen Kohlenhydrate und der Proteine." *Ber.* 59: 2973–2981.

———. 1930. "Emil Fischer." In *Das Buch der grossen Chemiker,* vol. 2, edited by G. Bugge 1157–1201. Berlin: Verlag Chemie, 1930.

Berthelot, Marcellin. 1866a. "Sur les caractères de la benzine et du styrolène, comparés avec ceux des autres d'hydrogène." *Bull. Soc. chim.* 6: 289–296.

———. 1866b. "Sur la présence du styrolène dans les huiles de goudron de huille." *Bull. Soc. chim.* 6: 296–298.

Berzelius, Jöns Jacob. 1832. *Jahres-Bericht Über die Fortschritte der physischen Wissenschaften von Jacob Berzelius.* Vol. 11. Translated by F. Wöhler. Tubingen: Heinrich Laupp.

———. 1833. *Jahres-Bericht Über die Fortschritte der physischen Wissenschaften von Jacob Berzelius.* Vol. 12. Translated by F. Wöhler. Tubingen: Heinrich Laupp.

"Better Late Than Never." 1953. *Newsweek* 42 (November 16): 86–87.

Beyerchen, Alan D. 1977. *Scientists Under Hitler: Politics and Physics Community in the Third Reich.* New Haven, Conn. and London: Yale University Press.

Billmeyer, Fred W., Jr. 1957. *Textbook of Polymer Chemistry.* New York: Interscience Publishers.

Blain, Daniel. 1980. "Twentieth Century Psychiatry—Living History in the Life of Kenneth E. Appel, M.D., 1896–1979." *Transactions & Studies of the College of Physicians of Philadelphia* 2, no. 2: 144–154.

Bolton, Elmer K. 1942a. "Development of Nylon." *Ind. Eng. Chem.* 34: 53–58.

———. 1942b. "Nylon." *Chem. Eng. News* 20: 1365–1366.

Bouchardat, F. Gustave. 1875. "Sur les produits de la distillation sèche du caoutchouc." *Bull. Soc. chim.* 24: 111–14.

Boyer, Raymond F. 1986a. "The Early Days at Dow." In Alfrey, 1986. Pp. 23–26.

———. 1986b. "The Plastics Department Research Laboratory (1955–1981)." In Alfrey, 1986. Pp. 27–28.

Braceland, Francis J. 1980. "Kenneth Ellmaker Appel, 1896–1979." *American Journal of Psychiatry* 137: 501–503.

Brady, John Paul. 1980. "Memoir of Kenneth Ellmaker Appel, 1896–1979." *Transactions & Studies of the College of Physicians of Philadelphia* 2, no. 2: 155–156.

Brock, William H. 1992. *The Fontana History of Chemistry*. London: Fontana Press. Also published as *The Norton History of Chemistry*. New York: W. W. Norton, 1993.

Brocke, Bernhard vom. 1990. "Die Kaiser-Wilhelm-Gesellschaft in der Weimarer Republik: Ausbau zu einer gesamtdeutschen Forschungsorganisation (1918–1933)." In Vierhaus and Brocke, eds., 1990. Pp. 197–355.

Brønsted, J. N. and G. Hevesy. 1922. "On the Separation of the Isotopes of Mercury." *Phil. Mag.* 43: 31–49.

Brown, Horace T. and G. Harris Morris. 1888. "The Determination of the Molecular Weights of the Carbohydrates." *J. Chem. Soc.* 53: 610–621.

——. 1889. "The Determination of the Molecular Weights of the Carbohydrates. Part II." *J. Chem. Soc.* 55: 462–74.

"Bruson, Dr. Herman A(lexander)." 1965. *American Men of Science,* 11th edition. New York: Bowker.

Bugarsky, Stefan and Leo Liebermann. 1898. "Ueber das Bindungsvermögen eiweissartiger Körper für Salzsäure, Natriumhydroxd und Kochsalz." *Pflüger's Arch. Physiol.* 72: 51–74.

Burch, C. R. 1929. "Some Experiments on Vacuum Distillation." *Proc. Roy. Soc.* 123: 271–284.

Burgdorf, C. C. 1926. "Artificial Rubber During the War in Germany." *Ind. Eng. Chem.* 18: 1172–1173.

Burk, R. E., Herman F. Mark, and G. Stafford Whitby. 1940. "Introduction to 'High Polymers.'" In Carothers, 1940. Pp. v–vii.

Burk, R. E. and Oliver Grummit, eds. 1943. *The Chemistry of Large Molecules.* New York: Interscience Publishers.

Buzágh, Aladar von. 1937. *Colloid Systems: A Survey of the Phenomena of Modern Colloid Physics and Chemistry.* Translated by Otto B. Darbishire, edited by William Clayton, with foreword by Wolfgang Ostwald. London: Technical Press.

Carothers, Wallace H. 1923. "The Isosterism of Phenylisocyanate and Diazobenzeneimide." *J. Amer. Chem. Soc.* 45: 1734–1738.

——. 1924a. "The Catalytic Reduction of Aldehydes with a Platinum Catalyst, Together with a Study of the Effects of Numerous Substances on the Platinum Catalysis of the Hydrogenation of Benzaldehyde." Ph.D. dissertation, University of Illinois, WHCF.

——. 1924b. "The Double Bond." *J. Amer. Chem. Soc.* 46: 2226–2236.

——. 1929. "An Introduction to the General Theory of Condensation Polymers." *J. Amer. Chem. Soc.* 51: 2548–2559.

——. 1931. "Polymerization." *Chem. Reviews* 8: 353–426.

——. 1932. Review of *Die hochmolekulare organischen Verbindungen: Kautschuk und Cellulose* by Hermann Staudinger. *J. Amer. Chem. Soc.* 54: 4469–4471.

——. 1936. "Polymers and Polyfunctionality." *Trans. Faraday Soc.* 32, Pt. 1: 39–49.

——. 1940. *Collected Papers of Wallace Hume Carothers on High Polymeric Substances.* High Polymers, 1. Edited by Herman F. Mark and G. Stafford Whitby. New York: Interscience Publishers.

Carothers, Wallace H. and Roger Adams, 1923. "Platinum Oxide as a Catalyst in the Reduction of Organic Compounds. II. Reduction of Aldehydes. Activation of the Catalyst by the Salts of Certain Metals." *J. Amer. Chem. Soc.* 45: 1071–1086.

Carothers, Wallace H. and Roger Adams. 1924. "Platinum Oxide as a Catalyst in the Reduction of Organic Compounds. V. The Preparation of Primary Alcohols by the Catalytic Hydrogenation of Aldehydes." *J. Amer. Chem. Soc.* 46: 1675–1683.

———. 1925. "Platinum Oxide as a Catalyst in the Reduction of Organic Compounds. VII. A Study of the Effects of Numerous Substances on the Platinum Catalysis of the Reduction of Benzaldehyde." *J. Amer. Chem. Soc.* 47: 1047–1063.

Carothers, Wallace H., and G. L. Dorough. 1930. "Ethylene Succinates." *J. Amer. Chem. Soc.* 52: 711–721.

Carothers, Wallace H., Ira Williams, Arnold M. Collins, and James E. Kirby. 1931. "A New Synthetic Rubber: Chloroprene and Its Polymers." *J. Amer. Chem. Soc.* 53: 4203–4225.

Carothers, Wallace H. and Julian W. Hill. 1931. "Artificial Fibers from Synthetic Linear Condensation Superpolymers." Abstract of paper presented at the Buffalo Meeting of the American Chemical Society, September 1. HML, Acc. 903, *Rutledge Scrapbook*, vol. 598.

———. 1932a. "The Use of Molecular Evaporation as a Means for Propagating Chemical Reactions." *J. Amer. Chem. Soc.* 54: 1557–1559.

———. 1932b. "Linear Superpolyesters." *J. Amer. Chem. Soc.* 54: 1559–1566.

———. 1932c. "Polyamides and Mixed Polyester-Polyamides." *J. Amer. Chem. Soc.* 54: 1566–1569.

———. 1932d. "A Linear Superpolyanhydride and a Cyclic Dimeric Anhydride from Sebastic Acid." *J. Amer. Chem. Soc.* 54: 1569–1579.

———. 1932e. "Artificial Fibers from Synthetic Linear Condensation Superpolymers." *J. Amer. Chem. Soc.* 54: 1579–1587.

Carothers, Wallace H. and F. J. van Natta. 1933. "Polyesters from ω-Hydroxydecanoic Acid." *J. Amer. Chem. Soc.* 55: 4714–4719.

Carra, Sergio, Federico Parisi, Italo Pasquon, and Piero Pino, eds. 1982. *Giulio Natta: Present Significance of His Contribution*. Milan: Editrice di Chimica.

Carraher, Charles E., Jr. 1981. "Polymer Education and the Mark Connection." In Stahl, ed., 1981. Pp. 123–142.

———. 1982. "History of Polymer Education — USA." In Seymour, ed., 1982. Pp. 173–197.

Carroll, P. Thomas. 1982. "Academic Chemistry in America, 1876–1976: Diversification, Growth, and Change." Ph.D. dissertation, University of Pennsylvania.

Cerruti, Luigi. 1990. "Natta, Giulio." *DSB*, vol. 18.

Chayut, Michael. 1993. "New Site for Scientific Change: Paul Flory's Initiation into Polymer Chemistry." *His. Stud. Phys. Sci.* 23: 193–218.

———. 1994. "The Hybridisation of Scientific Roles and Ideas in the Context of Centers and Peripheries." *Minerva* 32: 297–308.

Chemical Society of Japan, ed. 1978. *Nihon no kagaku hyaku-nen-shi: Kagaku to kagaku-kogyo no ayumi* (A Hundred Years of Japanese Chemistry: The Development of Chemistry and the Chemical Industry). Tokyo: Tokyo Kagaku Dojin.

"Chemists Produce Synthetic 'Silk.' " 1931. *New York Times*, September 2.

Clarke, Hans T. 1945. "Max Bergmann, 1886–1944." *Science* 102: 168–170.

"Coal and Castor Oil Challenge Silkworm." 1938. *Washington (D.C.) News*, September 22.

Cochrane, Rexmond C. 1978. *The National Academy of Sciences: The First Hundred Years, 1863–1963*. Washington, D.C.: National Academy of Sciences.

Cofman, Victor. 1927. "Colloid Dynamics." *Chem. Reviews* 4: 1–49.

Cohn, Edwin J., and James B. Conant. 1926. "The Molecular Weights of Proteins in Phenol." *Proceedings of the National Academy of Sciences* 12: 433–438.

Collins, Arnold M. 1973. "The Discovery of Polychloroprene: Charles Goodyear Medal Address — 1973." *Rubber Chem. Tech.* 46: 648–652.

"Colloquium im Kaiser Wilhelm-Institut für physikalische und Elektrochemie." 1929. *Z. angew. Chem.* 42: 52–53.

Commission on Macromolecular Nomenclature, Macromolecular Division, International Union of Pure and Applied Chemistry. 1991. *Compendium of Macromolecular Nomenclature.* Cambridge, Mass.: Blackwell Scientific Publications.

———. 1996. "Glossary of Basic Terms in Polymer Science." *Pure and Applied Chemistry* 68: 2287–2311.

Conant, James B. 1928. *Organic Chemistry: A Brief Introductory Course.* New York: Macmillan.

———. 1933. *The Chemistry of Organic Compounds: A Year's Course of Organic Chemistry.* New York: Macmillan.

———. 1970. *My Several Lives: Memoirs of a Social Inventor.* New York and London: Harper and Row.

Corey, E. J. 1977. "Roger Adams." In Milligan, ed., 1977. Pp. 204–228.

Couper, Archibald Scott. 1858a. "On a New Chemical Theory." *Phil. Mag.* ser. 4, 16: 104–116.

———. 1858b. "Sur une nouvelle théorie chimique." *Compt. rend.* 46: 1157–1160.

Crawford, Elizabeth, J. L. Heilbron, and Rebecca Ullrich. 1987. *The Nobel Population, 1901–1937: A Census of the Nominators and Nominees for the Prizes in Physics and Chemistry.* Berkeley: Office for History of Science and Technology, University of California.

Dalton, John. 1836. "Observations on Certain Liquids Obtained from Caoutchouc by Distillation." *Phil. Mag.* ser. 3, 9: 479–483.

Dawson, T. R. 1952. "Rubber Literature: An Annotated Bibliography." In Schidrowitz and Dawson, eds., 1952. Pp. 372–390.

Debye, Peter. 1944. "Light Scattering in Solutions." *J. Appl. Phys.* 15: 338–342.

Debye, Peter and Arthur M. Bueche. 1948. "Intrinsic Viscosity, Diffusion, and Sedimentation Rate of Polymers in Solution." *J. Chem. Phys.* 16: 573–579.

De Milt, Clara. 1948. "Carl Weltzien and the Congress at Karlsruhe." *Chymia* 1: 153–169.

Dinsmore, R. P. 1951. "Rubber Chemistry." In American Chemical Society, ed., 1951. Pp. 118–126.

Dolby, R. G. A. 1977. "The Transmission of Two New Scientific Disciplines from Europe to North America in the Late Nineteenth Century." *Ann. Sci.* 34: 287–310.

Donnan, Frederick G. 1942. "Herbert Freundlich, 1880–1941." *J. Chem. Soc.* (1942): 646–654.

———. 1951. "Ostwald Memorial Lecture." In *Memorial Lectures Delivered Before the Chemical Society, 1933–1942,* vol. 4, 1–17. London: The Chemical Society, 1951.

———. 1953. "Some Personal Reminiscences." *Trans. Faraday Soc.* 49: 511–514.

Dostal, H. and Herman F. Mark. 1936. "The Mechanism of Polymerisation." *Trans. Faraday Soc.* 32, Pt. 1: 51–69.

"Dr. Carothers Suicide by Poison." 1937. *Philadelphia Evening Bulletin,* April 30.

"Dr. Samuel S. Pickles." 1951. *Transactions of the Institution of the Rubber Industry* 27: 147.

"Dr. W. H. Carothers, Chemist, Found Dead." 1937. *New York Times,* April 30.

Drakeley, T. J. 1952. "Institutions and Associations." In Schidrowitz and Dawson, eds., 1952. Pp. 364–371.

"Du Pont Scientist Ends Life by Poison." 1937. *Philadelphia Inquirer,* April 30.

DuBois, J. Harry. 1972. *Plastics History U.S.A.* Boston: Cahners Books.

Duncan, Robert K. 1905. *The New Knowledge: A Popular Account of the New Physics and the New Chemistry in Their Relation to the New Theory of Matter.* New York: A. S. Barnes.

――――. 1907. *The Chemistry of Commerce: A Simple Interpretation of Some New Chemistry in Its Relation to Modern Industry.* New York and London: Harper and Brothers.

Dunstan, Wyndham R. 1906. "Some Imperial Aspects of Applied Chemistry." *West Indian Bulletin* 7: 263–277.

Ede, Andrew G. 1993a. "Colloid Chemistry in North America, 1900–1935: The Neglected Dimension." Ph.D. dissertation, University of Toronto.

――――. 1993b. "When Is a Tool Not a Tool? Understanding the Role of Laboratory Equipment in the Early Colloidal Chemistry Laboratory." *Ambix* 40: 11–24.

――――. 1996. "Colloids and Quantification: The Ultracentrifuge and Its Transformation of Colloid Chemistry." *Ambix* 43: 32–45.

"Editor's Outlook." 1925. *J. Chem. Educ.* 2: 518.

"Editorial." 1969. *Journal of the Oil and Colour Association* 52: 244–245.

Edsall, John T. 1962. "Proteins as Macromolecules: An Essay on the Development of the Macromolecule Concept and Some of Its Vicissitudes." *Archives of Biochemistry and Biophysics Supplement* 1: 12–20.

Eidgenössische Technische Hochschule. 1955. *Eidgenössische Technische Hochschule, 1855–1955.* Zürich: Buchverlag der Neuen Züricher Zeitung.

Elliott, Eric. 1989a. "Neglected Dimensions: The Life of a Colloid Chemist." Paper presented at the 25th Joint Atlantic Seminar in the History of Biology at Yale University, April 1.

――――. 1989b. "The IG Farbenindustrie: Is There Science Here for the Historian of Science?" Paper read at the Symposium, "Die IG Farben und der Staat," at the XVIII International Congress of the History of Science, Hamburg, August 5.

――――. 1993. "Wolfgang Ostwald and Martin Fischer: Two Lives Through Letters, 1903–1943." Paper presented at Brown Bag Luncheon, Chemical Heritage Foundation, March 5.

――――. n.d. "The Fall and Rise of the Kaiser Wilhelm Institute for Colloid Chemistry." Paper presented to the University of Pennsylvania.

Elzen, Boelie. 1993. "The Failure of a Successful Artifact: The Svedberg Ultracentrifuge." In *Center on the Periphery: Historical Aspects of 20th-Century Swedish Physics,* edited by Svante Lindqvist, 347–377. Canton, Mass.: Science History Publications.

Engel, Michael. 1984. *Geschichte Dahlems.* Berlin: Verlag Arno Spitz.

Engler, Carl. 1897. "Zur Frage der Entstehung des Erdöls und über die Selbstpolymerisation der Kohlenwasserstoffe." *Ber.* 30: 2358–2365.

Ewald, P. P., ed. 1962. *Fifty Years of X-ray Diffraction.* Utrecht: International Union of Crystallography.

Faraday, Michael. 1826. "On Pure Caoutchouc, and the Substances by Which It Is Accompanied in the State of Sap, or Juice." *Quart. J. Sci. Arts* 21: 19–28.

Farber, Eduard. 1969. *The Evolution of Chemistry: A History of Its Ideas, Methods, and Materials.* 2nd edition. New York: Ronald Press.

Farias, Victor. 1987. *Heidegger et Nazisme: Morale et politique.* Lagrasse: Verdier.

Farrar, W. V. 1974. "Meyer, Kurt Heinrich." *DSB,* vol. 9.

Feldman, Gerald D. 1973. "A German Scientist Between Illusion and Reality: Emil Fischer, 1909–1919." In *Deutschland in der Weltpolitik des 19. und 20. Jahrhunderts,* edited by Immanuel Geiss and Bernd Jürgen Wendt, 341–362. Düsseldorf: Bertesmann Universitätsverlag, 1973.

Findlay, Alexander. 1953. "Wilder Dwight Bancroft, 1867–1953." *J. Chem. Soc.* 56: 2506–2514.

———. 1965. *A Hundred Years of Chemistry.* 3rd edition. Revised by Trevor I. Williams. London: Gerald Duckworth & Co.

Fischer, Emil. 1907a. "Proteine und Polypeptide." *Z. angew. Chem.* 20: 913–917.

———. 1907b. "Synthese von Polypeptiden, XVII." *Ber.* 40: 1754–1767.

———. 1913a. "Über das Tannin und Synthese ähnlicher Stoffe. III. Hochmolekulare Verbindungen." *Ber.* 46: 1116–1138.

———. 1913b. "Synthese von Depsiden, Flechtenstoffen und Gerbstoffen." *Ber.* 46: 3253–3289.

———. 1914. "Synthesis of Depsides, Lichen-Substances, and Tannins." Translated from E. Fischer, 1913b, by Frank R. Elder. *J. Amer. Chem. Soc.* 36: 1170–1201.

———. 1916. "Isomerie der Polypeptide." *Sitzungsber. Preuss. Akad. Wiss. Berlin.* Halbbd. 2: 990–1008.

———. 1923. *Emil Fischer gesammelte Werke: Untersuchungen über Aminosäuren, Polypeptide und Proteine II.* Edited by Max Bergmann. Berlin: Verlag von Julius Springer.

Fisher, Harry L. 1944. "The Origin and Development of Synthetic Rubber." In *Symposium on the Applications of Synthetic Rubbers,* 3–16. Philadelphia: American Society for Testing Materials, 1944.

Florkin, Marcel. 1972. *A History of Biochemistry.* Comprehensive Biochemistry, 30. Amsterdam, London, and New York: Elsevier.

"Flory on Carothers and Du Pont." 1983. *CHOC News* 1, no. 2: 9–10.

Flory, Paul J. 1934. "The Photo-Decomposition of Nitric Oxide." Ph.D. dissertation, Ohio State University. PJFP, Additions Box 1.

———. 1936. "Molecular Size Distribution in Linear Condensation Polymers." *J. Amer. Chem. Soc.* 58: 1877–1895.

———. 1937. "The Mechanism of Vinyl Polymerizations." *J. Amer. Chem. Soc.* 59: 241–270.

———. 1939. "Kinetics of Polyesterification: A Study of the Effects of Molecular Weight and Viscosity on Reaction Rate." *J. Amer. Chem. Soc.* 61: 3334–3340.

———. 1941a. "Molecular Size Distribution in Three Dimensional Polymers. I. Gelation." *J. Amer. Chem. Soc.* 63: 3083–3090.

———. 1941b. "Molecular Size Distribution in Three Dimensional Polymers. II. Trifunctional Branching Units." *J. Amer. Chem. Soc.* 63: 3091–3096.

———. 1941c. "Molecular Size Distribution in Three Dimensional Polymers. III. Tetrafunctional Branching Units." *J. Amer. Chem. Soc.* 63: 3096–3100.

———. 1942a. "Constitution of Three-Dimensional Polymers and the Theory of Gelation." *J. Phys. Chem.* 46: 132–140.

———. 1942b. "Thermodynamics of High Polymer Solutions." *J. Chem. Phys.* 10: 51–61.

———. 1949. "The Configuration of Real Polymer Chains." *J. Chem. Phys.* 17: 303–310.

———. 1953a. *Principles of Polymer Chemistry.* Ithaca, N.Y.: Cornell University Press.

———. 1953b. "Reminiscences of Dr. Carothers" (in Japanese). Translated by Kotaro Tanemura from an unpublished lecture. *Toray Jiho* (December): 4–5.

———. 1973. "Macromolecules Vis-a-Vis the Traditions of Chemistry." *J. Chem. Educ.* 50: 732–735.

———. 1974. "The Science of Molecules." *Chem. Eng. News* 52, no. 30: 23–25.

———. 1978. "P. J. Flory." In Ulrich, ed., 1978. Pp. 69–72.

————. 1982. "Innovation and Polymer Science." Paper read at the Goodyear Innovation Conference, held in Akron, Ohio, April 20. PJFP, Additions Box 6.

————. 1985a. "Concepts in Polymer Science: A Half Century in Retrospect." Paper read at 30th IUPAC International Symposium of Macromolecules. PJFP, Additions, Box 6.

————. 1985b. *Selected Works of Paul J. Flory.* 3 vols. Edited by Leo Mandelkern, James E. Mark, Ulrich W. Suter, and Do Y. Yoon. Stanford, Calif.: Stanford University Press.

Fredga, Arne. 1953. "Chemistry 1953: Presentation Speech." In Nobel Foundation, ed., 1964. Pp. 393–396.

Freeth, F. A. 1957. "Frederick George Donnan." *Biog. Mem. FRS* 3: 23–39.

Freudenberg, Karl. 1921. "Zur Kenntnis der Cellulose." *Ber.* 54: 767–772.

Freundlich, Herbert. 1926. *Colloid and Capillary Chemistry.* Translated by H. Stafford Hatfield. London: Methuen.

Frey-Wyssling, Albert F. 1964. "Frügeschichte und Ergebnisse der submikroskopischen Morphologie." *Mikroskopie* 19: 2-12.

Friedel, Robert. 1983. *Pioneer Plastic: The Making and Selling of Celluloid.* Madison: University of Wisconsin Press.

Fruton, Joseph S. 1970. "Bergmann, Max." *DSB,* vol. 2.

————. 1972. *Molecules and Life: Historical Essays on the Interplay of Chemistry and Biology.* New York, London, Sydney, and Toronto: Wiley-Interscience.

————. 1976. "The Emergence of Biochemistry." *Science* 192: 327–334.

————. 1990. *Contrasts in Scientific Style: Research Groups in the Chemical and Biochemical Sciences.* Memoirs of the American Philosophical Society 191, Philadelphia: American Philosophical Society.

Fry, Harry S. 1928. "A Pragmatic System of Notation for Electronic Valence Conceptions in Chemical Formulas." *Chem. Reviews* 5: 557–569.

Furukawa, Yasu. 1982. "Staudinger and the Emergence of the Macromolecular Concept." *Historia Scientiarum* no. 22: 1–18.

————. 1987. "Scientists in Industry: the Emergence of American Industrial Science" (in Japanese). In *Kagakushi* (History of Science), edited by Chikara Sasaki, 241–272. Tokyo: Kobundo, 1987.

————. 1989. "Wallace Carothers's Suicide and the Context of Depression" (in Japanese). Paper read at the History of Science Conversazione, University of Tokyo, December 19.

————. 1993. "Staudinger's Scientific Activities and Political Struggles" (in Japanese). *Kagakushi* 20: 1–19.

————. 1993-1994. "Staudinger, Polymers, and Political Struggles." *Chemical Heritage* 10, no. 1: 4–6.

————. 1994. "Americanism in Chemical Research: The Genesis of American Polymer Chemistry" (in Japanese). In *Seimitsu kagaku no shiso* (Ideas in Exact Sciences), edited by Yoshihiro Nitta et al., 291–315. Tokyo: Iwanami Shoten, 1994.

————. 1997. "Polymer Chemistry." In *Science in the Twentieth Century,* edited by John Krige and Dominique Pestre, 547–563. Amsterdam: Harwood Academic Publishers.

Geer, William C. 1927. "Rubber." In Hale, ed., 1927. Pp. 171–176.

Gehman, S. D. and J. E. Field. 1937. "Colloidal Structure of Rubber in Solution." *Ind. Eng. Chem.* 29: 793–799.

Geison, Gerald L. 1978. *Michael Foster and the Cambridge School of Physiology: The Scientific Enterprise in Late Victorian Society.* Princeton, N.J.: Princeton University Press.

———. 1981. "Scientific Change, Emerging Specialties, and Research Schools." *History of Science* 19: 20–40.

Geison, Gerald L. and Frederic L. Holmes, eds. 1993. *Research Schools: Historical Reappraisals. Osiris* 2nd ser. 8.

Gladstone, J. H. and Walter Hibbert. 1889. "On the Molecular Weight of Caoutchouc and Other Colloid Bodies." *Phil. Mag.* ser. 5, 28: 38–42.

Gmelin, Leopold. 1848–1860. *Hand-Book of Chemistry.* 14 vols. Translated by Henry Watts. London: Cavendish Society.

Go, Yukichi. 1957. "Visiting Acquaintances to Think of the Development and the Future of Polymer Chemistry: Profiles of Polymer Chemists 1–7" (in Japanese). *Gomu* 4: 3–5, 69–71, 201–203, 267–270, 399–402, 527–530, 656–659.

Goto, Ryozo, and Kazuhiro Maruyama. 1983. "Professors Nodzu and Staudinger" (in Japanese). *Kobunshi* 32: 48–51.

Graham, Thomas. 1861. "Liquid Diffusion Applied to Analysis." *Phil. Trans.* 151: 183–224.

Gray, H. LeB. 1927. "Cellulose." In Hale, ed., 1927. Pp. 177–181.

Guth, Eugene. 1979. "Birth and Rise of Polymer Science — Myth and Truth." *J. Appl. Poly. Sci.* 35: 1–12.

Guth, Eugene and Herman F. Mark, 1934. "Zur innermolekularen Statistik, insbesondere bei Kettenmolekülen I." *Monatsh.* 65: 93–121.

Guth, Eugene, and Hubert M. James. 1941. "Elastic and Thermoelastic Properties of Rubberlike Materials." *Ind. Eng. Chem.* 33: 624–629.

Haber, L. F. 1971. *The Chemical Industry, 1900–1930: International Growth and Technological Change.* Oxford: Oxford University Press.

———. 1986. *The Poisonous Cloud: Chemical Warfare in the First World War.* Oxford: Clarendon Press.

Hahn, Otto. 1975. *Erlebnisse und Erkenntnisse.* Edited by Dietrich Hahn. Düsseldorf and Wien: Econ Verlag.

Hale, William J., ed. 1927. *A Survey of American Chemistry.* Vol. 1: *July 1, 1925–July 1, 1926.* New York: Chemical Catalog Co. for National Research Council.

Hannaway, Owen. 1976. "The German Model of Chemical Education in America: Ira Remsen at Johns Hopkins (1876–1913)." *Ambix* 23: 145–164.

Hardy, Kenneth R. 1974. "Social Origins of American Scientists and Scholars." *Science* 185: 497–506.

Harington, C. R. 1945. "Max Bergmann, 1886–1944." *J. Chem. Soc.* (1945): 716–718.

Harries, Carl D. 1901. "Über das Verhalten des Kautschuks gegen salpetrige Säure." *Ber.* 34: 2991–2992.

———. 1904. "Über den Abbau des Parakautschuks vermittelst Ozon." *Ber.* 37: 2708–2711.

———. 1905. "Zur Kenntnis der Kautschukarten: Ueber Abbau und Constitution des Parakautschuks." *Ber.* 38: 1195–1203.

———. 1910. "Ueber den gegenwärtigen Stand der Chemie des Kautschuks." *Gummi-Ztg.* 24: 850–854.

———. 1911. "Über Kohlenwasserstoffe der Butadienreihe und über einige aus ihnen darstellbare künstliche Kautschukarten." *Liebigs Ann. Chem.* 383: 157–227.

———. 1913a. "Über die Hydrohalogenide der künstlichen und näturlichen Kautschukarten und die daraus regenerierbaren kautschukähnlichen Stoffe." *Ber.* 46: 733–743.

———. 1913b. "Über die Künstlichen Kautschuk-Arten." *Liebigs Ann. Chem.* 395: 211–264.

———. 1914a. "Beiträge zur Kenntnis der Konstitution des Kautschuks und verwandter Verbindungen." *Liebigs Ann. Chem.* 406: 173–226.

———. 1914b. "Über Diacetyl-propan(1,5-Heptandion) aus Kautschuk." *Ber.* 47: 784–791.

———. 1919. *Untersuchungen ueber die näturlichen und künstlichen Kautschukarten.* Berlin: Verlag von Justus Springer.

———. 1923. "Über Aggregation und Desaggregation: Hydrolyse des Schellackharzes. Hydrierung des Kautschuks." *Ber.* 56: 1048–1051.

Hauser, Ernst A. 1939. *Colloidal Phenomena: An Introduction to the Science of Colloids.* New York and London: McGraw-Hill.

———. 1955. "The History of Colloid Science: In Memory of Wolfgang Ostwald." *J. Chem. Educ.* 32: 2–9.

Haworth, Walter N. 1928. "Aliphatic Division." *Annual Reports on the Progress of Chemistry for 1927* 21: 61–105.

Haworth, Walter N. and E. L. Hirst. 1930. "Aliphatic Division." *Annual Reports on the Progress of Chemistry for 1929* 26: 105–110.

Haynes, William. 1949. *American Chemical Industry.* Vol. 6: *The Chemical Companies.* Toronto, New York, and London: D. Van Nostrand.

———. 1953. *Cellulose: The Chemical That Grows.* Garden City, N.Y.: Doubleday.

Heidegger, Martin. 1933. *Die Selbstbehauptung der deutschen Universität.* Breslau: Wilh. Gottl. Korn Verlag.

———. 1991. " 'Only a God Can Save Us': *Der Spiegel*'s Interview with Martin Heidegger." Translated by Maria P. Alter and John D. Caputo. In *The Heidegger Controversy: A Critical Reader,* edited by Richard Wolin, 91–116. New York: Columbia University Press, 1991.

Helferich, Burckhardt. 1969. "Max Bergmann, 1886–1944." *Chem. Ber.* 102: i–xxvi.

Hengstenberg, Josef. 1927. "Röntgenuntersuchungen der Struktur der Polymerisations-produckte des Formaldehyds." *Annalen der Physik* 4, no. 84: 245–278.

Hermann, Armin. 1973. "Kuhn, Werner." *DSB,* vol. 7.

Hermes, Matthew E. 1996. *Enough for One Lifetime: Wallace Carothers, Inventor of Nylon.* Washington, D.C.: American Chemical Society and Chemical Heritage Foundation.

Hershberg, James G. 1993. *James B. Conant: Harvard to Hiroshima and the Making of the Nuclear Age.* New York: Alfred A. Knopf.

Herzfeld, Karl P. 1935. "Reginald Oliver Herzog." *Science* 81: 607–608.

Herzog, Reginald O. 1923. "Einige Arbeiten aus dem Kaiser-Wilhelm-Institut für Faserstoffchemie." *Naturwiss.* 10: 172–180.

———. 1925. "Zur Erkenntnis der Cellulose-Faser." *Ber.* 58: 1254–1262.

———. 1930. "Zur Deformation hochmolekularer Verbindungen." *Kolloid-Z.* 53: 46–51.

Herzog, Reginald O., and W. Janke. 1921. "Röntgenspektrographische Untersuchungen hochmolekularer organischer Verbindungen." *Z. angew. Chem.* 34: 385–387.

Herzog, Reginald O., and M. Kobel. 1924. "Protein Studien. II. Versuche zur Molekulargewichtesbestimmung an Seiden-fibroin." *Z. physiol. Chem.* 134: 269–299.

Hess, Kurt. 1920. "Über die Konstitution der Cellulose: Bemerkung zu einer Arbeit von Hrn. P. Karrer." *Helv. Chim. Acta* 3: 866–869.

——. 1924. "Über Cellulose." *Liebigs Ann. Chem.* 435: 1–114.

——. 1928. *Die Chemie der Zellulose und ihrer Begleiter.* Leipzig: Akademische Verlagsgesellschaft.

——. 1930. "Über alte und neue Auffassungen der Zellulosekonstitution und ihre experimentellen Grundlagen." *Kolloid-Z.* 53: 61–75.

Hess, Kurt, and Walter Wittelsbach. 1920. "Über die Konstitution der Zellulose. I: Die Acetolyse der Äthyzellulose." *Z. Elektrochem.* 26: 232–251.

Hiebert, Erwin N., and Hans-Günther Körber. 1978. "Ostwald, Friedrich Wilhelm." *DSB*, vol. 15.

Hill, Julian W. 1971. "Carothers, Wallace Hume." *DSB*, vol. 3.

——. 1977. "Wallace Hume Carothers." In Milligan, ed., 1977. Pp. 232–251.

Hochheiser, Sheldon. 1986. *Rohm and Haas: History of a Chemical Company.* Philadelphia: University of Pennsylvania Press.

Höcker, H. 1994. "A 50 Years' History of the First Journal in the Field of Macromolecular Chemistry." *Macromol. Chem. Phys.* 195 (1994): 1–6.

Hofmann, Fritz. 1912. "Der synthetische Kautschuk: Vom Standpunkt der Technik." *Z. angew. Chem.* 29: 1462–1467.

——. 1936. "Wie es zur Synthese des Kautschuks kam?" *Chemiker-Ztg.* 60: 693–696.

Holleman, Arnold F. 1915. *A Text-Book of Organic Chemistry.* 4th English edition. Edited by Owen E. Mott. New York: John Wiley & Sons; London: Chapman and Hall.

Holmes, Harry N. 1954. "The Growth of Colloid Chemistry in the United States." *J. Chem. Educ.* 31: 600–602.

Homburg, Ernst. 1992. "The Emergence of Research Laboratories in the Dyestuffs Industry, 1870–1900." *BJHS* 25: 91–111.

Hooke, Robert. 1665. *Micrographia: Or Some Physiological Descriptions of Minute Bodies Made by Magnifying Glasses with Observations and Inquiries Thereupon.* London: Martyn and Allestry.

Hopff, Heinrich. 1969. "Hermann Staudinger, 1881–1965." *Chem. Ber.* 102, no. 5: xli–xlviii.

Hoshino, Kohei. 1977. "The Technological Development of Synthetic Fibers as Viewed from Nylon 6" (in Japanese). *Kobunshi* 26: 831–843.

Hoshino, Toshio. 1952. "Exploring Macromolecular Compounds: from a Viewpoint of an Organic Chemist" (in Japanese). *Kobunshi* 1: 14–18.

Hounshell, David A., and John K. Smith. 1988. *Science and Corporate Strategy: Du Pont R&D, 1902–1980.* Cambridge and New York: Cambridge University Press.

Houwink, Roelof. 1940. "Zusammenhang zwischen viscosimetrisch und osmotisch bestimmten Polymerisationsgraden bei Hochpolymeren." *J. prakt. Chem.* 157: 15–18.

——. 1964. "Dr. L. H. Baekeland." *Chem. Ind.* (January): 50–51.

Hudson, John. 1992. *The History of Chemistry.* Houndmills and London: Macmillan.

Huggins, Maurice L. 1938. "The Viscosity of Dilute Solutions of Long-Chain Molecules. I." *J. Phys. Chem.* 42: 911–920.

Hunt, Morton M. 1958. "Profiles: Polymers Everywhere — II." *New Yorker* (September 20): 46–79.

Ihde, Aaron J. 1964. *The Development of Modern Chemistry.* New York: Harper and Row.

———. 1990. *Chemistry, as Viewed from Bascom's Hill: A History of the Chemistry Department at the University of Wisconsin in Madison.* Madison: Department of Chemistry, University of Wisconsin.

International Union of Pure and Applied Chemistry. 1952. "Report on Nomenclature in the Field of Macromolecules." *J. Poly. Sci.* 8: 257–277.

Iwabuchi, Susumu. 1985. "W. Kern" (in Japanese). *Kobunshi* 34: 472–475, 586–589, 656–659.

Iwakura, Yoshio. 1974. "Various Reminiscences" (in Japanese). In *Kobunshi to tomoni* (Along with Macromolecules), edited by Kuniharu Nagakubo, 3–85. Tokyo: Professor Yoshio Iwakura Retirement Memorial Committee.

———. 1995. "The History and Role of the Society of Polymer Science" (in Japanese). *Kagakushi* 22: 53–55.

James, Hubert M., and Eugene Guth. 1943. "Theory of the Elastic Properties of Rubber." *J. Chem. Phys.* 11: 455–481.

James, Laylin K. "Hermann Staudinger." In James, ed., 1993. Pp. 359–367.

———, ed. 1993. *Nobel Laureates in Chemistry, 1901–1992.* Washington, D.C.: American Chemical Society and Chemical Heritage Foundation.

Japanese Union of Fiber Research, ed. 1940. *Sen'i kenkyu to sen'i kokusaku* (Fiber Research and National Policy of Fibers). Tokyo: Japanese Union of Fiber Research.

Jeanloz, R. W. 1956. "Kurt Heinrich Meyer, 1883–1952." *Advances in Carbohydrate Chemistry* 11: xiii–xviii.

Johnson, Jeffrey A. 1985. "Academic Chemistry in Imperial Germany." *Isis* 76: 500–524.

———. 1990. *The Kaiser's Chemists: Science and Modernization in Imperial Germany.* Chapel Hill and London: University of North Carolina Press.

Johnson, John R. 1940. "Wallace Hume Carothers, 1896–1937." *J. Chem. Soc.* 143: 100–102.

———. 1958. "Carothers, Wallace Hume." *Dictionary of American Biography* Supplement 2.

Karrer, Paul. 1920. "Zur Kenntnis der Polysaccharide I. Methylierung der Stärke." *Helv. Chim. Acta* 3: 620–625.

———. 1925. *Polymere Kohlenhydrate.* Leipzig: Akademische Verlaggesellschaft.

Katz, Johan R. 1925. "Was sind die Ursachen der eigentümlichen Dehnbarkeit des Kautschuk?" *Kolloid-Z.* 36: 300–307; 37: 19–22.

———. 1936. "X-ray Spectrography of Polymers and in Particular Those Having a Rubber-like Extensibility." *Trans. Faraday Soc.* 32: 77–94.

Kauffman, George B. 1967. "Alfred Werner's Habilitationsschrift." *Chymia* 12: 183–187.

———. 1990a. "Marvel, Carl Shipp ('Speed')." *DSB*, vol. 18.

———. 1990b. "Roy H. Kienle (1896–1957), Polymer Pioneer." In Seymour and Mark, eds., 1990. Pp. 149–162.

Kay, Lily E. 1993. *The Molecular Vision of Life: Caltech, the Rockefeller Foundation, and the Rise of the New Biology.* New York and Oxford: Oxford University Press.

Kekulé, August. 1858. "Ueber die Constitution und die Metamorphosen der chemischen Verbindungen und über die chemische Natur des Kohlenstoffs." *Liebigs Ann. Chem.* 106: 129–159.

———. 1878. "The Scientific Aims and Achievements of Chemistry." Address delivered on assuming the Rectorate of the Rhenish Friedrich-Wilhelms University of Bonn, October 18, 1877. *Nature* 18: 210–213. The original German text was reprinted in Anschütz, 1929. Pp. 903–917.

Kenyon, W. O. 1951. "Cellulose Chemistry." In American Chemical Society, ed., 1951. Pp. 143–153.

Kettering, Charles F. 1947. "Biographical Memoir of Leo Hendrik Baekeland." *Biog. Mem. NAS* 24: 281–302.

Kienle, Roy H. 1930. "Observations as to the Formation of Synthetic Resins." *Ind. Eng. Chem.* 22: 590–594.

Kienle, Roy H. and A. G. Hovey. 1929. "The Polyhydric Alcohol-Polybasic Acid Reaction. I. Glycerol-Phthalic Anhydride." *J. Amer. Chem. Soc.* 51: 509–519.

Kienle, R. H. and C. S. Ferguson. 1929. "Alkyd Resins as Film-Forming Materials." *Ind. Eng. Chem.* 21: 352–348.

Kleeberg, W. 1891. "Ueber die Einwirkung des Formaldehyds auf Phenole." *Liebigs Ann. Chem.* 263: 283–351.

Kohler, Robert E., Jr. 1975. "Lewis-Langmuir Theory of Valence and the Chemical Community, 1920–1928." *His. Stud. Phys. Sci.* 6: 431–468.

———. 1978. "Adams, Roger." *DSB.* Vol. 15.

———. 1982. *From Medical Chemistry to Biochemistry: The Making of a Biomedical Discipline.* Cambridge and New York: Cambridge University Press.

Körber, Hans-Günther. 1975. "Ostwald, Carl Wilhelm Wolfgang." *DSB*, vol. 10.

Kovac, Jeffrey. 1993. "Paul J. Flory, 1910–1985." In James, ed., 1993. Pp. 564–570.

Kraemer, Elmer O. 1943a. "The Colloidal Behavior of Organic Macromolecular Materials." In Burk and Grummit, eds., 1943. Pp. 73–93.

———. 1943b. "The Ultracentrifuge and Its Application to the Study of Organic Macromolecules." In Burk and Grummit, eds., 1943. Pp. 95–124.

Kraft, F., and A. Stern. 1894. "Ueber das Verhalten der fettsauren Alkalien und der Seifen in Gegenwart von Wasser. I." *Ber.* 27: 1747–1754.

Kraft, F. and A. Strutz. 1896. "Ueber das Verhalten seifenähnlicher Substanzen gegen Wasser." *Ber.* 29: 1328–1334.

Kronstein, A. 1902. "Zur Kenntnis der Polymerisation." *Ber.* 35: 4150–4153, 4153–4157.

Krüll, Claudia. 1976. "Historische Experimente (um 1905): 'Denk-Experimente' zum Benzolring—aus Manuskripten Hermann Staudingers." *Chemie-Experiment und Didaktik* 2: 441–448.

———. 1978. "Hermann Staudinger—Aufbruch ins Zeitalter der Makromoleküle." *Kultur und Technik* 2, no. 3: 44–49.

Kuhn, Hans. 1964. "Werner Kuhn (1899–1963)." *Helv. Chim. Acta* 47: 689–695.

———. 1984. "Leben und Werk von Werner Kuhn, 1899–1963." *Chimia* 38: 191–211.

Kuhn, Werner. 1934. "Über die Gestalt fadenförmiger Moleküle in Lösungen." *Kolloid-Z.* 68: 2–15.

———. 1936. "Beziehungen zwischen Molekülgröße, statistischer Molekülgestalt und elastischen Eigenschaften hochpolymerer Stoffe." *Kolloid-Z.* 76: 258–271.

"Ladies' Hose and History." *Chicago Tribune*, September 28, 1938.

Langmuir, Irving. 1919a. "The Arrangement of Electrons in Atoms and Molecules." *J. Amer. Chem. Soc.* 41: 868–934.

———. 1919b. "Isomorphism, Isosterism, and Covalence." *J. Amer. Chem. Soc.* 41: 1543–1559.

———. 1920. "The Octet Theory of Valence and Its Applications with Special Reference to Organic Compounds." *J. Amer. Chem. Soc.* 42: 274–292.

Laszlo, Pierre. 1986. *A History of Biochemistry: Molecular Correlates of Biological Concepts.* Comprehensive Biochemistry, vol. 34A. Amsterdam and New York: Elsevier Science Publishers.

Leicester, Henry M. 1951. *The Historical Background of Chemistry.* New York: Dover.

Leonard, Nelson J. 1969. "Roger Adams." *J. Amer. Chem. Soc.* 91: a–d.

Levene, Phoebus A. 1927. "Proteins." In Hale, ed., 1927. Pp. 159–166.

Lewis, Gilbert N. 1916. "The Atom and the Molecule." *J. Amer. Chem. Soc.* 38: 762–785.

Liebig, Justus von. 1835. "Bemerkung über die Methoden der Darstellung flüchtiger, durch trockene Destillation organischer Materien erhaltene Produkte." *Liebigs Ann. Chem.* 16: 61–62.

Liesegang, R. E. 1933. Review of Hermann Staudinger, *Die hochmolekularen organischen Verbindungen. Kolloid-Z.* 62: 244.

———. 1940. Review of Hermann Staudinger, *Organische Kolloidchemie, Kolloid-Z.* 91: 93–94.

Linter, C. J. and G. Düll. 1893. "Ueber den Abbau der Stärke unter dem Einflusse der Diastasewirkung." *Ber.* 26: 2533–2547.

Löser, Bettina. 1983. "Der Einfluß der Arbeiten zur technischen Kautschuksynthese auf die Herausbildung der makromolekulare Chemie." *NTM* 20: 45–55.

———. 1991. "Zur Gründungsgeschichte des Kaiser-Wilhelm-Institut für Faserstoffchemie in Berlin-Dahlem." *NTM* 28: 73–93.

Lottermoser, Alfred. 1943. "Wolfgang Ostwald 60 Jahre alt." *Kolloid-Z.* 103, no. 2: 89–94.

Macrakis, Kristie. 1993. *Surviving the Swastika: Scientific Research in Nazi Germany.* New York and Oxford: Oxford University Press.

Marckwald, E. and F. Frank. 1906. "Dem Andenken Carl Otto Weber." *Kolloid-Z.* 1: 4–10.

Mark, Herman F. 1926. "Über die röntgenographische Ermittlung der Struktur organischer besonders hochmolekularer Substanzen." *Ber.* 59: 2982–3000. 530.

———. 1930. "Über das Verhalten der Hochpolymeren in Lösung." *Kolloid-Z.* 53: 32–41.

———. 1938a. "Aspects of High Polymeric Chemistry." *Nature* 142: 937–939.

———. 1938b. *Der feste Körper.* Leipzig: Verlag Hirzel.

———. 1940. *Physical Chemistry of High Polymeric Systems.* High Polymers, 2. New York: Interscience Publishers.

———. 1952. "Kurt Heinrich Meyer, 1883–1952." *Z. angew. Chem.* 64: 521–523.

———. 1962. "Recollections of Dahlem and Ludwigshafen." In Ewald, ed., 1962. Pp. 603–607.

———. 1967. "Polymers—Past, Present, Future." In Milligan, ed., 1967. Pp. 19–43.

———. 1973. "The Early Days of Polymer Science." *J. Chem. Educ.* 50: 757–760.

———. 1976a. "An Architect of Polymer Science." *J. Poly. Sci., Symposium* no. 54: 1–2.

———. 1976b. "Polymer Chemistry: The Past 100 Years." *Chem. Eng. News* 54 (April 6): 176–189.

———. 1981. "Polymer Chemistry in Europe and America—How It All Began." *J. Chem. Educ.* 58: 527–534.

———. 1982. "Coming to an Age of Polymers in Science and Technology." In Seymour, ed., 1982. Pp. 1–9.

———. 1984a. "The Contribution of Carl (Speed) Marvel to Polymer Science." *J. Macromol. Sci.—Chemistry* A 21 (13 & 14): 1567–1606.

———. 1984b. "The Development of Plastics." *American Scientist* 72: 156–162.

———. 1987. "From Revolution to Evolution." *J. Chem. Educ.* 64: 858–861.

———. 1988a. "C. G. Overberger: Co-Founder and Permanent Standard Bearer of the Journal of Polymer Science." *J. Poly. Sci., Part A: Polymer Chemistry* 26: xi–xiii.

———. 1988b. "Interscience Publishers, Inc.: Dr. Maurits Dekker and Dr. Eric S. Proskauer." *J. Poly. Sci., Part A: Polymer Chemistry* 26: xv–xvi.

———. 1989. "Hermann Staudinger: Father of Modern Polymer Chemistry." In Seymour, ed., 1989. Pp. 93–109.

———. 1993. *From Small Organic Molecules to Large: A Century of Progress.* Profiles, Pathways, and Dreams: Autobiographies of Eminent Chemists. Washington, D.C.: American Chemical Society.

Mark, Herman F., and G. Stafford Whitby, 1940. "Introduction to 'Collected Papers of Wallace Hume Carothers.' " In Carothers, 1940. Pp. viii–x.

Mark, Herman F., and A. Raff, 1941. *High Polymeric Reactions: Their Theory and Practice.* High Polymers, 3. New York: Interscience Publishers.

Mark, Herman F., and Herman A. Bruson, 1956. "Hermann Staudinger." *J. Poly. Sci.* 19: 387–388.

Marvel, Carl S. 1978. "C. S. Marvel." In Ulrich, ed., 1978. Pp. 133–141.

———. 1981. "The Development of Polymer Chemistry in America—The Early Days." *J. Chem. Educ.* 58: 535–539.

Mattmüller, Markus. 1968. *Leonhard Ragaz und der religiöse Sozialismus.* Vol. 2: *Die Zeit des ersten Weltkriegs und der Revolutionen.* Basel and Stuttgart: Verlag von Helbing & Lichtenhahn.

Mayo, Frank R. and Frederick M. Lewis. 1944. "Copolymerization. I. A Basis for Comparing the Behavior of Monomers in Copolymerization: The Copolymerization of Styrene and Methylmethacrylate." *J. Amer. Chem. Soc.* 66: 1594–1601.

McBain, James W. 1929. "Structure in Amorphous and Colloidal Matter." *J. Chem. Educ.* 6: 2115–2127.

McBain, James W. and C. S. Salmon. 1920. "Colloidal Electrolytes: Soap Solutions and Their Constitution." *J. Amer. Chem. Soc.* 42: 426–461.

McGrew, Frank C. 1958. "Structure of Synthetic High Polymers." *J. Chem. Educ.* 35: 178–186.

McMillan, Frank M. 1979. *The Chain Straighteners.* London: Macmillan.

Meikle, Jeffrey I. 1995. *American Plastic: A Cultural History.* New Brunswick, N.J.: Rutgers University Press.

Meinel, Christoph. 1990a. "Freudenberg, Karl Johann." *DSB,* vol. 17.

———. 1990b. "Ziegler, Karl Waldemar." *DSB,* vol. 18.

Melville, Harry W. 1953. "High Polymers." *Trans. Faraday Soc.* 49: 565–570.

Meyer, Kurt H. 1929. "Bemerkungen zu den Arbeiten von H. Staudinger." *Z. angew. Chem.* 42: 76–77.

———. 1930. "Räumliche Vorstellungen über den Bau der Kohlenstoffverbindungen und ihre Verwendeng in der Chemie der Hochpolymeren." *Kolloid-Z.* 53: 8–17.

———. 1942. *Natural and Synthetic High Polymers.* High Polymers, 4. New York: Interscience Publishers.

Meyer, Kurt H. and Herman F. Mark. 1928a. "Über den Bau des kristallisierten Anteils der Cellulose." *Ber.* 61: 593–614.

———. 1928b. "Über den Kautschuk." *Ber.* 61: 1939–1949.

———. 1930a. *Aufbau der hochpolymeren Substanzen.* Berlin: Hirschwaldsche Buchhandlung.

———. 1930b. *Der Aufbau der hochpolymeren organischen Naturstoffe.* Leipzig: Akademische Verlagsgesellschaft.

———. 1937. *Hochpolymere Chemie.* Leipzig: Akademische Verlagsgesellschaft.

Meyer, Kurt H., Georg von Susich, and E. Valkó. 1932. "Die elastischen Eigenschaften der organischen Hochpolymeren und ihre kinetische Deutung." *Kolloid-Z.* 59: 208–216.

Meyer-Thurow, Georg. 1982. "The Industrialization of Invention: A Case Study from the German Chemical Industry." *Isis* 73: 363–381.

Miller, Jane A. 1993. "Paul Karrer, 1889–1971." In James, ed., 1993. Pp. 242–247.

Milligan, W. O., ed. 1967. *Polymers.* Proceedings of the Robert A. Welch Foundation Conferences on Chemical Research, X. Houston: Robert A. Welch Foundation.

———, ed. 1977. *American Chemistry—Bicentennial.* Proceedings of the Robert A. Welch Foundation Conferences on Chemical Research, XX. Houston: Robert A. Welch Foundation.

Milligan, W. O., J. W. Williams, and E. J. Miller. 1951. "Colloid Chemistry." In American Chemical Society, ed., 1951. Pp. 169–172.

Morawetz, Herbert. 1985. *Polymers: Origins and Growth of a Science.* New York: John Wiley & Sons.

Morrell, J. B. 1972. "The Chemist Breeders: The Research Schools of Liebig and Thomas Thomson." *Ambix* 19: 1–46.

Morris, Peter J. T. 1982. "The Development of Acetylene Chemistry and Synthetic Rubber by I. G. Farbenindustrie Aktiengesellschaft: 1926–1945." Ph.D. thesis, Oxford University.

———. 1986. *Polymer Pioneers: A Popular History of the Science and Technology of Large Molecules.* Philadelphia: Center for History of Chemistry.

———. 1989. *The American Synthetic Rubber Research Program.* Philadelphia: University of Pennsylvania Press.

———. 1994. "Synthetic Rubber: Autarky and War." In Mossman and Morris, eds., 1994. Pp. 54–69.

Morton, Maurice. 1982. "History of Synthetic Rubber." In Seymour, ed., 1982. Pp. 225–238.

———. 1989. "From Rubber Chemistry to Polymers: A History of Polymer Science at the University of Akron." *Rubber Chem. Tech.* 62: G19–G37.

Mosley, Leonard. 1980. *Blood Relations: The Rise and Fall of the du Ponts of Delaware.* New York: Atheneum, 1980.

Mossman, S. T. I. "Parkesine and Celluloid." In Mossman and Morris, eds., 1994. Pp. 10–25.

Mossman, S. T. I. and Peter J. T. Morris, eds. 1994. *The Development of Plastics.* Cambridge: Royal Society of Chemistry.

Moy, Timothy D. 1989. "Emil Fischer as 'Chemical Mediator': Science, Industry, and Government in World War One." *Ambix* 36: 109–120.

Mulvaney, J. E. 1976. "Interview with Carl S. Marvel." *J. Chem. Educ.* 53: 609–913.

Nägeli, Carl W. 1877. *Das Mikroskop: Theorie und Anwendung desselben.* 2nd edition. Leipzig: W. Engelmann.

Nairon (Nylon). Osaka: Boshoku Zasshi-sha, 1939.

Nastukoff, A. 1900. "Ueber einige Oxycellulosen und über das Molekulargewicht der Cellulose." *Ber.* 33: 2237–2243.

Natta, Giulio. 1963. "Macromolecular Chemistry: Nobel Lecture, December 12, 1963." *Science* 147: 261–272.

Naunton, W. J. S. 1952. "Synthetic Rubber." In Schidrowitz and Dawson, eds., 1952. Pp. 100–109.

Neuberger, A. 1945. "Dr. Max Bergmann." *Nature* 155: 419–420.

"New Fiber Called Superior to Silk for Making Stockings." 1938. *New York World-Telegram*, September 21.

"New Silk Made on Chemical Base Rivals Quality of Natural Product." 1938. *New York Times*, September 22.

Nichols, James B. 1924. "Nitrocellulose and Camphor." *J. Phys. Chem.* 28: 769–771.

Nicolai, Georg Friedrich. 1918. *The Biology of War.* Translated by Constance A. Grande and Julian Grande. New York: Century Co.

Nishikawa, S. and S. Ono. 1913. "Transmission of X-rays through Fibrous, Lamellar and Granular Substances." *Proceedings of Tokyo Mathematics-Physics Society* II, 7: 131–138.

Nishikawa Commemorative Committee, ed. 1982. *Nishikawa Shoji seisei: hito to gyoseki* (Professor Shoji Nishikawa: The Man and the Work). Tokyo: Kodansha.

"Nobel Prize to German, Hollander." 1953. *Chem. Eng. News* 31: 4760–4761.

Nobel Foundation, ed. 1964. *Nobel Lectures: Chemistry, 1942–1962.* Amsterdam, London, and New York: Elsevier.

———, ed. 1966. *Nobel Lectures: Chemistry, 1922–1941.* Amsterdam, London, and New York: Elsevier.

Nodzu, Ryuzaburo. 1930. "High-molecular Organic Compounds and Professor Staudinger" (in Japanese). *Warera no Kagaku* 3: 43–52.

———. 1954. "My Impression of Dr. Staudinger" (in Japanese). *Kobunshi* 8: 375–377.

———. 1957. "On Professor H. Staudinger's Visit to Japan" (in Japanese). *Kinki Kagaku Kogyo-kai Kaiho* no. 47: 3.

Noyes, Arthur A. 1905. "The Preparation and Properties of Colloidal Mixtures." *J. Amer. Chem. Soc.* 27: 85–104.

Nye, Mary Jo. 1972. *Molecular Reality: A Perspective on the Scientific Work of Jean Perrin.* London: MacDonald; New York: Elsevier.

———. 1993. *From Chemical Philosophy to Theoretical Chemistry: Dynamics of Matter and Dynamics of Disciplines, 1800–1950.* Berkeley and London: University of California Press.

"Nylon." 1940. *Fortune 22*, no. 1 (July): 57–60, 114, 116.

Ochiai, Eiji. 1953. "Accomplishments of Dr. Staudinger" (in Japanese). *Kobunshi* 6: 161–162.

———. 1954. "Profile of Dr. Staudinger" (in Japanese). *Kobunshi* 3: 223–226.

———. 1966. "Memoirs of Professor Staudinger" (in Japanese). *Kobunshi* 15: 409–411.

Oesper, Ralph E. 1945. "Wolfgang Ostwald (1883–1943)." *J. Chem. Educ.* 22: 263–264.

———. 1951. "Rudolf Pummerer." *J. Chem. Educ.* 28: 243–244.

Ogata, Naoya. 1982. "H. F. Mark" (in Japanese). *Kobunshi* 31: 698–701, 792–795, 864–867.

Ohtsu, Takayuki. 1983. "Carl S. Marvel" (in Japanese). *Kobunshi* 32: 448–451, 506–509, 604–607.

Okamura, Seizo. 1995. "Ichiro Sakurada and Japanese Polymer Chemistry: A Perspective of the History of Science" (in Japanese). *Kagakushi* 22: 56–58.

Olby, Robert. 1970. "The Macromolecular Concept and the Origins of Molecular Biology." *J. Chem. Educ.* 47: 168–171.

———. 1974. *The Path to the Double Helix.* Seattle: University of Washington Press. Reprinted as *The Path to the Double Helix: The Discovery of DNA.* New York: Dover, 1994.

———. 1976. "Staudinger, Hermann." *DSB,* vol. 13.

———. 1979. "The Significance of the Macromolecules in the Historiography of Molecular Biology." *History and Philosophy of the Life Sciences* 1: 185–198.

Ostwald, Wilhelm. 1926–1927. *Lebenslinien: Eine Selbstbiographie.* 3 vols. Berlin: Klasing.

Ostwald, Wolfgang. 1909. *Grundriss der Kolloidchemie.* Dresden: Theodor Steinkopff.

———. 1915. *Die Welt der vernachlässigten Dimensionen.* Dresden and Leipzig: Verlag von Theodor Steinkopff.

———. 1917. *An Introduction to Theoretical and Applied Colloid Chemistry: The World of Neglected Dimensions.* Translated from Wolfgang Ostwald, 1915, by Martin H. Fischer. New York: John Wiley and Sons; London: Chapman and Hall.

———. 1919. *A Handbook of Colloid-Chemistry: The Recognition of Colloids, the Theory of Colloids, and Their General Physico-Chemical Properties.* 2nd English edition. Translated from Wolfgang Ostwald, 1909, by Martin H. Fischer. London: J. & A. Churchill.

———. 1920. *Kleines Praktikum der Kolloidchemie.* Dresden and Leipzig: Verlag von Theodor Steinkopff.

———. 1924. *Practical Colloid Chemistry.* Translated from Wolfgang Ostwald, 1920, by J. Newton Kugelmass. London: Methuen.

———. 1940. Review of W. Röhrs, H. Staudinger, and R. Vieweg, eds., *Fortschritte der Chemie, Physik, und Technik der makromolekularen Stoffe* (1939). *Kolloid-Z.* 90: 370–375.

Ott, Emil. 1926a. "Röntgenmetrische Untersuchungen an hochpolymeren organischen Substanzen zum Zwecke einer Abgrenzung des Molekulargewichts derselben." *Z. Physik* 27: 174–177.

———. 1926b. "Untersuchungen an Cellulose und Licheinin mit Hilf der Roentgenspektren." *Helv. chim. Acta* 9: 31–32.

Ott, Hugo. 1988. *Martin Heidegger: Unterwegs zu seiner Biographie.* Frankfurt and New York: Campus Verlag.

Pardee, Arthur M. 1916. "A Study of the Conductivity of Certain Organic Salts in Absolute Alcohol at 15°, 25°, and 35°C." Ph.D. dissertation, Johns Hopkins University.

Partington, J. R. 1961–1970. *A History of Chemistry.* 4 vols. London: Macmillan; New York: St. Martins Press.

Pauly, Philip J. 1987. *Controlling Life: Jacques Loeb and the Engineering Ideal in Biology.* New York and Oxford: Oxford University Press.

Pedersen, Kai O. 1940. "The Protein Molecule." In Svedberg and Pedersen, eds., 1940. Pp. 406–415.

Pennsylvania Hospital. 1937. *186th Annual Report of the Pennsylvania Hospital, Philadelphia.* Philadelphia: Member Welfare Federation of Philadelphia.

Perkin, William H., Jr. 1912. "The Production and Polymerisation of Butadiene, Isoprene, and Their Homologues." *J. Soc. Chem. Ind.* 31: 616–624.

Perrin, Jean B. 1916. *Atoms.* Translated from *Les Atomes* (1914) by D. Ll. Hammik. London: Constable.

Pflüger, Eduard F. W. 1875. "Ueber die physiologische Verbrennung in den lebendigen Organismen." *Pflüger's Arch. Physiol.* 10: 251–367.

"Physical Chemical Research Discussed by Kienle in Mattiello Memorial Lecture." 1949. *Chem. Eng. News* 27: 3788–3789.

Picken, L. E. R. 1952. "Prof. Kurt H. Meyer." *Nature* 169: 820.

Pickles, Samuel S. 1910. "The Constitution and Synthesis of Caoutchouc." *J. Chem. Soc.* 97: 1085–1090.

———. 1951. "The Chemical Constitution of the Rubber Molecule." *Transactions of the Institution of the Rubber Industry* 27: 148–165.

"Pickles, Samuel Shrowder." 1953. *Who's Who in British Science, 1953.* London: Leonard Hill.

Planck, Max, ed. 1936. *25 Jahre Kaiser Wilhelm-Gesellschaft zur Förderung der Wissenschaften.* 3 vols. Berlin: Verlag Julius Springer.

Polanyi, Michael. 1921a. "Das Röntgen Faserdiagramm." *Z. Physik* 7: 149–180.

———. 1921b. "Faserstruktur im Röntgenlichte." *Naturwiss.* 9: 337–340.

———. 1962. "My Time with X-rays and Crystals." In Ewald, ed., 1962. Pp. 629–636.

Prelog, Vladimir, and O. Jeger. 1980. "Leopold Ružička." *Biog. Mem. FRS* 26: 411–501.

Priesner, Claus. 1980. *H. Staudinger, H. Mark und K. H. Meyer: Thesen zur Grösse und Struktur der Makromoleküle.* Weinheim; Deerfield Beach, Fl.; and Basel: Verlag Chemie.

———. 1987. "Hermann Staudinger und die Makromolekulare Chemie in Freiburg: Dokumente zur Hochschulpolitik 1925–1955." *Chemie in unserer Zeit* 21: 151–160.

Pringsheim, Hans. 1925. "Über die Chemie Complexer Naturstoffe." *Naturwiss.* 13: 1084–1090.

———. 1926. "Abbau und Aufbau der Polysaccharide." *Ber.* 59: 3008–3018.

Pritykin, L. M. 1981. "The Role of Concepts of Structure in the Development of the Physical Chemistry of Polymers." *Isis* 72: 446–156.

Proskauer, Eric S. 1988. "A Tribute to Dr. Mark." *J. Poly. Sci., Part A: Polymer Chemistry* 26: vii–ix.

Pummerer, Rudolf. 1930. "Zur Konstitution des Kautschuks." *Kolloid-Z.* 53: 75–78.

Pummerer, Rudolf, and Peter A. Burkard. 1922. "Über Kautschuk." *Ber.* 55: 3458–3472.

Pummerer, Rudolf and Albert Koch. 1924. "Über einen Krystallisierten Kautschuk und über Hydro-Kautschuk." *Liebigs Ann. Chem.* 438: 294–313.

Pummerer, Rudolf, Hilde Nielsen, and Wolfgang Gündel. 1927. "Kryoskopische Molekulargewichts-Bestimmungen des Kautschuks." *Ber.* 60: 2167–2175.

Pummerer, Rudolf, and Wolfgang Gündel. 1928. "Über Darstellung und Molekulargrösse des Isokautschuknitrons." *Ber.* 61: 1591–1596.

Purves, Clifford B. 1954. "Historical Survey." In *Cellulose and Cellulose Derivatives.* 2nd edition. Part 1, edited by Emil Ott, Harold M. Spurlin, and Mildred W. Grafflin, 29–53. New York: Interscience Publishers.

Quarles, Willem. 1951. "Hermann Staudinger: Thirty Years of Macromolecules." *J. Chem. Educ.* 28: 120–122.

Ragaz, Leonhard. 1982. *Leonhard Ragaz in seinen Briefen.* Vol. 2: *1914–1932.* Edited by Christine Ragaz, Markus Mattmüller, and Arthur Rich. Zurich: Theologischer Verlag.

———. 1984. *Signs of the Kingdom: A Ragaz Reader.* Edited and translated by Paul Bock. Grand Rapids, Mich.: Eerdmans.

Raoult, François-Marie. 1882. "Loi générale de congélation des dissolvants." *Compt. rend.* 95: 1030–1033.

———. 1887. "Loi générale des tensions de vapeur des dissolvants." *Compt. rend.* 104: 1430–1433.

Reich, Leonard. 1985. *The Making of American Industrial Research: Science and Business at GE and Bell, 1876–1926.* Cambridge: Cambridge University Press.

Reis, A. J. 1920. "Zur Kenntnis der kristallgitter." *Z. Physik* 1: 204–220.

Remane, Horst and Frank Weise. 1993. "Das Chemiehistorische Dokument: Der Studienplan des Begründers der makromolekularen Chemie—Hermann Staudinger (1881–1965)." *J. prakt. Chem.* 334: 211–213.

Rhees, David J. 1907. "The Chemists' Crusade: The Rise of an Industrial Science in Modern America, 1907–1922." Ph.D. dissertation, University of Pennsylvania.

Ridgway, David W. 1977. "Interview with Paul J. Flory." *J. Chem. Educ.* 54: 341–344.

Rockmore, Tom. 1992. *On Heidegger's Nazism and Philosophy.* Berkeley: University of California Press.

Rodewald, H., and A. Kattein. 1900. "Über natürliche und künstliche Stärkekörper." *Z. physik. Chem.* 33: 579–592.

Rogers, H. 1952. "Development of Manufacturing Activities." In Schidrowitz and Dawson, eds., 1952. Pp. 40–63.

Rolland, Romain. 1952. *Journal des années de guerre, 1914–1919.* Edited by Marie Romain Rolland. Paris: Éditions Albin Michel.

Ronge, Grete. 1972. "Hess, Kurt." *Neue Deutsche Biographie,* vol. 9.

Ross, Sydney. 1978. "Freundlich, Herbert Max Finlay." *DSB,* vol. 15.

Rossiter, Margaret W. 1975. *The Emergence of Agricultural Science: Justus Liebig and the Americans, 1840–1880.* New Haven, Conn., and London: Yale University Press.

Russell, Colin A. 1971. *The History of Valency.* Leicester: Leicester University Press.

Rutkoff, Peter M., and William B. Scott. 1981. "Biographical Afterword: Hans Staudinger, 1889–1980." In Hans Staudinger, 1981. Pp. 137–153.

Sabanijeff, A. P. and N. A. Alexandrov. 1891. "Cryoscopic Investigations of Colloids . . . III. On the Molecular Weight of Egg Albumin" (in Russian). *J. Russ. Phys. Chem. Soc.* 23: 7–9.

Sachsse, Hans. 1984. "Ein Chemiker zur Friedensdiskussion: Hermann Staudinger zu Technik und Politik." *Nachrichten aus Chemie, Technik, und Laboratorium* 32: 974–976.

Sakurada, Ichiro. 1931. "The Size and Structure of the Cellulose Molecule (I): the Cellulose Molecule from Chemical Viewpoint" (in Japanese). *Warera no Kagaku* 4: 262–287.

———. 1938. "Future Perspectives of Artificial Fibers" (in Japanese). *Kogyo Kagaku Zassi* 41: 478–482.

———. 1940. "The Shape of Threadlike Molecules in Solution and the Relationship Between the Solution Viscosity and the Molecular Weight" (in Japanese). *Kasen Kohen-shu* 5: 33–44.

———. 1969. *Kokunshi kagaku to tomoni* (Along with Polymer Chemistry). Tokyo: Kinokuniya-shoten.

———. 1972. "The Viscosity and Molecular Weight of Macromolecular Solutions" (in Japanese). *Kagaku* 27: 867–873.

———. 1983. "Recollections of Polemic with Staudinger" (in Japanese). *Kobunshi Kako* 32: 506–508.

Sakurada, Ichiro, Hiroshi Sobue, Yukichi Go, Keikichi Arai, Masahide Yazawa, Kohei Hoshino, and Yoshio Iwakura. 1965. "Discussion: The Background and Future of Polymer Science" (in Japanese). *Kobunshi* 14: 1038–1057.

Scatchard, George. 1973. "Half a Century as a Part-time Colloid Chemist." In *Twenty Years of Colloid and Surface Chemistry: The Kendall Award Addresses,* edited by Karol J. Mysels, Carlos M. Samour, and John H. Hollister, 103–106. Washington, D.C.: American Chemical Society, 1973.

Scheraga, Harold A. 1980. "Paul J. Flory on His 70th Birthday." *Macromolecules* 13, no. 3: 8A–10A.

Schidrowitz, Philip and T. R. Dawson, eds. 1952. *History of the Rubber Industry.* Cambridge: W. Heffer & Sons.

Schneeberger, Guido. 1962. *Nachlese zu Heidegger: Dokumente zu seinem Leben und Denken.* Bern: Buchdruckerei.

Schroeter, Georg. 1916. "Über die Beziehungen zwischen den polymeren Ketenen und dem Cyclobutan-1, 3-dion und seinen Derivaten." *Ber.* 49: 2697–2745.

Schulz, Günther V. 1935. "Über die Beziehung zwischen Reaktionsgeschwindidkeit und Zusammensetzung das Reaktionsprodukties bei Makropolymerisationsvorgängen." *Z. physik. Chem.* B 30: 379–3978.

———. 1936a. "Über die Verteilung der Molekulargewichte in hochpolymeren Gemischen und die Bestimmung des mittleren Molekulargewichtes." *Z. physik. Chem.* B 32: 27–45.

———. 1936b. "Osmotische Molekulargewichts-bestimmungen in polymerhomologen Reihen hochmolekularer Stoffe." *Z. physik. Chem.* A 176: 317–337.

Schwartz, A. Truman. 1981. "Importance of Good Teaching: The Influence of Arthur Pardee on Wallace Carothers." *Journal of College Science Teaching* 10: 218–221.

Schwarzhaupt, Elizabeth. 1937. "Dr. Magda Staudinger's 80th Birthday." Paper presented to the International Federation of University Women.

"Scientist Ends Life by Poison in Hotel Room." 1937. *Philadelphia Daily News,* April 30.

Seltzer, Richard J. 1985. "Paul Flory: A Giant Who Excelled in Many Roles." *Chem. Eng. News* (December 23): 27–30.

Servos, John W. 1990. *Physical Chemistry from Ostwald to Pauling: The Making of a Science in America.* Princeton, N.J.: Princeton University Press.

Seymour, Raymond B. 1980. "Polymer Science Pioneers: Herman Alexander Bruson." *Polymer News* 6: 268–269.

———, ed. 1982. *History of Polymer Science and Technology.* New York and Basel: Marcel Dekker.

———, ed. 1989. *Pioneers in Polymer Science.* Dordrecht: Kluwer Academic Publishers.

Seymour, Raymond B. and Tai Cheng, eds. 1986. *History of Polyolefins: The World's Most Widely Used Polymers.* Dordrecht and Boston: R. Reidel.

Seymour, Raymond B. and Gerald S. Kirshenbaum, eds. 1986. *High Performance Polymers: Their Origin and Development.* New York, Amsterdam, and London: Elsevier.

Seymour, Raymond B. and Herman F. Mark, eds. 1990. *Organic Coatings: Their Origin and Development.* New York, Amsterdam, and London: Elsevier.

Seymour, Raymond B. and Roger S. Porter, eds. 1993. *Manmade Fibers: Their Origin and Development.* London and New York: Elsevier Applied Science.

Simon, Mansfred. 1980. "Kränzlein, Georg." *Neue Deutsche Biographie.* Vol. 12.

Skolnik, Herman and Kenneth M. Reese, eds. 1976. *A Century of Chemistry: The Role of Chemists and the American Chemical Society.* Washington, D.C.: American Chemical Society.

Smith, John K. 1985. "The Ten-Year Invention: Neoprene and Du Pont Research, 1930–1939." *Tech. Cult.* 26: 34–55.

Smith, Watson. 1901. "A New Glyceride: Glycerol Phthalate." *J. Soc. Chem. Ind.* (November 30): 1075–1076.

Soma, Jun-ichi. 1976. "The Dawn of Polymer Research in Our Country 6" (in Japanese). *Kobunshi Kako* (February, 1976): 13–20.

Sørensen, Søren P. L. 1917. "Proteinstudien." *Compt. rend. trav. Lab. Carlsberg* 12: 1–364.

Sponsler, Olenus L. and Walter H. Dore. 1926. "The Structure of Ramie Cellulose as Derived from X-ray Data." *Colloid Symp. Monogr.* 4: 174–202.

Stahl, G. Allan. 1979. "Interview with Herman F. Mark." *J. Chem. Educ.* 56: 83–86.

———, ed. 1981. *Polymer Science Overview: A Tribute to Herman F. Mark.* Washington, D.C.: American Chemical Society.

———. 1981a. "Herman F. Mark: The Early Years, 1895–1926." In Stahl, ed., 1981. Pp. 5–19.

———. 1981b. "Herman F. Mark: The Geheimrat." In Stahl, ed., 1981. Pp. 61–88.

———. 1981c. "Herman F. Mark: The Continuing Invasion." In Stahl, ed., 1981. Pp. 105–121.

Staudinger, Franz. 1907. *Wirtschaftliche Grundlagen der Moral.* Darmstadt: E. Roether.

Staudinger, Hans. 1981. *The Inner Nazi: A Critical Analysis of Mein Kampf.* Edited by Peter M. Rutkoff and William B. Scott. Baton Rouge and London: Louisiana State University Press.

Staudinger, Hermann. 1903. "Anlagerung des Malonesters an ungestättigte Verbindungen." Dissertation, Universität Halle, HSP, A IV 1.

———. 1905. "Einwirkung von Natriummalonester auf Äthoxybernsteinsäureester und Athoxybenzylmalonester." *Liebigs Ann. Chem.* 341: 99–117.

———. 1906. "Cinnamyliden-acetophenon und Natriummalonester." *Liebigs Ann. Chem.* 345: 217–226.

———. 1912. *Die Ketene.* Stuttgart: Verlag Enke.

———. 1917a. "Über Kautschuksynthese." Paper read at the 36th general meeting of Schweizerischen Gesellschaft für chemische Industrie, held October 7. HSP, B I 157. Reprinted under the title, "Über Isopren und Kautschuk: Kautschuk-Synthese," in Staudinger, 1969–1976, vol. 1: *Arbeiten über Isopren, Kautschuk und Balata* (1969). Pp. 22–39.

———. 1917b. "Technik und Krieg." *Die Friedens-Warte* 19 (July): 196.

———. 1919a. "La technique moderne et la guerre." *Revue internationale de la Croix-rouge* 1 (15 May): 508–515.

———. 1919b. "Über Isopren und Kautschuk: Kautschuk-Synthese." *Schweizerische Chemiker-Zeitung* 1919: 1–5, 28–33, 60–64. Reprinted in Staudinger, 1969–1976, vol. 1: *Arbeiten über Isopren, Kautschuk und Balata* (1969). Pp. 40–60.

———. 1920a. "Über Polymerisation." *Ber.* 53: 1073–1085.

———. 1920b. "Die drei Nobelpreisträger Adolf von Baeyer, Emil Fischer und Alfred Werner." *Schweizerische Chemiker-Zeitung* Heft 11/17: 1–12.

———. 1924. "Über die Konstitution des Kautschuks." *Ber.* 57: 1203–1208.

———. 1925a. "Rapport technique sur la guerre chimique de M. H. Staudinger." *Revue internationale de la Croix-Rouge* 7: 17–45.

———. 1925b. "Zur Chemie des Kautschuk und der Guttapercha." *Kautschuk* (August): 5–6; (September): 8–10.

———. 1926. "Die Chemie der hochmolukularen organischen Stoffe im Sinne der Kekuléschen Strukturlehre." *Ber.* 59: 3019–3043.

———. 1929a. "Die Chemie der hochmolukularen organischen Stoffe im Sinne der Kekuléschen Strukturlehre. I." *Z. angew. Chem.* 42: 37–40, 67–73.

———. 1929b. "Schlusswort (zu den Bemerkungen von K. H. Meyer)." *Z. angew. Chem.* 42: 77.

———. 1930. "Über hochpolymere Verbindungen: Organische Chemie und Kolloidchemie." *Kolloid-Z.* 53: 19–30.

———. 1931a. "Sur la Structure des Composés à poids moléculaire élevé." Paper read at the ninth International Solvay Congress, Brussels, April. Reprinted in Staudinger, 1969–1976, vol. 5: *Arbeiten allgemeiner Richtung von Hermann Staudinger* (1973). Pp. 50–125.

———. 1931b. "Sur la Constitution des Colloïdes moléculaires." *Bull. Soc. chim.* 49: 1267–1279.

———. 1932. *Die hochmolekularen organischen Verbindungen: Kautschuk und Cellulose.* Berlin: Verlag von Julius Springer.

———. 1933. "Viscosity Investigations for the Examination of the Constitution of Natural Products of High Molecular Weight and of Rubber and Cellulose." *Trans. Faraday Soc.* 29, Pt. 1: 18–32.

———. 1934a. "Die Bedeutung der Chemie für das deutsche Volk." *Völkische Zeitung,* February 25. Reprinted as "Die Bedeutung der Chemie für die Existenzmöglichkeit des deutschen Volkes." *Chemiker-Ztg.* 59 (1935): 201–202.

———. 1934b. "Die neuere Entwicklung der organischen Kolloidchemie." Paper read at the IXth International Congress for Pure and Applied Chemistry, Madrid, April 10. *Trabajos del IX Congreso Internacional de Quimica Pura y Aplicada,* vol. IV: 9–47. Reprinted in Staudinger, 1969–1976, vol. 4: *Physikalisch-chemische Untersuchungen an makromolekularen Stoffen* (1975). Pp. 45–83.

———. 1934c. "Die Poly-oxymethylene als Modell der Cellulose (Bemerkungen zu einer Arbeit von K. Hess und Mitarbeitern)." *Ber.* 67: 475–479.

———. 1935. "Über die Einteilung der Kolloide." *Ber.* 68: 1682–1691.

———. 1936a. "The Formation of High Polymers of Unsaturated Substances." *Trans. Faraday Soc.* 32, Pt. 1: 97–115.

———. 1936b. "Zur Entwicklung der makromolekularen Chemie. Zugleich Antwort auf die Entgegnung von K. H. Meyer und A. van der Wyk." *Ber.* 69: 1168–1185.

———. 1936c. "Über die makromolekulare Chemie." *Z. angew. Chem.* 49: 801–813.

———. 1937. "Über Cellulose, Stärke und Glycogen." *Naturwiss.* 25: 173–681.

———. 1940a. *Organische Kolloidchemie.* Braunschweig: Verlag Vieweg & Sohn.

———. 1940b. "Über niedermolekulare und makromolekulare Chemie." *J. prakt. Chem.* 155: 1–12.

———. 1942. "Kolloidik und makromolekulare Chemie." *J. prakt. Chemie* 160: 245–280.

———. 1947. *Makromolekulare Chemie und Biologie.* Basel: Verlag Wepf & Co.

———. 1953. "Macromolecular Chemistry: Nobel Lecture, December 11, 1953." In Nobel Foundation, ed., 1964. Pp. 397–419.

———. 1955. "Über die Entwicklung der makromolekularen Chemie in den Jahren 1920 bis 1926." In *Festgabe der GEP zur Hundertjahrfeier der Eidgenössischen Technischen Hochschule Zürich* (1955): 399–411. Reprinted in Staudinger, 1969–1976, vol. 5: *Arbeiten allgemeiner Richtung von Hermann Staudinger* (1975). Pp. 3–15.

———. 1958. "Über die Entwicklung der makromolekularen Chemie." Paper read at the symposium, "Organic Polymer Chemistry: Prospects and Retrospects," at the Polytechnic Institute of Brooklyn, September 20. HSP, B I 143.

———. 1961. *Arbeitserinnerungen.* Heidelberg: Dr. Alfred Hüthig Verlag.

———. 1966. *Kenkyu Kaiko* (Memoirs of My Research). Translated from Staudinger, 1961, into Japanese by Yoshio Kobayashi. Tokyo: Iwanami Shoten.

———. 1969–1976. *Das wissenschaftliche Werk von Hermann Staudinger: Gesammelte Arbeiten nach Sachgebieten geordnet.* 7 vols. Edited by Magda Staudinger, Heinrich Hopff, and Werner Kern. Basel and Heidelberg: Hüthig & Wepf Verlag.

———. 1970. *From Organic Chemistry to Macromolecules: A Scientific Autobiography on My Original Papers.* Translated from Staudinger, 1961, by Jerome Fock and Michael Fried. New York, London, Sydney, and Toronto: Wiley-Interscience Publishers.

Staudinger, Hermann and Helmut W. Klever. 1911. "Über die Darstellung von Isopren aus Terpenkohlenwasserstoffen." *Ber.* 44: 2212–2215.

Staudinger, Hermann and Jakob Fritschi. 1922. "Über Isopren und Kautschuk: Über die Hydrierung des Kautschuks und über seine Konstitution." *Helv. Chim. Acta* 5: 785–806.

Staudinger, Hermann and Herman A. Bruson. 1926a. "Über das Dicyclopentadien und weitere polymere Cyclopentadiene." *Liebigs Ann. Chem.* 447: 97–110.

———. 1926b. "Über die Polymerisation des Cyclopentadiens." *Liebigs Ann. Chem.* 447: 110–122.

Staudinger, Hermann, H. Johner, R. Signer, G. Mie, and J. Hengstenberg. 1927. "Der polymere Formaldehyd, ein Modell der Cellulose." *Z. physik. Chem.* 126: 425–448.

Staudinger, Hermann, A. A. Ashdown, M. Brunner, H. A. Bruson, and S. Wehrli. 1929. "Über die Konstitution des Poly-indens." *Helv. Chim. Acta* 12: 934–957.

Staudinger, Hermann and Rudolf Signer, 1929. "Über den Kristallbau hochmolekularer Verbindungen." *Zeitschrift für Kristallographie* 70: 193–210.

Staudinger, Hermann and Werner Heuer. 1930. "Beziehungen zwischen Viskosität und Molekulargewicht bei Polystyrolen." *Ber.* 63: 222–234.

Staudinger, Hermann and Avery A. Ashdown. 1930. "Über Poly-a-phenylbutadien." *Ber.* 63: 717–721.

Staudinger, Hermann, and Ryuzaburo Nodzu. 1930. "Viskositätsuntersuchungen an Paraffin-Lösungen." *Ber.* 63: 721–724.

Staudinger, Hermann and Eiji Ochiai. 1932. "Viskositätsmessungen an Lösungen von Fadenmolekülen." *Z. physik. Chem.* (A), 158: 35–55.

Staudinger, Hermann and Georg Kränzlein. 1966. *Zur Strukturaufklärung der Makromoleküle: Ein Briefwechsel zwischen Prof. Staudinger und Dr. Kränzlein.* Dokumente aus Hoechster Archiven: Beiträge zur Geschichte der Chemie, 15. Edited by Farbwerke Hoechst AG. Reprint. Frankfurt(M)-Hoechst: Farbwerke Hoechst AG, 1972.

Staudinger, Magda. 1982. "Zur Geschichte der Zeitschrift 'Die Makromolekulare Chemie.'" *Makromol. Chem.* 183: 1829–1831.

———. 1987. "Hermann Staudinger — der Mensch und der Forscher." In *Makromolekulare Chemie: Das Werk Hermann Staudingers in seiner heutigen Bedeutung,* edited by Ernst Jostkleigrewe, 9–29. Munich and Zürich: Verlag Schnell & Steiner. 1987.

Staverman, A. J. 1975. "The Gaussian Chain in the Theory of Rubber Elasticity." *J. Poly. Sci., Symposium,* no. 51: 45–56.

Steinkopff, Theodor, Dietrich Steinkopff, and A. Lottermoser. 1943. "Wolfgang Ostwald." *Kolloid-Z.* 105, no. 3: front pages.

Stine, Charles M. A. 1932. "Chemical Research: A Factor of Prime Importance in American Industry." *J. Chem. Educ.* 9: 2032–2039.

———. 1934. "The Approach to Chemical Research Based on a Specific Example." *Journal of the Franklin Institute* 218: 397–410.

———. 1936. "The Place of the Fundamental Research in an Industrial Research Organization." *Trans. Amer. Inst. Chem. Eng.* 32: 127–137.
Stockmayer, Walter H. 1974. "The 1974 Nobel Prize for Chemistry." *Science* 186: 724–726.
Stockmayer, Walter H., and Bruno H. Zimm. 1984. "When Polymer Science Looked Easy." *Ann. Rev. Phys. Chem.* 35: 1–21.
Stoltzenberg, Dietrich. 1994. *Fritz Haber: Chemiker, Nobelpreisträger, Deutscher, Jude: Eine Biographie.* Weinheim: Verlag Chemie.
Strube, Irene. 1987. "War Hermann Staudinger der Initiator des ersten Appels gegen die Anwendung chemischer Kampfstoffe im Jahre 1918?" *NTM* 24: 87–92.
Sturchio, Jeffrey L. 1981. "Chemists and Industry in Modern America: Studies in the Historical Application of Science Indicators." Ph.D. dissertation, University of Pennsylvania.
Sühnel, Klaus. 1989. "80 Jahre Kolloidchemie: Leben und Werk Wolfgang Ostwalds." *NTM* 26: 31–45.
Sutton, Leslie, and Mansel Davies. 1996. *The History of the Faraday Society.* Cambridge: Royal Society of Chemistry.
Svedberg, The, and Robin Fåhraeus. 1926. "A New Method for the Determination of the Molecular Weight of the Proteins." *J. Amer. Chem. Soc.* 48: 430–438.
Svedberg, The, and Kai O. Pedersen, eds. 1940. *The Ultracentrifuge.* Oxford: Oxford University Press.
Szabadváary, Ferenc. 1990. "Polyáni, Mihály (Michael)." *DSB*, vol. 18.
Takei, Muneo. 1947. "Death of Wolfgang Ostwald" (in Japanese). *Kagaku no Ryoiki* 1: 27–29.
———. 1948. "Ostwald" (in Japanese). *Kagakuken* 3 (December): 59–63.
Tamamushi, Bun'ichi. 1978. *Ichi kagakusha no kaiso* (Memoirs of a Chemist). Tokyo: Chuo Kohron-sha.
Tanaka, Atsushi. 1992–1993. "H. Staudinger's Research and the Birth of the Polymer Industry in Germany" (in Japanese). *Kagakushi* 19: 172–187, 247–261; 20: 243–258.
Tanaka, Ryukichi. 1930. "Wolfgang Ostwald" (in Japanese). *Warera no Kagaku* 3: 396–399.
Tarbell, D. Stanley and Ann T. Tarbell. 1981. *Roger Adams: Scientist and Statesman.* Washington, D.C.: American Chemical Society.
———. 1991. "Carl S. Marvel at Illinois Wesleyan, 1911–1915." *J. Chem. Educ.* 68: 539–542.
Tarbell, D. Stanley, Ann T. Tarbell, and R. M. Joyce. 1980. "The Students of Ira Remsen and Roger Adams." *Isis* 71: 620–626.
Thackray, Arnold, Jeffrey L. Sturchio, P. Thomas Carroll, and Robert Bud. 1985. *Chemistry in America, 1876–1976: Historical Indicators.* Dordrecht: Reidel.
Thiele, Johannes. 1899. "Zur Kenntnis der ungesättigten Verbindungen." *Liebigs Ann. Chem.* 306: 87–142.
Thomas, Arthur W. 1925. "The Modern Trend in Colloid Chemistry." *J. Chem. Educ.* 2: 323–340.
Tilden, William A. 1884. "On the Decomposition of Terpenes by Heat." *J. Chem. Soc.* 45: 410–420.
———. 1908. "Synthetic Rubber." *India-Rubber J.* 36: 321–322.
Törnqvist, Erik G. M. 1968. "The Historical Background of Synthetic Elastomers with Particular Emphasis on the Early Period." In *Polymer Chemistry of Synthetic*

Elastomers, Part I, edited by Joseph P. Kennedy and Erik G. M. Törnqvist, 21–94. New York, London, and Sydney: Interscience Publishers, 1968.

Travis, Anthony S. 1993. *The Rainbow Makers: The Origins of the Synthetic Dyestuffs Industry in Western Europe.* Bethlehem, Penn.: Lehigh University Press; London and Toronto: Associated University Press.

Twiss, Summer B., ed. 1945. *Advancing Fronts in Chemistry.* Vol. 1: *High Polymers.* New York: Reinhold.

Ulrich, Robert D. 1978. "The History of ACS Division of Polymer Chemistry, Inc." In Ulrich, ed., 1978. Pp. 1–30.

———, ed. 1978. *Contemporary Topics in Polymer Science.* Vol. 1: *Macromolecular Science: Retrospect and Prospect.* New York and London: Plenum Press.

Van der Wyk, A. J. A. 1952. "Kurt Heinrich Meyer." *Helv. Chim. Acta* 35: 1418–1422.

Van't Hoff, J. H. 1887. "Die Rolle des osmotischen Druckes in der Analogie zwischen Lösungen und Gasen." *Z. physik. Chem.* 1: 481–508.

———. 1888. "The Function of Osmotic Pressure in the Analogy Between Solutions and Gases." *Phil. Mag.* ser. 5, 26: 81–105.

Vierhaus, Rudolf and Bernhard vom Brocke, eds. 1990. *Forschung im Spannungsfeld von Politik und Gesellschaft: Geschichte und Struktur der Kaiser-Wilhelm-/Max-Planck-Gesellschaft.* Stuttgart: Deutsche Verlags-Anstalt.

Vorländer, Daniel. 1894. "Aethylenester zweibasischer Säuren und Phenole." *Liebigs Ann. Chem.* 280: 167–206.

Vorländer, Daniel, and Hermann Staudinger. 1903a. "Über Zwischenprodukte bei Additions- und Kondensationsreaktionen des Malonesters." *Zeitschrift für Naturwissenschaften* (Halle) 75: 385–432.

———. 1903b. "Über die Anlagerung des Malonesters an das System CH=CH−CH=CH−C=O." *Zeitschrift für Naturwissenschaften* (Halle) 75: 433–454.

Wada, Takeshi. 1987. "The Harries-Pickles Controversy on the Theory of Molecular Structure of Rubber" (in Japanese). *Kagakushi* 14: 16–28, 49–60.

Wadano, Motoi. 1952. "Recollections of the Kaiser Wilhelm Institute" (in Japanese). *Kobunshi* 1, no. 7: 43–47.

Waldschmidt-Leitz, Ernst. 1926. "Zur Struktur der Proteine." *Ber.* 59: 3000–3007.

Wallace, G. L. 1952. "Economic and Social Aspects of the Industry." In Schidrowitz and Dawson, eds., 1952. Pp. 327–342.

Wallick, Merritt. 1988. "Nylon's Story Is One of Success, Tragedy." *Sunday News Journal*, Wilmington, Del., January 17.

Washburn, Edward W. 1929. "Molecular Stills." *Bureau of Standards Journal of Research* 2: 476–483.

Watts, Henry. 1863–1868. *A Dictionary of Chemistry and Applied Branches of Other Sciences.* 5 vols. London: Longmans, Green.

Weber, Carl O. 1900a. "The Nature of India-Rubber." *J. Soc. Chem. Ind.* 19: 215–221.

———. 1900b. "Über die Natur des Kautschuks." *Ber.* 33: 779–796.

———. 1902. *Chemistry of India Rubber.* London: Charles Griffin.

———. 1906. "Grundzüge einer Theorie der Kautschukvulkanisation." *Kolloid-Z.* 1: 33–38, 65–71.

Weimarn, P. P. von. 1907. "Zur Lehre von den Kolloiden, amorphen und kristallinischen Zuständen." *Kolloid-Z.* 2: 76–83.

Weissenberg, K. 1925. "Kristallbau und chemische Konstitution, 1–3." *Z. Physik* 34: 406–419, 420–432, 433–452.

Werner, Alfred. 1891. "Beiträge zur Theorie der Affinität und Valenz: Über Stereochemie des Stickstoffs in der Benzhydroxamsäurereihe." *Vierteljahrsschrift der Züricher Naturforscher Gesellschaft* 36: 129–169.

———. 1902. "Ueber Haupt- und Nebenvalenzen und die Constitution der Ammoniumverbindungen." *Liebigs Ann. Chem.* 322: 261–296.

"What Does Nylon Teach Us?" (in Japanese). 1939. *Osaka Asahi Shinbun,* April 17.

Whinfield, John Rex. 1946. "Chemistry of Terilen." *Nature* 153: 930–931.

Whitby, G. Stafford. 1921. "Recent Work of Harries on Caoutchouc." *India-Rubber J.* 617 (February 12): 313–315.

———. 1940. "Acetylene Polymers and Their Derivatives. I. Introduction." In Carothers, 1940. Pp. 273–281.

———. 1967. "Hermann Staudinger (1881–1965)." *Rubber Chem. Tech.* 40 (June): xvi–xxiii.

———. 1989a. "Thomas Hancock (1786–1865)." In Zimmerman, ed., 1989. Pp. 3–11.

———. 1989b. "Carl Dietrich Harries (1866–1923)." In Zimmerman, ed., 1989. Pp. 20–22.

———. 1989c. "Johan Rudolph Katz (1880–1938)." In Zimmerman, ed., 1989. Pp. 47–59.

Whitby, G. Stafford and Morris Katz. 1928. "The Polymerization of Indene, Cinnamal Fluorene and Some Derivatives of Indene." *J. Amer. Chem. Soc.* 50: 1160–1171.

———. 1933. "Synthetic Rubber." *Ind. Eng. Chem.* 25: 1204–1211, 1338–1348.

Wigner, Eugene P. and R. A. Hodgkin. 1977. "Michael Polanyi, 12 March 1891–22 February 1976." *Biog. Mem. FRS* 23: 413–448.

Williams, C. Greville. 1860. "On Isoprene and Caoutchine." *Phil. Trans.* 150: 241–255.

Williams, J. W. 1979. "The Development of the Ultracentrifuge and Its Contributions." In *The Origins of Modern Biochemistry: A Retrospect on Proteins,* edited by P. R. Srinivasan, Joseph S. Fruton, and John T. Edsall, 77–91. *Annals of the New York Academy of Sciences* 325. New York: New York Academy of Sciences.

Willstätter, Richard. 1965. *From My Life: The Memoirs of Richard Willstätter.* Translated by Lilli S. Hornig. New York and Amsterdam: W. A. Benjamin.

Wise, George. 1985. *Willis R. Whitney, General Electric, and the Origins of U.S. Industrial Research.* New York: Columbia University Press.

Wise, Louis E. 1960. "Emil Heuser, 1882–1953." *Advances in Carbohydrate Chemistry* 15: 1–9.

Witkop, Bernhard. 1992. "Remembering Heinrich Wieland (1877–1957): Portrait of an Organic Chemist and Founder of Modern Biochemistry." *Medicinal Research Reviews* 12: 195–274.

Yarsley, V. E. 1967. "Hermann Staudinger — His Life and Work: Memorial Lecture." *Chem. Ind.* 7 (February 18): 250–271.

Zandvoort, H. 1988. "Macromolecules, Dogmatism, and Scientific Change: The Prehistory of Polymer Chemistry as Testing Ground for Philosophy of Science." *Studies in History and Philosophy of Science* 19: 489–515.

Zimmerman, Barbara Nute, ed. 1989. *Vignettes from the International Rubber Science Hall of Fame (1958–1988): 36 Major Contributors to Rubber Science — A Biographical Collection.* Akron, Ohio: Rubber Division, American Chemical Society.

Zinoffsky, Oscar. 1886. "Ueber die Grösse des Hämoglobinmoleküls." *Z. physiol. Chem.* 10: 16–34.

Zsigmondy, Richard. 1909. *Colloids and the Ultramicroscope: A Manual of Colloid Chemistry and Ultramicroscopy.* Translated by J. Alexander. New York: John Wiley & Sons.

———. 1925. "Properties of Colloids." In Nobel Foundation, ed., 1966. Pp. 45–66.

Zsigmondy, Richard and W. Bachmann. 1912. "Ueber Gallerten: Ultramikro-skopische Studien an Seifenlösungen und -gallerten." *Kolloid-Z.* 11: 145–157.

Interviews

Adams, Roger. Interviews with John B. Mellecker, November 1964; February, March, July, October 1965, RAP, Box 9.

Berchet, Gerard J. Interview with Adeline B. C. Strange, n.d. HML.

———. Interview with the author, March 13, 1982.

Bolton, Elmer K. Interview with Alfred D. Chandler, Richard D. Williams, and Norman B. Wilkinson, September 14, 1961, HML, Acc. 1689.

Dykstra, H. and H. Cupery. Interview with Adeline B. C. Strange, August 2, 1978, HML.

Flory, Paul J. Interview with Charles G. Overberger. Eminent Chemists Video-tapes Series, ACS, 1982.

Go, Yukichi. Interview with the author, November 18, 1985.

Hanford, William E. Interview with the author, April 29, 1993.

Hill, Julian W. Interviews with the author, February 22, November 29, 1982.

Labovsky, Joseph. Interview with the author and James J. Bohning, May 6, 1993.

Mark, Herman F. Interview with the author, March 19, 1982.

———. Interviews with James J. Bohning and Jeffrey L. Sturchio, February 3, March 17, June 20, 1986. CHF Oral Histories, CHF.

Marvel, Carl S. Interview with Leon Gorter and Charles Price, July 13, 1983. CHF Oral Histories, CHF.

Okamura, Seizo. Interview with the author, June 20, 1994.

Sakurada, Yutaka. Interview with the author, June 23, 1994.

Signer, Rudolf. Interview with Tonja Koeppel, September 30, 1986. CHF Oral Histories, CHF.

———. Interview with the author, July 29, 1989.

Staudinger, Hans. Interview with Jehuda Riemer, July 25, 1978. HWSP, Box 1, Folder 2.

Staudinger, Magda. Interviews with the author, March 13, 1987; July 30, 31, 1989.

Index

SHENANDOAH UNIVERSITY LIBRARY
WINCHESTER, VA 22601